# Relativity Demystified

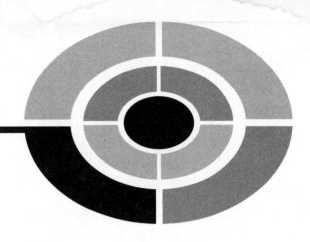

# Relativity Demystified

**DAVID McMAHON**

**McGRAW-HILL**

New York Chicago San Francisco Lisbon London Madrid
Mexico City Milan New Delhi San Juan Seoul
Singapore Sydney Toronto

The McGraw·Hill Companies

Library of Congress Cataloging-in-Publication Data

McMahon, David (David M.)
  Relativity demystified / David McMahon.
    p. cm.—(Demystified series)
  Includes index.
  ISBN 0-07-145545-0 (acid-free paper)
  1. Relativity (Physics)—Popular works.   I. Title.   II. McGraw-Hill "Demystified" series.

QC173.57.M36   2006
530.11—dc22

                                                                    2005054738

2 3 4 5 6 7 8 9 0   DOC/DOC   0 1 0 9 8 7 6

ISBN 0-07-145545-0

*The sponsoring editor for this book was Judy Bass and the production supervisor was Pamela A. Pelton. It was set in Times Roman by TechBooks. The art director for the cover was Margaret Webster-Shapiro.*

*Printed and bound by RR Donnelley.*

This book was printed on recycled, acid-free paper containing a minimum of 50% recycled, de-inked fiber.

McGraw-Hill books are available at special quantity discounts to use as premiums and sales promotions, or for use in corporate training programs. For more information, please write to the Director of Special Sales, McGraw-Hill Professional, Two Penn Plaza, New York, NY 10121-2298. Or contact your local bookstore.

# CONTENTS

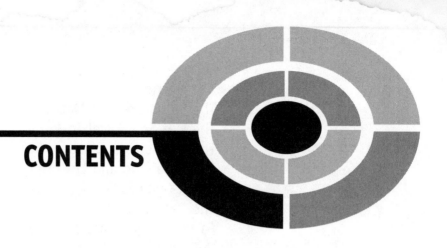

# CONTENTS

# CONTENTS

# PREFACE

The theory of relativity stands out as one of the greatest achievements in science. The "special theory", which did not include gravity, was put forward by Einstein in 1905 to explain many troubling facts that had arisen in the study of electricity and magnetism. In particular, his postulate that the speed of light in vacuum is the same constant seen by all observers forced scientists to throw away many closely held commonsense assumptions, such as the absolute nature of the passage of time. In short, the theory of relativity challenges our notions of what reality is, and this is one of the reasons why the theory is so interesting.

Einstein published the "general" theory of relativity, which is a theory about gravity, about a decade later. This theory is far more mathematically daunting, and perhaps this is why it took Einstein so long to come up with it. This theory is more fundamental than the special theory of relativity; it is a theory of space and time itself, and it not only describes, it *explains* gravity. Gravity is the distortion of the structure of spacetime as caused by the presence of matter and energy, while the paths followed by matter and energy (think of bending of passing light rays by the sun) in spacetime are governed by the structure of spacetime. This great feedback loop is described by Einstein's field equations.

This is a book about general relativity. There is no getting around the fact that general relativity is mathematically challenging, so we cannot hope to learn the theory without mastering the mathematics. Our hope with this book is to "demystify" that mathematics so that relativity is easier to learn and more accessible to a wider audience than ever before. In this book we will not skip any of the math that relativity requires, but we will present it in what we hope to be a clear fashion and illustrate how to use it with many explicitly solved examples. Our goal is to make relativity more accessible to everyone. Therefore we hope that engineers, chemists, and mathematicians or anyone who has had basic mathematical training at the college level will find this book useful. And of course the book is aimed at physicists and astronomers who want to learn the theory.

The truth is that relativity looks much harder than it is. There is a lot to learn, but once you get comfortable with the new math and new notation, you will actually find it a bit easier than many other technical areas you have studied in the past.

This book is meant to be a self-study guide or a supplement, and not a full-blown textbook. As a result we may not cover every detail and will not provide lengthly derivations or detailed physical explanations. Those can be found in any number of fine textbooks on the market. Our focus here is also in "demystifying" the mathematical framework of relativity, and so we will not include lengthly descriptions of physical arguments. At the end of the book we provide a listing of references used in the development of this manuscript, and you can select books from that list to find the details we are leaving out. In much of the material, we take the approach in this book of stating theorems and results, and then applying them in solved problems. Similar problems in the end-of chapter quiz help you try things out yourself.

So if you are taking a relativity course, you might want to use this book to help you gain better understanding of your main textbook, or help you to see how to accomplish certain tasks. If you are interested in self-study, this book will help you get started in your own mastery of the subject and make it easier for you to read more advanced books.

While this book is taking a lighter approach than the textbooks in the field, we are not going to cut corners on using advanced mathematics. The bottom line is you are going to need some mathematical background to find this book useful. Calculus is a must, studies of differential equations, vector analysis and linear algebra are helpful. A background in basic physics is also helpful.

Relativity can be done in different ways using a coordinate-based approach or differential forms and Cartan's equations. We much prefer the latter approach and will use it extensively. Again, it looks intimidating at first because there are lots of Greek characters and fancy symbols, and it is a new way of doing things. When doing calculations it does require a bit of attention to detail. But after a bit of practice, you will find that its not really so hard. So we hope that readers will invest the effort necessary to master this nice mathematical way of solving physics problems.

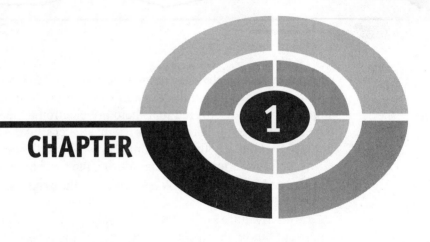

# A Quick Review of Special Relativity

Fundamentally, our commonsense intuition about how the universe works is tied up in notions about space and time. In 1905, Einstein stunned the physics world with the special theory of relativity, a theory of space and time that challenges many of these closely held commonsense assumptions about how the world works. By accepting that the speed of light in vacuum is the same constant value for all observers, regardless of their state of motion, we are forced to throw away basic ideas about the passage of time and the lengths of rigid objects.

This book is about the general theory of relativity, Einstein's theory of gravity. Therefore our discussion of special relativity will be a quick overview of concepts needed to understand the general theory. For a detailed discussion of special relativity, please see our list of references and suggested reading at the back of the book.

The theory of special relativity has its origins in a set of paradoxes that were discovered in the study of electromagnetic phenomena during the nineteenth century. In 1865, a physicist named James Clerk Maxwell published his famous set of results we now call *Maxwell's equations*. Through theoretical studies alone, Maxwell discovered that there are electromagnetic waves and that they travel at one speed—the speed of light $c$. Let's take a quick detour to get a glimpse into the way this idea came about. We will work in SI units.

In careful experimental studies, during the first half of the nineteenth century, Ampere deduced that a steady current $\vec{J}$ and the magnetic field $\vec{B}$ were related by

$$\nabla \times \vec{B} = \mu_0 \vec{J} \tag{1.1}$$

However, this law cannot be strictly correct based on simple mathematical arguments alone. It is a fundamental result of vector calculus that the divergence of any curl vanishes; that is,

$$\nabla \cdot \left( \nabla \times \vec{A} \right) = 0 \tag{1.2}$$

for any vector $\vec{A}$. So it must be true that

$$\nabla \cdot \left( \nabla \times \vec{B} \right) = 0 \tag{1.3}$$

However, when we apply the divergence operator to the right-hand side, we run into a problem. The problem is that the continuity equation, which is the mathematical description of the conservation of charge, tells us that

$$\frac{\partial \rho}{\partial t} + \nabla \cdot \vec{J} = 0 \tag{1.4}$$

where $\rho$ is the current density. Therefore, when we apply the divergence operator to the right-hand side of (1.4), we obtain

$$\nabla \cdot \left( \mu_0 \vec{J} \right) = \mu_0 \nabla \cdot \vec{J} = -\mu_0 \frac{\partial \rho}{\partial t} \tag{1.5}$$

We can take this process even further. Gauss's law tells us how to relate the charge density to the electric field. In SI units, this law states

$$\nabla \cdot \vec{E} = \frac{1}{\varepsilon_0} \rho \tag{1.6}$$

This allows us to rewrite (1.5) as

$$-\mu_0 \frac{\partial \rho}{\partial t} = -\mu_0 \varepsilon_0 \frac{\partial}{\partial t} \left( \nabla \cdot \vec{E} \right) = -\nabla \cdot \left( \mu_0 \varepsilon_0 \frac{\partial \vec{E}}{\partial t} \right) \qquad (1.7)$$

Putting our results together, we've found that

$$\nabla \cdot \left( \nabla \times \vec{B} \right) = -\nabla \cdot \left( \mu_0 \varepsilon_0 \frac{\partial \vec{E}}{\partial t} \right) \qquad (1.8)$$

when in fact it *must* be zero. Considerations like these led Maxwell to "fix up" Ampere's law. In modern form, we write it as

$$\nabla \times \vec{B} = \mu_0 \vec{J} + \mu_0 \varepsilon_0 \frac{\partial \vec{E}}{\partial t} \qquad (1.9)$$

The extra term $\mu_0 \varepsilon_0 \frac{\partial \vec{E}}{\partial t}$ is called the displacement current and its presence led to one of Maxwell's most dramatic discoveries. Using simple vector calculus, one can show that the electric and magnetic fields satisfy the following wave equations:

$$\nabla^2 \vec{E} = \mu_0 \varepsilon_0 \frac{\partial^2 \vec{E}}{\partial t^2} \quad \text{and} \quad \nabla^2 \vec{B} = \mu_0 \varepsilon_0 \frac{\partial^2 \vec{B}}{\partial t^2}$$

Now, the *wave equation* is

$$\nabla^2 f = \frac{1}{v^2} \frac{\partial^2 f}{\partial t^2}$$

where $v$ is the velocity of the wave. Comparison of these equations shows that electromagnetic waves in vacuum travel at speed

$$v = \frac{1}{\sqrt{\mu_0 \varepsilon_0}} = 3 \times 10^8 \text{ m/s } = c$$

where $c$ is nothing more than the speed of light.

The key insight to gain from this derivation is that electromagnetic waves (light) *always* travel at one and the same speed in vacuum. It doesn't matter who you are or what your state of motion is, this is the speed you are going to find. It took many years for this insight to sink in—and it was Einstein who simply accepted this result at face value.

We can give a quick heuristic insight as to why this result leads to the "para-doxes" of relativity. What is *speed* anyway (our argument here is qualitative, so we are going to be a bit sloppy here)? It is distance covered per unit time:

$$v = \frac{\Delta x}{\Delta t}$$

The commonsense viewpoint, which is formalized mathematically in prerel-ativistic Newtonian physics, is that distances and times are fixed—thus, how could you possibly have a constant velocity that is the same for all observers? That wouldn't make any sense. However, the theoretical result that the speed of light in vacuum is the same for all observers is an experimental fact con-firmed many times over. If $v$ is the constant speed of light seen by all observers regardless of their state of motion

$$c = \frac{\Delta x}{\Delta t}$$

then distances and time intervals must be different for different observers. We will explore this in detail below.

In many treatments of special relativity, you will see a detailed discussion of the Michelson-Morley experiment. In a nutshell, the idea was that waves need a medium to travel through, so physicists at the time made the completely reasonable assumption that there was a medium that filled all of space, called the *luminiferous ether*. It was thought that the ether provided the medium nec-essary for the propagation of electromagnetic waves. The Michelson-Morley experiment was designed to detect the motion of the earth with respect to the ether—but it found nothing. This is one of the most famous "null" results in the history of experimental physics.

This experiment is a crucial result in the history of physics, but the record seems to indicate that Einstein based his derivations on an acceptance of Maxwell's equations and what they tell us more so than on the Michelson-Morley experiment (in fact Einstein may not have even known very much, if anything, about the experiment at the time of his derivations). As a result, while the experiment is interesting and very important, we are going to skip it and move on to the theoretical framework of special relativity.

Interestingly, other researchers, Lorentz, Fitzgerald, and Poincare, indepen-dently derived the Lorentz transformations in order to explain the null results of the Michelson-Morley experiment. The gist of these equations is that clocks slow down and the lengths of rigid bodies contract, making it impossible to construct an experimental apparatus of any kind to detect motion with respect to the ether.

In addition to the arguments we have described here, Einstein used results that came out of Faraday's law and the studies of electromagnetic induction to come up with ideas about relative motion. We won't discuss those here, but the interested reader is encouraged to explore the references for details.

The history of the discovery of the special theory of relativity is a good lesson in the way science works—for it demonstrates the crucial interplay between theory and experiment. Careful experiments within the limits of technology available at the time led to Ampere's law and Faraday's law. Later, purely mathematical arguments and theoretical considerations were used to show that Ampere's law was only approximately correct and that electromagnetic waves travel through space at the speed of light. More theoretical considerations were put forward to explain how those waves traveled through space, and then the dramatic experimental result found by Michelson and Morley forced those ideas to be thrown away. Then Einstein came along and once again used mostly theoretical arguments to derive special relativity. The bottom line is this: Physics is a science that depends on two legs—theory and experiment—and it cannot stand on either alone.

We now turn to a quick tour of the basics of special relativity, and begin with some definitions.

# Frame of Reference

A frame of reference is more or less a fancy way of saying *coordinate system*. In our thought experiments, however, we do more than think in mathematical terms and would like to imagine a way that a frame of reference could really be constructed. This is done by physically constructing a coordinate system from measuring rods and clocks. Local clocks are positioned everywhere within the frame and can be used to read off the time of an event that occurs at that location. You might imagine that you have 1-m long measuring rods joined together to form a lattice, and that there is a clock positioned at each point where rods are joined together.

# Clock Synchronization

One problem that arises in such a construction is that it is necessary to synchronize clocks that are physically separated in space. We can perform the synchronization using light rays. We can illustrate this graphically with a simple spacetime diagram (more on these below) where we represent space on the horizontal axis and time on the vertical axis. This allows us to plot the motion of

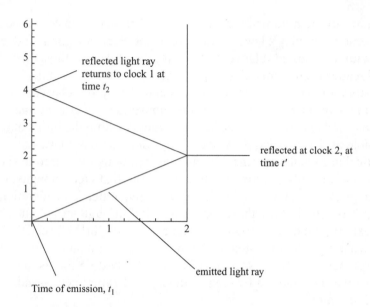

**Fig. 1-1.** Clock synchronization. At time $t_{\bar{1}}$, a light beam is emitted from a clock at the origin. At time $t'$, it reaches the position of clock 2 and is reflected back. At time $t_{\bar{2}}$, the reflected beam reaches the position of clock 1. If $t'$ is halfway between the times $t_{\bar{1}}$ and $t_{\bar{2}}$, then the two clocks are synchronized.

objects in space and time (we are of course only considering one-dimensional motion). Imagine two clocks, clock 1 located at the origin and clock 2 located at some position we label $x_1$ (Fig. 1-1).

To see if these clocks are synchronized, at time $t_1$ we send a beam of light from clock 1 to clock 2. The beam of light is reflected back from clock 2 to clock 1 at time $t_2$, and the reflected beam arrives at the location of clock 1 at time $t_1'$. If we find that

$$t' = \frac{1}{2}(t_1 + t_2)$$

then the clocks are synchronized. This process is illustrated in Fig. 1-1. As we'll see later, light rays travel on the straight lines $x = t$ in a spacetime diagram.

# Inertial Frames

An *inertial frame* is a frame of reference that is moving at constant velocity. In an inertial frame, Newton's first law holds. In case you've forgotten, Newton's first

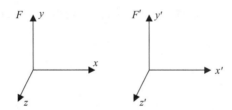

**Fig. 1-2.** Two frames in *standard configuration*. The primed frame $(F')$ moves at velocity $v$ relative to the unprimed frame $F$ along the $x$-axis. In prerelativity physics, time flows at the same rate for all observers.

law states that a body at rest or in uniform motion will remain at rest or in uniform motion unless acted upon by a force. Any other frame that is moving uniformly (with constant velocity) with respect to an inertial frame is also an inertial frame.

# Galilean Transformations

The study of relativity involves the study of how various physical phenomena appear to different observers. In prerelativity physics, this type of analysis is accomplished using a *Galilean transformation*. This is a simple mathematical approach that provides a transformation from one inertial frame to another. To study how the laws of physics look to observers in relative motion, we imagine two inertial frames, which we designate $F$ and $F'$. We assume that they are in the *standard configuration*. By this we mean the frame $F'$ is moving in the $x$ direction at constant velocity $v$ relative to frame $F$. The $y$ and $z$ axes are the same for both observers (see Fig. 1-2). Moreover, in prerelativity physics, there is uniform passage of time throughout the universe for everyone everywhere. Therefore, we use the same time coordinate for observers in both frames.

The Galilean transformations are very simple algebraic formulas that tell us how to connect measurements in both frames. These are given by

$$t = t', \qquad x = x' + vt, \qquad y = y', \qquad z = z' \qquad (1.10)$$

# Events

An event is anything that can happen in spacetime. It could be two particles colliding, the emission of a flash of light, a particle just passing by, or just anything else that can be imagined. We characterize each event by its spatial location and the time at which it occurs. Idealistically, events happen at a single mathematical point. That is, we assign to each event $E$ a set of four coordinates $(t, x, y, z)$.

# The Interval

The spacetime interval gives the *distance* between two events in space and time. It is a generalization of the pythagorean theorem. You may recall that the distance between two points in cartesian coordinates is

$$d = \sqrt{(x_1 - x_2)^2 + (y_1 - y_2)^2 + (z_1 - z_2)^2} = \sqrt{(\Delta x)^2 + (\Delta y)^2 + (\Delta z)^2}$$

The interval generalizes this notion to the arena of special relativity, where we must consider distances in time together with distances in space. Consider an event that occurs at $E_1 = (ct_1, x_1, y_1, z_1)$ and a second event at $E_2 = (ct_2, x_2, y_2, z_2)$. The *spacetime interval*, which we denote by $(\Delta S)^2$, is given by

$$(\Delta S)^2 = c^2 (t_1 - t_2)^2 - (x_1 - x_2)^2 - (y_1 - y_2)^2 - (z_1 - z_2)^2$$

or more concisely by

$$(\Delta S)^2 = c^2 (\Delta t)^2 - (\Delta x)^2 - (\Delta y)^2 - (\Delta z)^2 \tag{1.11}$$

An interval can be designated *timelike*, *spacelike*, or *null* if $(\Delta S)^2 > 0$, $(\Delta S)^2 < 0$, or $(\Delta S)^2 = 0$, respectively. If the distance between two events is infinitesimal, i.e., $x_1 = x, x_2 = x + dx \Rightarrow \Delta x = x + dx - x = dx$, etc., then the interval is given by

$$ds^2 = c^2 dt^2 - dx^2 - dy^2 - dz^2 \tag{1.12}$$

The *proper time*, which is the time measured by an observer's own clock, is defined to be

$$d\tau^2 = -ds^2 = -c^2 dt^2 + dx^2 + dy^2 + dz^2 \tag{1.13}$$

This is all confusing enough, but to make matters worse different physicists use different sign conventions. Some write $ds^2 = -c^2 dt^2 + dx^2 + dy^2 + dz^2$, and in that case the sign designations for timelike and spacelike are reversed. Once you get familiar with this it is not such a big deal, just keep track of what the author is using to solve a particular problem.

The interval is important because it is an *invariant* quantity. The meaning of this is as follows: While observers in motion with respect to each other will assign different values to space and time differences, they *all* agree on the value of the interval.

# Postulates of Special Relativity

In a nutshell, special relativity is based on three simple postulates.

**Postulate 1:** *The principle of relativity.*
The laws of physics are the same in all inertial reference frames.

**Postulate 2:** *The speed of light is invariant.*
All observers in inertial frames will measure the same speed of light, regardless of their state of motion.

**Postulate 3:** *Uniform motion is invariant.*
A particle at rest or with constant velocity in one inertial frame will be at rest or have constant velocity in all inertial frames.

We now use these postulates to seek a replacement of the Galilean transformations with the caveat that the speed of light is invariant. Again, we consider two frames $F$ and $F'$ in the standard configuration (Fig. 1-2). The first step is to consider Postulate 3. Uniform motion is represented by straight lines, and what this postulate tells us is that straight lines in one frame should map into straight lines in another frame that is moving uniformly with respect to it. This is another way of saying that the transformation of coordinates must be linear. A linear transformation can be described using matrices. If we write the coordinates of frame $F$ as a column vector

$$\begin{bmatrix} ct \\ x \\ y \\ z \end{bmatrix}$$

then the coordinates of $F'$ are related to those of $F$ via a relationship of the form

$$\begin{bmatrix} ct' \\ x' \\ y' \\ z' \end{bmatrix} = L \begin{bmatrix} ct \\ x \\ y \\ z \end{bmatrix} \tag{1.14}$$

where $L$ is a $4 \times 4$ matrix. Given that the two frames are in standard configuration, the $y$ and $z$ axes are coincident, which means that

$$y' = y \quad \text{and} \quad z' = z$$

To get the form of the transformation, we rely on the invariance of the speed of light as described in postulate 2. Imagine that at time $t = 0$ a flash of light is emitted from the origin. The light moves outward from the origin as a spherical wavefront described by

$$c^2 t^2 = x^2 + y^2 + z^2 \tag{1.15}$$

Subtracting the spatial components from both sides, this becomes

$$c^2 t^2 - x^2 - y^2 - z^2 = 0$$

Invariance of the speed of light means that for an observer in a frame $F'$ moving at speed $v$ with respect to $F$, the flash of light is described as

$$c^2 t'^2 - x'^2 - y'^2 - z'^2 = 0$$

These are equal, and so

$$c^2 t^2 - x^2 - y^2 - z^2 = c^2 t'^2 - x'^2 - y'^2 - z'^2$$

Since $y' = y$ and $z' = z$, we can write

$$c^2 t^2 - x^2 = c^2 t'^2 - x'^2 \tag{1.16}$$

Now we use the fact that the transformation is linear while leaving $y$ and $z$ unchanged. The linearity of the transformation means it must have the form

$$\begin{aligned} x' &= Ax + Bct \\ ct' &= Cx + Dct \end{aligned} \tag{1.17}$$

We can implement this with the following matrix [see (1.14)]:

$$L = \begin{pmatrix} D & C & 0 & 0 \\ B & A & 0 & 0 \\ 0 & 0 & 1 & 0 \\ 0 & 0 & 0 & 1 \end{pmatrix}$$

Using (1.17), we rewrite the right side of (1.16) as follows:

$$x'^2 = (Ax + Bct)^2 = A^2x^2 + 2ABctx + B^2c^2t^2$$
$$c^2t'^2 = (Cx + Dct)^2 = C^2x^2 + 2CDctx + D^2c^2t^2$$

$$\Rightarrow c^2t'^2 - x'^2 = C^2x^2 + 2CDctx + D^2c^2t^2 - A^2x^2 - 2ABctx - B^2c^2t^2$$
$$= c^2\left(D^2 - B^2\right)t^2 - \left(A^2 - C^2\right)x^2 + 2\left(CD - AB\right)ctx$$

This must be equal to the left side of (1.16). Comparison leads us to conclude that

$$CD - AB = 0$$
$$\Rightarrow CD = AB$$

$$D^2 - B^2 = 1$$
$$A^2 - C^2 = 1$$

To obtain a solution, we recall that $\cosh^2 \phi - \sinh^2 \phi = 1$. Therefore we make the following identification:

$$A = D = \cosh \phi \qquad (1.18)$$

In some sense we would like to think of this transformation as a rotation. A rotation leads to a transformation of the form

$$x' = x \cos \phi - y \sin \phi$$
$$y' = -x \sin \phi + y \cos \phi$$

In order that (1.17) have a similar form, we take

$$B = C = -\sinh \phi \qquad (1.19)$$

With $A$, $B$, $C$, and $D$ determined, the transformation matrix is

$$L = \begin{pmatrix} \cos h\phi & -\sin h\phi & 0 & 0 \\ -\sin h\phi & \cos h\phi & 0 & 0 \\ 0 & 0 & 1 & 0 \\ 0 & 0 & 0 & 1 \end{pmatrix} \qquad (1.20)$$

Now we solve for the parameter $\phi$, which is called the *rapidity*. To find a solution, we note that when the origins of the two frames are coincident;

that is, when $x' = 0$, we have $x = vt$. Using this condition together with (1.17), (1.18), and (1.19), we obtain

$$x' = 0 = x \cosh \phi - ct \sinh \phi = vt \cosh \phi - ct \sinh \phi$$
$$= t (v \cosh \phi - c \sinh \phi)$$

and so we have $v \cosh \phi - c \sinh \phi = 0$, which means that

$$v \cosh \phi = c \sinh \phi$$
$$\Rightarrow \frac{\sinh \phi}{\cosh \phi} = \tanh \phi = \frac{v}{c} \qquad (1.21)$$

This result can be used to put the Lorentz transformations into the form shown in elementary textbooks. We have

$$x' = \cosh \phi x - \sinh \phi ct$$
$$ct' = - \sinh \phi x + \cosh \phi ct$$

Looking at the transformation equation for $t'$ first, we have

$$lct' = - \sinh \phi x + \cosh \phi ct = \cosh \phi \left( \frac{- \sinh \phi}{\cosh \phi} x + ct \right)$$
$$= \cosh \phi \left( - \tanh \phi x + ct \right)$$
$$= \cosh \phi \left( ct - \frac{v}{c} x \right)$$
$$= c \cosh \phi \left( t - \frac{v}{c^2} x \right)$$

$$\Rightarrow t' = \cosh \phi \left( t - \frac{v}{c^2} x \right)$$

We also find

$$x' = \cosh \phi x - \sinh \phi ct$$
$$= \cosh \phi \left( x - \tanh \phi ct \right)$$
$$= \cosh \phi \left( x - vt \right)$$

Now let's do a little trick using the hyperbolic cosine function, using

$$\cosh \phi = \frac{\cosh \phi}{1} = \frac{\cosh \phi}{\sqrt{1}} = \frac{\cosh \phi}{\sqrt{\cosh^2 \phi - \sinh^2 \phi}}$$

$$= \frac{1}{(1/\cosh \phi)} \frac{1}{\sqrt{\cosh^2 \phi - \sinh^2 \phi}}$$

$$= \frac{1}{\sqrt{(1/\cosh^2 \phi)(\cosh^2 \phi - \sinh^2 \phi)}}$$

$$= \frac{1}{\sqrt{1 - \tanh^2 \phi}} = \frac{1}{\sqrt{1 - v^2/c^2}}$$

This is none other than the definition used in elementary textbooks:

$$\gamma = \frac{1}{\sqrt{1 - v^2/c^2}} = \cosh \phi \qquad (1.22)$$

And so we can write the transformations in the familiar form:

$$t' = \gamma \left(t - vx/c^2\right), \qquad x' = \gamma \left(x - vt\right), \qquad y' = y, \qquad z' = z \qquad (1.23)$$

It is often common to see the notation $\beta = v/c$.

# Three Basic Physical Implications

There are three physical consequences that emerge immediately from the Lorentz transformations. These are time dilation, length contraction, and a new rule for composition of velocities.

## TIME DILATION

Imagine that two frames are in the standard configuration so that frame $F'$ moves at uniform velocity $v$ with respect to frame $F$. An interval of time $\Delta t'$

as measured by an observer in $F'$ is seen by $F$ to be

$$\Delta t = \frac{1}{\sqrt{1 - \beta^2}} \Delta t' = \gamma \Delta t'$$

that is, the clock of an observer whose frame is $F'$ runs slow relative to the clock of an observer whose frame is $F$ by a factor of $\sqrt{1 - \beta^2}$.

## LENGTH CONTRACTION

We again consider two frames in the standard configuration. At fixed time $t$, measured distances along the direction of motion are related by

$$\Delta x' = \frac{1}{\sqrt{1 - \beta^2}} \Delta x$$

that is, distances in $F'$ along the direction of motion appear to be shortened in the direction of motion by a factor of $\sqrt{1 - \beta^2}$.

## COMPOSITION OF VELOCITIES

Now imagine three frames of reference in the standard configuration. Frame $F'$ moves with velocity $v_1$ with respect to frame $F$, and frame $F''$ moves with velocity $v_2$ with respect to frame $F'$. Newtonian physics tells us that frame $F''$ moves with velocity $v_3 = v_1 + v_2$ with respect to frame $F$, a simple velocity addition law. However, if the velocities are significant fraction of the speed of light, this relation does not hold. To obtain the correct relation, we simply compose two Lorentz transformations.

**EXAMPLE 1-1**
Derive the relativistic velocity composition law.

**SOLUTION 1-1**
Using $\beta = v/c$, the matrix representation of a Lorentz transformation between $F$ and $F'$ is

$$L_1 = \begin{pmatrix} \frac{1}{\sqrt{1-\beta_1^2}} & \frac{-\beta_1}{\sqrt{1-\beta_1^2}} & 0 & 0 \\ \frac{-\beta_1}{\sqrt{1-\beta_1^2}} & \frac{1}{\sqrt{1-\beta_1^2}} & 0 & 0 \\ 0 & 0 & 1 & 0 \\ 0 & 0 & 0 & 1 \end{pmatrix} \tag{1.24}$$

The transformation between $F'$ and $F''$ is

$$L_2 = \begin{pmatrix} \dfrac{1}{\sqrt{1-\beta_2^2}} & \dfrac{-\beta_2}{\sqrt{1-\beta_2^2}} & 0 & 0 \\ \dfrac{-\beta_2}{\sqrt{1-\beta_2^2}} & \dfrac{1}{\sqrt{1-\beta_2^2}} & 0 & 0 \\ 0 & 0 & 1 & 0 \\ 0 & 0 & 0 & 1 \end{pmatrix} \qquad (1.25)$$

We can obtain the Lorentz transformation between $F$ and $F''$ by computing $L_2 L_1$ using (1.25) and (1.24). We find

$$\begin{pmatrix} \dfrac{1}{\sqrt{1-\beta_1^2}} & \dfrac{-\beta_1}{\sqrt{1-\beta_1^2}} & 0 & 0 \\ \dfrac{-\beta_1}{\sqrt{1-\beta_1^2}} & \dfrac{1}{\sqrt{1-\beta_1^2}} & 0 & 0 \\ 0 & 0 & 1 & 0 \\ 0 & 0 & 0 & 1 \end{pmatrix} \begin{pmatrix} \dfrac{1}{\sqrt{1-\beta_2^2}} & \dfrac{-\beta_2}{\sqrt{1-\beta_2^2}} & 0 & 0 \\ \dfrac{-\beta_2}{\sqrt{1-\beta_2^2}} & \dfrac{1}{\sqrt{1-\beta_2^2}} & 0 & 0 \\ 0 & 0 & 1 & 0 \\ 0 & 0 & 0 & 1 \end{pmatrix}$$

$$= \begin{pmatrix} \dfrac{1+\beta_1\beta_2}{\sqrt{(1-\beta_1^2)(1-\beta_2^2)}} & \dfrac{-(\beta_1+\beta_2)}{\sqrt{(1-\beta_1^2)(1-\beta_2^2)}} & 0 & 0 \\ \dfrac{-(\beta_1+\beta_2)}{\sqrt{(1-\beta_1^2)(1-\beta_2^2)}} & \dfrac{1+\beta_1\beta_2}{\sqrt{(1-\beta_1^2)(1-\beta_2^2)}} & 0 & 0 \\ 0 & 0 & 1 & 0 \\ 0 & 0 & 0 & 1 \end{pmatrix}$$

This matrix is itself a Lorentz transformation, and so must have the form

$$L_3 = \begin{pmatrix} \dfrac{1}{\sqrt{1-\beta_3^2}} & \dfrac{-\beta_3}{\sqrt{1-\beta_3^2}} & 0 & 0 \\ \dfrac{-\beta_3}{\sqrt{1-\beta_3^2}} & \dfrac{1}{\sqrt{1-\beta_3^2}} & 0 & 0 \\ 0 & 0 & 1 & 0 \\ 0 & 0 & 0 & 1 \end{pmatrix}$$

We can find $\beta_3$ by equating terms. We need to consider only one term, so pick the terms in the upper left corner of each matrix and set

$$\frac{1+\beta_1\beta_2}{\sqrt{(1-\beta_1^2)(1-\beta_2^2)}} = \frac{1}{\sqrt{1-\beta_3^2}}$$

Let's square both sides:

$$\frac{(1 + \beta_1\beta_2)^2}{\left(1 - \beta_1^2\right)\left(1 - \beta_2^2\right)} = \frac{1}{1 - \beta_3^2}$$

Inverting this, we get

$$\frac{\left(1 - \beta_1^2\right)\left(1 - \beta_2^2\right)}{(1 + \beta_1\beta_2)^2} = 1 - \beta_3^2$$

Now we isolate the desired term $\beta_3$:

$$\beta_3^2 = 1 - \frac{\left(1 - \beta_1^2\right)\left(1 - \beta_2^2\right)}{(1 + \beta_1\beta_2)^2}$$

On the right-hand side, we set $1 = \frac{(1+\beta_1\beta_2)^2}{(1+\beta_1\beta_2)^2}$, and rewrite the expression as

$$\beta_3^2 = \frac{(1 + \beta_1\beta_2)^2}{(1 + \beta_1\beta_2)^2} - \frac{\left(1 - \beta_1^2\right)\left(1 - \beta_2^2\right)}{(1 + \beta_1\beta_2)^2} = \frac{(1 + \beta_1\beta_2)^2 - \left(1 - \beta_1^2\right)\left(1 - \beta_2^2\right)}{(1 + \beta_1\beta_2)^2}$$

Now we expand the terms on the right to get

$$\beta_3^2 = \frac{(1 + \beta_1\beta_2)^2 - \left(1 - \beta_1^2\right)\left(1 - \beta_2^2\right)}{(1 + \beta_1\beta_2)^2}$$

$$= \frac{1 + 2\beta_1\beta_2 + \beta_1^2\beta_2^2 - \left(1 - \beta_1^2 - \beta_2^2 + \beta_1^2\beta_2^2\right)}{(1 + \beta_1\beta_2)^2}$$

This simplifies to

$$\beta_3^2 = \frac{2\beta_1\beta_2 + \beta_1^2 + \beta_2^2}{(1 + \beta_1\beta_2)^2} = \frac{(\beta_1 + \beta_2)^2}{(1 + \beta_1\beta_2)^2}$$

Taking square roots of both sides, we obtain

$$\beta_3 = \frac{\beta_1 + \beta_2}{1 + \beta_1\beta_2}$$

Now we use $\beta = v/c$ to write

$$\frac{v_3}{c} = \frac{v_1/c + v_2/c}{1 + v_1 v_2/c^2}$$

Multiplication of both sides by $c$ gives the velocity composition law

$$v_3 = \frac{v_1 + v_2}{1 + v_1 v_2/c^2} \tag{1.26}$$

# Light Cones and Spacetime Diagrams

It is often helpful to visualize spacetime by considering a flash of light emitted at the origin. As we discussed earlier, such a flash of light is described by a spherical wavefront. However, our minds cannot visualize four dimensions and it's not possible to draw it on paper. So we do the next best thing and suppress one or more of the spatial dimensions. Let's start with the simplest of all cases, where we suppress two spatial dimensions.

Doing so gives us a simple spacetime diagram (see Fig. 1- for the basic idea). In a spacetime diagram, the vertical axis represents time while one or two horizontal axes represent space. It is convenient to work in units where $c = 1$. The upper half plane where $t > 0$ represents events to the future of the origin. Past events are found in the lower half plane where $t < 0$. The motion of light in

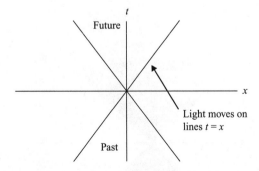

**Fig. 1-3.** The division of spacetime into future and past regions. Light rays move on the lines $t = x$ and $t = -x$. These lines define the light cone while the origin is some event $E$ in spacetime. The inside of the light cone in the lower half plane is the past of $E$, where we find all events in the past that could affect $E$. The future of $E$, which contains all events that can be causally affected by $E$, is inside the light cone defined in the upper half plane. Regions outside the light cone are called *spacelike*.

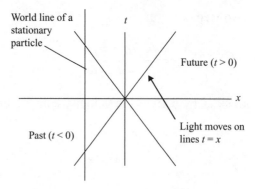

**Fig. 1-4.** The worldline of a stationary particle is a straight line.

such a diagram is then described by lines that make a 45°angle with the $x$-axis, i.e., lines that satisfy

$$t^2 = x^2$$

In the first quadrant, the paths of light rays are described by the lines $t = x$.

    The motion of a particle through spacetime as depicted in a spacetime diagram (Fig. 1-4) is called a *worldline*. The simplest of all particle motion is a particle just sitting somewhere. To indicate the worldline of a stationary particle on a spacetime diagram, we simply draw a straight vertical line that passes through the $x$-axis at the spatial location of the particle. This makes sense because the particle is located at some position $x$ that does not change, but time keeps marching forward.

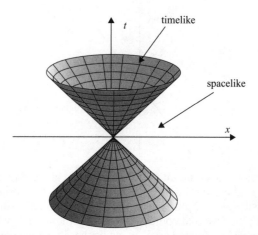

**Fig. 1-5.** A light cone with two spatial dimensions.

Special relativity tells us that a particle cannot move faster than the speed of light. On a spacetime diagram, this is indicated by the fact that particle motion is restricted to occur only *inside* the light cone. The region inside the light cone is called *timelike*. Regions outside the light cone, which are casually unrelated to the event $E$, are called *spacelike*.

More insight is gained when we draw a spacetime diagram showing two spatial dimensions. This is shown in Fig. 1-5.

# Four Vectors

In special relativity we work with a unified entity called *spacetime* rather than viewing space as an arena with time flowing in the background. As a result, a vector is going to have a time component in addition to the spatial components we are used to. This is called a *four vector*. There are a few four vectors that are important in relativity. The first is the four velocity, which is denoted by $\vec{u}$ and has components

$$\vec{u} = \left( \frac{dt}{d\tau}, \frac{dx}{d\tau}, \frac{dy}{d\tau}, \frac{dz}{d\tau} \right)$$

We can differentiate this expression again with respect to proper time to obtain the four acceleration $\vec{a}$. The norm or magnitude squared of $\vec{v} \cdot \vec{v}$ tells us if a vector is timelike, spacelike, or null. The definition will depend on the sign convention used for the line element. If we take $ds^2 = c^2 dt^2 - dx^2 - dy^2 - dz^2$, then if $\vec{v} \cdot \vec{v} > 0$ we say that $\vec{v}$ is timelike. If $\vec{v} \cdot \vec{v} < 0$, we say $\vec{v}$ is spacelike. When $\vec{v} \cdot \vec{v} = 0$, we say that $\vec{v}$ is null. The four velocity is always a timelike vector. Following this convention, we compute the dot product as

$$\vec{v} \cdot \vec{v} = (v_t)^2 - (v_x)^2 - \left(v_y\right)^2 - (v_z)^2$$

The dot product is an invariant, and so has the same value in all Lorentz frames. If a particle is moving in some frame $F$ with velocity $\vec{u}$, we find that energy and momentum conservation can be achieved if we take the energy to be $E = \gamma m_0 c^2$ and define the four momentum $\vec{p}$ using

$$\vec{p} = \gamma m_0 \vec{u} \tag{1.27}$$

where $m_0$ is the particle's rest mass and $\vec{u}$ is the velocity four vector. In a more familiar form, the momentum four vector is given by $\vec{p} = \left( E/c, \ p_x, \ p_y, \ p_z \right)$.

For two frames in the standard configuration, components of the momentum four vector transform as

$$p'_x = \gamma p_x - \beta \gamma E/c$$
$$p'_y = p_y$$
$$p'_z = p_z \tag{1.28}$$
$$E' = \gamma E - \beta \gamma c p_x$$

Using our sign convention for the dot product, we find

$$\vec{p} \cdot \vec{p} = E^2/c^2 - p_x^2 - p_y^2 - p_z^2 = E^2/c^2 - p^2$$

Remember, the dot product is a Lorentz invariant. So we can find its value by calculating it in any frame we choose. In the rest frame of the particle, the momentum is zero (and so $p^2 = 0$) and the energy is given by Einstein's famous formula $E = m_0 c^2$. Therefore

$$\vec{p} \cdot \vec{p} = m_0^2 c^2$$

Putting these results together, we obtain

$$E^2 - p^2 c^2 = m_0^2 c^4 \tag{1.29}$$

# Relativistic Mass and Energy

The *rest mass* of a particle is the mass of the particle as measured in the instantaneous rest frame of that particle. We designate the rest mass by $m_0$. If a particle is moving with respect to an observer $O$ with velocity $v$ then $O$ measures the mass of the particle as

$$m = \frac{m_0}{\sqrt{1 - v^2/c^2}} = \gamma m_0 \tag{1.30}$$

Now consider the binomial expansion, which is valid for $|x| < 1$,

$$(1 + x)^n \approx 1 + nx$$

For $n = -1/2$,

$$(1 - x)^{-1/2} \approx 1 + \frac{1}{2}x$$

Setting $x = v^2/c^2$ in (1.30), we obtain

$$m = \frac{m_0}{\sqrt{1 - v^2/c^2}} \approx m_0 \left( 1 + \frac{1}{2}\frac{v^2}{c^2} \right) = m_0 + \frac{1}{2}m_0\frac{v^2}{c^2}$$

Multiplying through by $c^2$ we obtain an expression that relates the relativistic energy to the rest mass energy plus the Newtonian kinetic energy of the particle:

$$mc^2 = m_0c^2 + \frac{1}{2}m_0v^2$$

# Quiz

1. An inertial frame is best described by
   (a) one that moves with constant acceleration
   (b) a frame that is subject to constant forces
   (c) a frame that moves with constant velocity
   (d) a frame that is subject to galilean transformations

2. The proper time $d\tau^2$ is related to the interval via
   (a) $d\tau^2 = -ds^2$
   (b) $d\tau^2 = ds^2$
   (c) $d\tau^2 = -c^2 ds^2$
   (d) $\frac{d\tau^2 = -ds^2}{c^2}$

3. The principle of relativity can be best stated as
   (a) The laws of physics differ only by a constant in all reference frames differing by a constant acceleration.
   (b) The laws of physics change from one inertial reference frames to another.
   (c) The laws of physics are the same in all inertial reference frames.

4. Rapidity is defined using which of the following relationships?
   (a) $\tanh \phi = \frac{v}{c}$
   (b) $\tan \phi = \frac{v}{c}$
   (c) $\tanh \phi = -\frac{v}{c}$
   (d) $v \tanh \phi = c$

5. Consider two frames in standard configuration. The phenomenon of length contraction can be described by saying that distances are shortened by a factor of

   (a) $\sqrt{1 + \beta^2}$

   (b) $\sqrt{1 - \beta^2}$

   (c) $-\dfrac{\sqrt{1 + \beta^2}}{c^2}$

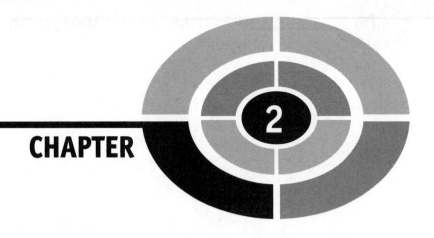

# Vectors, One Forms, and the Metric

In this chapter we describe some of the basic objects that we will encounter in our study of relativity. While you are no doubt already familiar with vectors from studies of basic physics or calculus, we are going to be dealing with vectors in a slightly different light. We will also encounter some mysterious objects called *one forms,* which themselves form a vector space. Finally, we will learn how a geometry is described by the *metric*.

## Vectors

A vector is a quantity that has both magnitude and direction. Graphically, a vector is drawn as a directed line segment with an arrow head. The length of the arrow is a graphic representation of its magnitude. (See Figure 2-1).

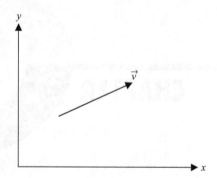

**Fig. 2-1.** Your basic vector, a directed line segment drawn in the $x$–$y$ plane.

The reader is no doubt familiar with the graphical methods of vector addition, scalar multiplication, and vector subtraction. We will not review these methods here because we will be looking at vectors in a more abstract kind of way. For our purposes, it is more convenient to examine vectors in terms of their *components*. In the plane or in ordinary three-dimensional space, the components of a vector are the projections of the vector onto the coordinate axes. In Fig. 2-2, we show a vector in the $x$–$y$ plane and its projections onto the $x$ and $y$ axes.

The components of a vector are numbers. They can be arranged as a list. For example, in 3 dimensions, the components of a vector $A$ can be written as $A = (A_x, A_y, A_z)$. More often, one sees a vector written as an expansion in terms of a set of basis vectors. A basis vector has unit length and points along the direction of a coordinate axis. Elementary physics books typically denote the basis for cartesian coordinates by $(\hat{i}, \hat{j}, \hat{k})$, and so in ordinary three-dimensional

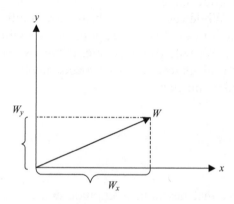

**Fig. 2-2.** A vector $W$ in the $x$–$y$ plane, resolved into its components $W_x$ and $W_y$. These are the projections of $W$ onto the $x$ and $y$ axes.

cartesian space, we can write the vector $A$ as

$$\vec{A} = A_x \hat{i} + A_y \hat{j} + A_z \hat{k}$$

In more advanced texts a different notation is used:

$$\vec{A} = A_x \hat{x} + A_y \hat{y} + A_z \hat{z}$$

This has some advantages. First of all, it clearly indicates which basis vector points along which direction (the use of $(\hat{i}, \hat{j}, \hat{k})$ may be somewhat mysterious to some readers). Furthermore, it provides a nice notation that allows us to define a vector in a different coordinate system. After all, we could write the same vector in spherical coordinates:

$$\vec{A} = A_r \hat{r} + A_\theta \hat{\theta} + A_\phi \hat{\phi}$$

There are two important things to note here. The first is that the vector $A$ is a geometric object that exists independent of coordinate system. To get its components we have to choose a coordinate system that we want to use to represent the vector. Second, the numbers that represent the vector in a given coordinate system, the components of the vector, are in general different depending on what coordinate system we use to represent the vector. So for the example we have been using so far $(A_x, A_y, A_z) \neq (A_r, A_\theta, A_\phi)$.

# New Notation

We are now going to use a different notation that will turn out to be a bit more convenient for calculation. First, we will identify the coordinates by a set of labeled indices. The letter $x$ is going to be used to represent all coordinates, but we will write it with a superscript to indicate which particular coordinate we are referring to. For example, we will write $y$ as $x^2$. It is important to recognize that the "2" used here is just a label and is not an exponent. In other words,

$$y^2 = \left(x^2\right)^2$$

and so on. For the entire set of cartesian coordinates, we make the following identification:

$$(x, y, z) \rightarrow \left(x^1, x^2, x^3\right)$$

With this identification, when we write $x^1$ we mean $x$, while $x^2$ means $y$ and so forth. This identification is entirely general, and we can use it to represent another coordinate system such as cylindrical coordinates. What represents what will be made clear from the context. Also, it will be convenient to move back and forth between this representation and the ones you are used to.

When using more than one coordinate system, or more importantly when considering transformations between coordinate system, we need a way to apply this representation to two different sets of coordinates. One way to do so is to put primes on the indices. As an example, suppose we are considering cartesian and spherical coordinates within the context of the same problem. If we label cartesian coordinates with $(x, y, z) \rightarrow \left(x^1, x^2, x^3\right)$, then we add primes to the labels used for spherical coordinates and write

$$(r, \theta, \phi) \rightarrow \left(x^{1'}, x^{2'}, x^{3'}\right)$$

We will label the components of a vector in the same way, with a raise index. In the current example, the components of a vector $A$ in cartesian coordinates would be given by

$$A = \left(A^1, A^2, A^3\right)$$

While primed coordinates would represent the same vector in spherical coordinates

$$A = \left(A^{1'}, A^{2'}, A^{3'}\right)$$

another useful notation change is to write partial derivatives in a succinct way. We write

$$\frac{\partial f}{\partial x} = \partial_x f$$

or, using indices, we write

$$\frac{\partial}{\partial x^a} \rightarrow \partial_a$$

There are two reasons for writing things in this apparently obscure fashion. The first is that when you do relativity a great deal of writing is required. Hey! Anything that can cut back on that is a good thing. But we will see that the placement of the indices (as shown in $\partial_a$) will prove to be more enlightening and useful. Unfortunately, at this point we're not ready to say why, so you'll just have to take my word for it and just keep in mind what the shorthand symbols mean.

# Four Vectors

In many cases, we are not going to work with specific components of an objectlike vector, but will rather work with a general component. In this book we will label components of objectlike vectors with lowercase Latin letters. For example, we can refer to the vector $A$ by

$$A^a$$

As we get involved with relativity, a vector will have space and time components (it will be a *four vector*). In that case, the *time component* of the vector will be labeled by the index 0, and so the components of a four vector $V$ will be given by

$$V = \left(V^0, V^1, V^2, V^3\right)$$

Vector addition can be described in the following way:

$$\vec{A} + \vec{B} = \left(A^0 + B^0, A^1 + B^1, A^2 + B^2, A^3 + B^3\right)$$

while scalar multiplication can be written as

$$\alpha \vec{A} = \left(\alpha A^0, \alpha A^1, \alpha A^2, \alpha A^3\right)$$

Keep in mind that some authors prefer to use (1, 2, 3, 4) as their indices, using 4 to label time. We will stick to using 0 to label the time coordinate, however.

Many authors prefer to use the following labeling convention. When all four components (space and time) in an expression are used, Greek letters are used for indices. So if an author writes

$$T^{\mu}{}_{\nu}{}^{\gamma}$$

the indices $\mu, \nu, \gamma$ can range over (0, 1, 2, 3). In this context, Latin indices are reserved for spatial components only, and so in the expression

$$S^i{}_j$$

the indices $i, j$ will range only over $(1, 2, 3)$. Typing a lot of Greek symbols is a bit of extra work, so we will stick to using Latin indices all the time. When possible, we will use early letters $(a, b, c, \ldots)$ to range over all possible values $(0, 1, 2, 3)$ and use letters from the middle of the alphabet such as $i$, $j$ to range only over the spatial components $(1, 2, 3)$.

# The Einstein Summation Convention

The *Einstein summation convention* is a way to write sums in a shorthand format. When the same index appears twice in an expression, once raised and once lowered, a sum is implied. As a specific example,

$$\sum_{i=1}^{3} A_i B^i \rightarrow A_i B^i = A_1 B^1 + A_2 B^2 + A_3 B^3$$

Another example is that $S^a T_{ab}$ is shorthand for the expression

$$\sum_{a=0}^{3} S^a T_{ab}$$

An index that is summed over is called a *dummy index*, and can be replaced by another label if it is convenient. For example,

$$S^a T_{ab} = S^c T_{cb}$$

The index $b$ in the previous expressions is not involved in the sum operations. Such an index is known as a *free* index. A free index will typically appear on both sides of an expression. For example, consider the following equation:

$$A^{a'} = \Lambda^{a'}{}_b A^b$$

In this expression, $b$ is once again a dummy index. The sum implied here means that

$$A^{a'} = \Lambda^{a'}{}_b A^b = \Lambda^{a'}{}_0 A^0 + \Lambda^{a'}{}_1 A^1 + \Lambda^{a'}{}_2 A^2 + \Lambda^{a'}{}_3 A^3$$

The other index found in this expression, $a'$, is a free index. If we elect to change the free index, it must be changed on both sides of the equation. Therefore it would be valid to make the change $a' \to b'$, provided that we make this change on both sides; i.e.,

$$A^{b'} = \Lambda^{b'}{}_b A^b$$

# Tangent Vectors, One Forms, and the Coordinate Basis

We will often label basis vectors with the notation $e_a$. Using the Einstein summation convention, a vector $V$ can be written in terms of some basis as

$$V = V^a e_a$$

In this context the notation $e_a$ makes sense, because we can use it in the summation convention (this would not be possible with the cumbersome $(\hat{i}, \hat{j}, \hat{k})$ for example).

In a given coordinate system, the basis vectors $e_a$ are tangent to the coordinate lines. (See Fig. 2-3 and Fig. 2-4) This is the reason why we can write basis vectors as partial derivatives in a particular coordinate direction (for an explanation, see Carroll, 2004). In other words, we take

$$e_a = \partial_a = \frac{\partial}{\partial x^a}$$

This type of basis is called a *coordinate basis*. This allows us to think of a vector as an operator, one that maps a function into a new function that is related to its derivative. In particular,

$$Vf = (V^a e_a) = V^a \partial_a f$$

A vector $V$ can be represented with *covariant* components $V_a$. This type of vector is called a *one form*. Basis one forms have raised indices and are often denoted by $\omega^a$. So we can write

$$\tilde{V} = V_a \omega^a$$

We have used a tilde to note that this is a one form and not an *ordinary* vector (but it is the same object, in a different representation). Later, we will see how to move between the two representations by raising and lowering indices with the

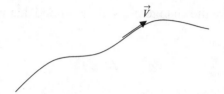

**Fig. 2-3.** A tangent vector to a curve.

metric. The basis one forms form a dual vector space to ordinary vectors, in the sense that the one forms constitute a vector space in their own right and the basis one forms map basis vectors to a number, the Kronecker delta function; i.e.,

$$\omega^a(e_b) = \delta^a_b \tag{2.1}$$

where

$$\delta^a_b = \begin{cases} 1 & a = b \\ 0 & \text{otherwise} \end{cases}$$

In a coordinate representation, the basis one forms are given by

$$\omega^a = \mathrm{d}x^a \tag{2.2}$$

With this representation, it is easy to see why (2.1) holds.

An arbitrary one form $\sigma_a$ maps a vector $V^a$ to a number via the scalar product

$$\sigma \cdot V = \sigma_a V^a$$

We can think of this either way: we can visualize vectors as maps that take one forms to the real numbers via the scalar product. More generally, we can define a $(p, q)$ *tensor* as a function that takes $p$ one forms and $q$ vectors as input and

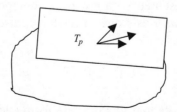

**Fig. 2-4.** An admittedly crude representation. The blob represents a manifold (basically a space of points). $T_p$ is the tangent space at a point $p$. The tangent vectors live here.

maps them to the real numbers. We can write a general tensor in the following way:

$$T = T_{abc\cdots}{}^{lmn\cdots} \omega^a \otimes \omega^b \otimes \omega^c \cdots e_l \otimes e_m \otimes e_n \cdots$$

We will have more to say about one forms, basis vectors, and tensors later.

# Coordinate Transformations

In relativity it is often necessary to change from one coordinate system to another, or from one frame to another. A transformation of this kind is implemented with a transformation matrix that we denote by $\Lambda^a{}_{b'}$. The placement of the indices and where we put the prime notation will depend on the particular transformation. In a coordinate transformation, the components of $\Lambda^a{}_{b'}$ are formed by taking the partial derivative of one coordinate with respect to the other. More specifically,

$$\Lambda^a{}_{b'} = \frac{\partial x^a}{\partial x^{b'}} \tag{2.3}$$

The easiest way to get a handle on how to apply this is to simply write the formulas down and apply them in a few cases. Basis vectors transform as

$$e_{a'} = \Lambda^b{}_{a'} e_b \tag{2.4}$$

We can do this because, as you know, it is possible to expand any vector in terms of some basis. What this relation gives us is an expansion of the basis vector $e_{a'}$ in terms of the old basis $e_b$. The components of $e_{a'}$ in the $e_b$ basis are given by $\Lambda^a{}_{b'}$. Note that we are denoting the new coordinates by primes and the old coordinates are unprimed indices.

**EXAMPLE 2-1**
Plane polar coordinates are related to cartesian coordinates by

$$x = r \cos\theta \quad \text{and} \quad y = r \sin\theta$$

Describe the transformation matrix that maps cartesian coordinates to polar coordinates, and write down the polar coordinate basis vectors in terms of the basis vectors of cartesian coordinates.

**SOLUTION 2-1**

Using $\Lambda^a{}_{b'} = \frac{\partial x^a}{\partial x^{b'}}$, the components of the transformation matrix are

$$\Lambda^x{}_r = \frac{\partial x}{\partial r} = \cos\theta \quad \text{and} \quad \Lambda^y{}_r = \frac{\partial y}{\partial r} = \sin\theta$$

$$\Lambda^x{}_\theta = \frac{\partial x}{\partial\theta} = -r\sin\theta \quad \text{and} \quad \Lambda^y{}_\theta = \frac{\partial y}{\partial\theta} = r\cos\theta$$

Using (2.4), we can write down the basis vectors in polar coordinates. We obtain

$$e_r = \Lambda^x{}_r e_x + \Lambda^y{}_r e_y = \cos\theta e_x + \sin\theta e_y$$
$$e_\theta = \Lambda^x{}_\theta e_x + \Lambda^y{}_\theta e_y = -r\sin\theta e_x + r\cos\theta e_y$$

The components of a vector transform in the opposite manner to that of a basis vector (this is why, an ordinary vector is sometimes called *contravariant*; it transforms contrary to the basis vectors). This isn't so surprising given the placement of the indices. In particular,

$$V^{a'} = \Lambda^{a'}{}_b V^b = \frac{\partial x^{a'}}{\partial x^b} V^b \tag{2.5}$$

The components of a one form transform as

$$\sigma_{a'} = \Lambda^b{}_{a'} \sigma_b \tag{2.6}$$

Basis one forms transform as

$$\omega^{a'} = dx^{a'} = \Lambda^{a'}{}_b dx^b \tag{2.7}$$

To find the way an arbitrary tensor transforms, you just use the basic rules for vectors and one forms to transform each index (OK we aren't transforming the indices, but you get the drift). Basically, you add an appropriate $\Lambda$ for each index. For example, the metric tensor, which we cover in the next section, transforms as

$$g_{a'b'} = \Lambda^c{}_{a'} \Lambda^d{}_{b'} g_{cd}$$

# The Metric

At the most fundamental level, one could say that geometry is described by the pythagorean theorem, which gives the distance between two points (see

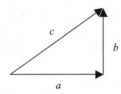

**Fig. 2-5.** The pythagorean theorem tells us that the lengths of *a, b, c* are related by
$$c = \sqrt{a^2 + b^2}.$$

Fig. 2-5). If we call $P_1 = (x_1, y_1)$ and $P_2 = (x_2, y_2)$, then the distance $d$ is given by

$$d = \sqrt{(x_1 - x_2)^2 + (y_1 - y_2)^2}$$

Graphically, of course, the pythagorean theorem gives the length of one side of a triangle in terms of the other two sides, as shown in Fig. 2-3.

As we have seen, this notion can be readily generalized to the flat spacetime of special relativity, where we must consider differences between spacetime events. If we label two events by $(t_1, x_1, y_1, z_1)$ and $(t_2, x_2, y_2, z_2)$, then we define the spacetime *interval* between the two events to be

$$(\Delta s)^2 = (t_1 - t_2)^2 - (x_1 - x_2)^2 - (y_1 - y_2)^2 - (z_1 - z_2)^2$$

Now imagine that the distance between the two events is infinitesimal. That is, if the first event is simply given by the coordinates $(t, x, y, z)$, then the second event is given by $(t + dt, x + dx, y + dy, z + dz)$. In this case, it is clear that the differences between each term will give us $(dt, dx, dy, dz)$. We write this infinitesimal interval as

$$ds^2 = dt^2 - dx^2 - dy^2 - dz^2$$

As we shall see, the form that the spacetime interval takes, which describes the geometry, is closely related to the gravitational field. Therefore it's going to become quite important to familiarize ourselves with the metric. In short, the interval $ds^2$, which often goes by the name the metric, contains information about how the given space (or spacetime) deviates from a flat space (or spacetime).

You are already somewhat familiar with the notion of a metric if you've studied calculus. In that kind of context, the quantity $ds^2$ is often called a *line element*. Let's quickly review some familiar line elements. The most familiar is that of ordinary cartesian coordinates. That one is given by

$$ds^2 = dx^2 + dy^2 + dz^2 \tag{2.8}$$

For spherical coordinates, we have

$$ds^2 = dr^2 + r^2 d\theta^2 + r^2 \sin\theta \, d\phi^2 \tag{2.9}$$

Meanwhile, the line element for cylindrical coordinates is

$$ds^2 = dr^2 + r^2 d\phi^2 + dz^2 \tag{2.10}$$

We can write these and other line elements in a succinct way by writing the coordinates with indices and summing. Generally, the line element is written as

$$ds^2 = g_{ab}(x) \, dx^a dx^b \tag{2.11}$$

where $g_{ab}(x)$ are the components of a second rank tensor (note that we can write these components as a matrix) called the metric. You can remember what this thing is by recalling that the components of the metric are given by the coefficient functions that multiply the differentials in the line element. For a metric describing ordinary three-dimensional space, these components are arranged into a matrix as follows:

$$g_{ij} = \begin{pmatrix} g_{11} & g_{12} & g_{13} \\ g_{21} & g_{22} & g_{23} \\ g_{31} & g_{32} & g_{33} \end{pmatrix}$$

For example, looking at (2.8), we see that for cartesian coordinates we can write

$$g_{ij} = \begin{pmatrix} 1 & 0 & 0 \\ 0 & 1 & 0 \\ 0 & 0 & 1 \end{pmatrix}$$

When dealing with spacetime, we take the convention that the time coordinate is labeled by $x^0$ and write the matrix representation of the metric as

$$g_{ab} = \begin{pmatrix} g_{00} & g_{01} & g_{02} & g_{03} \\ g_{10} & g_{11} & g_{12} & g_{13} \\ g_{20} & g_{21} & g_{22} & g_{23} \\ g_{30} & g_{31} & g_{32} & g_{33} \end{pmatrix}$$

For spherical coordinates, we make the identification $(r, \theta, \phi) \to (x^1, x^2, x^3)$ and using (2.9) write

$$g_{ij} = \begin{pmatrix} 1 & 0 & 0 \\ 0 & r^2 & 0 \\ 0 & 0 & r^2 \sin^2 \theta \end{pmatrix} \qquad (2.12)$$

For cylindrical coordinates, the matrix takes the form

$$g_{ij} = \begin{pmatrix} 1 & 0 & 0 \\ 0 & r^2 & 0 \\ 0 & 0 & 1 \end{pmatrix} \qquad (2.13)$$

In many cases, like the ones we have considered so far, the metric has components only along the diagonal. However, be aware that this is not always the case. For example, a metric can arise in the study of gravitational radiation that is called the *Bondi metric*. The coordinates used are $(u, r, \theta, \phi)$ and the line element can be written as

$$\begin{aligned} ds^2 &= \left( \frac{f}{r} e^{2\beta} - g^2 r^2 e^{2\alpha} \right) du^2 + 2e^{2\beta} du\, dr + 2gr^2 e^{2\alpha} du\, d\theta \\ &\quad - r^2 \left( e^{2\alpha} d\theta^2 + e^{-2\alpha} \sin^2 \theta d\phi^2 \right) \end{aligned} \qquad (2.14)$$

Here $f, g, \alpha, \beta$ are functions of the coordinates $(u, r, \theta, \phi)$. With these coordinates, we can write the matrix representation of the metric as

$$g_{ab} = \begin{pmatrix} g_{uu} & g_{ur} & g_{u\theta} & g_{u\phi} \\ g_{ru} & g_{rr} & g_{r\theta} & g_{r\phi} \\ g_{\theta u} & g_{\theta r} & g_{\theta\theta} & g_{\theta\phi} \\ g_{\phi u} & g_{\phi r} & g_{\phi\theta} & g_{\phi\phi} \end{pmatrix}$$

A good piece of information to keep in the back of your mind is that the metric is symmetric; i.e., $g_{ab} = g_{ba}$. This information is useful when writing down components of the metric associated with the mixed terms in the line element. For example, in this case we have

$$2e^{2\beta} du\, dr = e^{2\beta} du\, dr + e^{2\beta} dr\, du = g_{ur} du\, dr + g_{ru} dr\, du$$

$$2gr^2 e^{2\alpha} du\, d\theta = gr^2 e^{2\alpha} du\, d\theta + gr^2 e^{2\alpha} d\theta\, du = g_{u\theta} du\, d\theta + g_{\theta u} d\theta\, du$$

With this in mind, we write

$$
g_{ab} = \begin{pmatrix} \frac{f}{r}e^{2\beta} - g^2 r^2 e^{2\alpha} & e^{2\beta} & gr^2 e^{2\alpha} & 0 \\ e^{2\beta} & 0 & 0 & 0 \\ gr^2 e^{2\alpha} & 0 & -r^2 e^{2\alpha} & 0 \\ 0 & 0 & 0 & -r^2 e^{-2\alpha} \sin^2 \theta \end{pmatrix}
$$

The metric is a coordinate-dependent function, as can be seen from the examples discussed so far. Furthermore, recall that different sign conventions are used for space and time components. As an example, consider flat Minkowski space written with spherical coordinates. It is entirely appropriate to use

$$
ds^2 = dt^2 - dr^2 - r^2 d\theta^2 - r^2 \sin^2 \theta d\phi^2
$$

and it is equally appropriate to use

$$
ds^2 = -dt^2 + dr^2 + r^2 d\theta^2 + r^2 \sin^2 \theta d\phi^2
$$

The important thing is to make a choice at the beginning and stick with it for the problem being solved. When reading textbooks or research papers, be aware of the convention that the author is using. We will use both conventions from time to time so that you can get used to seeing both conventions.

# The Signature of a Metric

The sum of the diagonal elements in the metric is called the *signature*. If we have

$$
ds^2 = -dt^2 + dx^2 + dy^2 + dz^2
$$

then

$$
g_{ab} = \begin{pmatrix} -1 & 0 & 0 & 0 \\ 0 & 1 & 0 & 0 \\ 0 & 0 & 1 & 0 \\ 0 & 0 & 0 & 1 \end{pmatrix}
$$

The signature is found to be

$$
-1 + 1 + 1 + 1 = 2
$$

# The Flat Space Metric

By convention, the flat metric of Minkowski spacetime is denoted by $\eta_{ab}$. Therefore

$$\eta_{ab} = \begin{pmatrix} -1 & 0 & 0 & 0 \\ 0 & 1 & 0 & 0 \\ 0 & 0 & 1 & 0 \\ 0 & 0 & 0 & 1 \end{pmatrix}$$

provided that $ds^2 = -dt^2 + dx^2 + dy^2 + dz^2$.

# The Metric as a Tensor

So far we have casually viewed the metric as a collection of the coefficients found in a given line element. But as we mentioned earlier, the metric is a symmetric second rank tensor. Let's begin to think about it more in this light. In fact the metric **g**, which we sometimes loosely call the line element, is written formally as a sum over tensor products of basis one forms

$$\mathbf{g} = g_{ab} dx^a \otimes dx^b$$

First, note that the metric has an inverse, which is written with raised indices. The inverse is defined via the relationship

$$g_{ab}g^{bc} = \delta_a^c \tag{2.15}$$

where $\delta_a^c$ is the familiar (hopefully) Kronecker delta function.

When the metric is diagonal, this makes it easy to find the inverse. For example, looking at the metric for spherical coordinates (2.12), it is clear that all components $g_{ab} = 0$ when $a \neq b$. So using (2.15), we arrive at the following:

$$g^{rr}g_{rr} = 1 \;\Rightarrow\; g^{rr} = 1$$

$$g^{\theta\theta}g_{\theta\theta} = g^{\theta\theta}r^2 = 1 \;\Rightarrow\; g^{\theta\theta} = \frac{1}{r^2}$$

$$g^{\phi\phi}g_{\phi\phi} = g^{\phi\phi}r^2\sin^2\theta = 1 \;\Rightarrow\; g^{\phi\phi} = \frac{1}{r^2\sin^2\theta}$$

These components can be arranged in matrix form as

$$g^{ab} = \begin{pmatrix} 1 & 0 & 0 \\ 0 & \frac{1}{r^2} & 0 \\ 0 & 0 & \frac{1}{r^2 \sin^2 \theta} \end{pmatrix} \qquad (2.16)$$

# Index Raising and Lowering

In relativity it is often necessary to use the metric to manipulate expressions via index raising and lowering. That is, we can use the metric with lowered indices to lower an upper index present on another term in the expression, or use the metric with raised indices to raise a lower index present on another term. This probably sounds confusing if you have never done it before, so let's illustrate with an example. First, consider some vector $V^a$. We can use the metric to obtain the covariant components by writing

$$V_a = g_{ab} V^b \qquad (2.17)$$

Remember, the summation convention is in effect and so this expression is shorthand for

$$V_a = g_{ab} V^b = g_{a0} V^0 + g_{a1} V^1 + g_{a2} V^2 + g_{a3} V^3$$

Often, but not always, the metric will be diagonal and so only one of the terms in the sum will contribute. Indices can be raised in an analogous manner:

$$V^a = g^{ab} V_b \qquad (2.18)$$

Let's provide a simple illustration with an example.

**EXAMPLE 2-2**
Suppose we are working in spherical coordinates where a contravariant vector $X^a = (1, r, 0)$ and a covariant vector $Y_a = (0, -r^2, \cos^2 \theta)$. Find $X_a$ and $Y^a$.

**SOLUTION 2-2**
Earlier we showed that $g_{rr} = g^{rr} = 1$, $g_{\theta\theta} = r^2$, $g^{\theta\theta} = \frac{1}{r^2}$, $g_{\phi\phi} = r^2 \sin^2 \theta$, $g^{\phi\phi} = \frac{1}{r^2 \sin^2 \theta}$. Now

$$X_a = g_{ab} X^b$$

and so

$$X_r = g_{rr} X^r = (1)(1) = 1$$
$$X_\theta = g_{\theta\theta} X^\theta = (r^2)(r) = r^3$$
$$X_\phi = g_{\phi\phi} X^\phi = (r^2 \sin^2 \theta)(0) = 0$$

Therefore, we obtain $X_a = (1, r^3, 0)$. For the second case, we need to raise indices, so we write

$$Y^a = g^{ab} Y_b$$

This gives

$$Y^r = g^{rr} Y_r = (1)(0) = 0$$
$$Y^\theta = g^{\theta\theta} Y_\theta = \left(\frac{1}{r^2}\right)(-r^2) = -1$$
$$Y^\phi = g^{\phi\phi} Y_\phi = \left(\frac{1}{r^2 \sin^2 \theta}\right)(\cos^2 \theta) = \frac{\cos^2 \theta}{r^2 \sin^2 \theta} = \frac{\cot^2 \theta}{r^2}$$

and so $Y^a = \left(0, -1, \frac{\cot^2 \theta}{r^2}\right)$.

**EXAMPLE 2-3**

In this example, we consider a fictitious two-dimensional line element given by

$$ds^2 = x^2 dx^2 + 2dx\,dy - dy^2$$

Write down $g_{ab}$, $g^{ab}$ and then raise and lower indices on $V_a = (1, -1)$, $W^a = (0, 1)$.

**SOLUTION 2-3**

With lowered indices the metric can be written in matrix form as

$$g_{ab} = \begin{pmatrix} g_{xx} & g_{xy} \\ g_{yx} & g_{yy} \end{pmatrix}$$

We see immediately from the coefficients of $dx^2$ and $dy^2$ in the line element that

$$g_{xx} = x^2 \quad \text{and} \quad g_{yy} = -1$$

To obtain the terms with mixed indices, we use the symmetry of the metric to write $g_{xy} = g_{yx}$ and so

$$2dxdy = dxdy + dydx = g_{xy}dxdy + g_{yx}dydx$$

Therefore, we find that

$$g_{xy} = g_{yx} = 1$$

In matrix form, the metric is

$$g_{ab} = \begin{pmatrix} x^2 & 1 \\ 1 & -1 \end{pmatrix}$$

To obtain the inverse, we use the fact that multiplying the two matrices gives the identity matrix. In other words,

$$\begin{pmatrix} g^{xx} & g^{xy} \\ g^{yx} & g^{yy} \end{pmatrix} \begin{pmatrix} x^2 & 1 \\ 1 & -1 \end{pmatrix} = \begin{pmatrix} 1 & 0 \\ 0 & 1 \end{pmatrix}$$

Four equations can be obtained by carrying out ordinary matrix multiplication in the above expression. These are

$$g^{xx}x^2 + g^{xy} = 1$$
$$g^{xx} - g^{xy} = 0 \implies g^{xy} = g^{xx}$$
$$g^{xx}x^2 + g^{yy} = 0$$
$$g^{yy} = -x^2g^{xx}$$
$$g^{yx} - g^{yy} = 1$$

In addition, the symmetry of the metric provides the constraint $g^{xy} = g^{yx}$. Using $g^{xy} = g^{xx}$ in the first equation, we find

$$g^{xx}x^2 + g^{xx} = g^{xx}(1 + x^2) = 1$$
$$\implies g^{xx} = \frac{1}{1 + x^2}$$

and so

$$g^{xy} = g^{xx} = \frac{1}{1 + x^2} = g^{yx}$$

Finally, for the last term we get

$$g^{yy} = -x^2 g^{yx} = \frac{-x^2}{1+x^2}$$

Now with this information in place, we can raise and lower indices as desired. We find

$$V^a = g^{ab} V_b$$

$$\Rightarrow V^x = g^{xb} V_b = g^{xx} V_x + g^{xy} V_y = \frac{1}{1+x^2}(1) + \frac{1}{1+x^2}(-1) = 0$$

$$V^y = g^{yb} V_b = g^{yx} V_x + g^{yy} V_y = \frac{1}{1+x^2}(1) - \frac{x^2}{1+x^2}(-1) = \frac{1+x^2}{1+x^2} = 1$$

$$\Rightarrow V^a = (0, 1)$$

In the other case, we have

$$W_a = g_{ab} W^b$$

$$\Rightarrow W_x = g_{xb} W^b = g_{xx} W^x + g_{xy} W^y = \frac{1}{1+x^2}(0) + \frac{1}{1+x^2}(1) = \frac{1}{1+x^2}$$

$$W_y = g_{yb} W^b = g_{yx} W^x + g_{yy} W^y = \frac{1}{1+x^2}(0) + \left(\frac{-x^2}{1+x^2}\right)(1) = \frac{-x^2}{1+x^2}$$

$$\Rightarrow W_a = \left(\frac{1}{1+x^2}, \frac{-x^2}{1+x^2}\right)$$

# Index Gymnastics

Often we raise and lower indices with the metric in a more abstract fashion. The reason for doing this is to get equations in a more desirable form to derive some result, say. We will be seeing more of this later, so it will make more sense as we go along. Right now we will just provide a few examples that show how this works. We've already seen a bit of this with an ordinary vector

$$X^a = g^{ab} X_b$$

We can also apply this technique to a vector that is present in a more complicated expression, such as

$$X^a Y^c = g^{ab} X_b Y^c$$

or we can use it with higher rank tensors. Some examples are

$$S_b^a = g^{ac} S_{cb}$$
$$T^{ab} = g^{ac} T_c^b = g^{ac} g^{bd} T_{cd}$$
$$R_{abcd} = g_{ae} R_{bcd}^e$$

In flat spacetime, the metric $\eta_{ab}$ is used to raise and lower indices. When we begin to prove results involving tensors, this technique will be used frequently.

# The Dot Product

Earlier we briefly mentioned the scalar or dot product. The metric also tells us how to compute the dot or scalar product in a given geometry. In particular, the dot product is written as

$$V \cdot W = V_a W^a$$

Now we can use index raising and lowering to write the scalar product in a different way:

$$V \cdot W = V_a W^a = g_{ab} V^b W^a = g^{ab} V_a W_b$$

**EXAMPLE 2-4**
Consider the metric in plane polar coordinates with components given by

$$g_{ab} = \begin{pmatrix} 1 & 0 \\ 0 & r^2 \end{pmatrix} \quad \text{and} \quad g^{ab} = \begin{pmatrix} 1 & 0 \\ 0 & \frac{1}{r^2} \end{pmatrix}$$

Let $V^a = (1, 1)$ and $W_a = (0, 1)$. Find $V_a$, $W^a$, and $V \cdot W$.

**SOLUTION 2-4**
Proceeding in the usual manner, we find

$$V_a = g_{ab} V^b$$
$$\Rightarrow V_r = g_{rr} V^r = (1)(1) = 1$$
$$V_\theta = g_{\theta\theta} V^\theta = (r^2)(1) = r^2$$
$$\Rightarrow V_a = (1, r^2)$$

In an analogous manner, we obtain

$$W^a = g^{ab} W_b$$
$$\Rightarrow W^r = g^{rr} W_r = (1)(0) = 0$$
$$W^\theta = g^{\theta\theta} W_\theta = \left(\frac{1}{r^2}\right)(1) = \frac{1}{r^2}$$
$$\Rightarrow W^a = \left(0, \frac{1}{r^2}\right)$$

For the dot product, we find

$$V \cdot W = g_{ab} V^a W^b = g_{rr} V^r W^r + g_{\theta\theta} V^\theta W^\theta$$
$$= (1)(0) + \left(r^2\right)\left(\frac{1}{r^2}\right) = 0 + 1 = 1$$

As a check, we compute

$$V \cdot W = V^a W_a = V^r W_r + V^\theta W_\theta = (1)(0) + (1)(1) = 0 + 1 = 1$$

# Passing Arguments to the Metric

Thinking of a tensor as a map from vectors and one forms to the real numbers, we can think of the metric in different terms. Specifically, we can view the metric as a second rank tensor that accepts two vector arguments. The output is a real number, the dot product between the vectors

$$g(V, W) = V \cdot W$$

Looking at the metric tensor in this way, we see that the components of the metric tensor are found by passing the basis vectors as arguments. That is

$$g(e_a, e_b) = e_a \cdot e_b = g_{ab} \tag{2.19}$$

In flat space, we have $e_a \cdot e_b = \eta_{ab}$.

**EXAMPLE 2-5**
Given that the basis vectors in cartesian coordinates are orthnormal, i.e.,

$$\partial_x \cdot \partial_x = \partial_y \cdot \partial_y = \partial_z \cdot \partial_z = 1$$

with all other dot products vanishing, show that the dot products of the basis vectors in spherical polar coordinates give the components of the metric.

## SOLUTION 2-5

The basis vectors in spherical coordinates are written in terms of those of cartesian coordinates using the basis vector transformation law

$$e_{a'} = \Lambda^b{}_{a'} e_b$$

where the elements of the transformation matrix are given by

$$\Lambda^b{}_{a'} = \frac{\partial x^b}{\partial x^{a'}}$$

The coordinates are related in the familiar way:

$$x = r \sin\theta \cos\phi, \qquad y = r \sin\theta \sin\phi, \qquad z = r \cos\theta$$

We have

$$\partial_r = \frac{\partial r}{\partial x} \partial_x + \frac{\partial r}{\partial y} \partial_y + \frac{\partial r}{\partial z} \partial_z$$
$$= \sin\theta \cos\phi \partial_x + \sin\theta \sin\phi \partial_y + \cos\theta \partial_z$$

Therefore, the dot product is

$$g_{rr} = \partial_r \cdot \partial_r = \sin^2\theta \cos^2\phi + \sin^2\theta \sin^2\phi + \cos^2\theta$$
$$= \sin^2\theta \left(\cos^2\phi + \sin^2\phi\right) + \cos^2\theta$$
$$= \sin^2\theta + \cos^2\theta = 1$$

The basis vector $\partial_\theta$ is given by

$$\partial_\theta = r \cos\theta \cos\phi \partial_x + r \cos\theta \sin\phi \partial_y - r \sin\theta \partial_z$$

and the dot product is

$$g_{\theta\theta} = \partial_\theta \cdot \partial_\theta$$
$$= r^2 \cos^2\theta \cos^2\phi + r^2 \cos^2\theta \sin^2\phi + r^2 \sin^2\theta$$
$$= r^2 \left[\cos^2\theta \left(\cos^2\phi + \sin^2\phi\right) + \sin^2\theta\right] = r^2$$

The last basis vector is

$$\partial_\phi = -r \sin\theta \sin\phi \partial_x + r \sin\theta \cos\phi \partial_y$$

Therefore, the last component of the metric is

$$g_{\phi\phi} = \partial_\phi \cdot \partial_\phi = r^2 \sin^2\theta \sin^2\phi + r^2 \sin^2\theta \cos^2\phi$$
$$= r^2 \sin^2\theta \left(\sin^2\phi + \cos^2\phi\right) = r^2 \sin^2\theta$$

As an exercise, you can check to verify that all the other dot products vanish.

# Null Vectors

A *null vector* $V^a$ is one that satisfies

$$g_{ab}V^a V^b = 0 \qquad\qquad (2.20)$$

# The Metric Determinant

The determinant of the metric is used often. We write it as

$$g = \det(g_{ab}) \qquad\qquad (2.21)$$

# Quiz

1. The following is a valid expression involving tensors:
   (a) $S^a T_{ab} = S^c T_{ab}$
   (b) $S^a T_{ab} = S^a T_{ac}$
   (c) $S^a T_{ab} = S^c T_{cb}$

2. Cylindrical coordinates are related to cartesian coordinates via $x = r\cos\phi$, $y = r\sin\phi$, and $z = z$. This means that $\Lambda^z{}_z$ is given by
   (a) 1
   (b) −1
   (c) 0

3.  If $ds^2 = dr^2 + r^2 \, d\phi^2 + dz^2$ then
    (a) $g_{rr} = dr$, $g_{\phi\phi} = r \, d\phi$, and $g_{zz} = dz$
    (b) $g_{rr} = 1$, $g_{\phi\phi} = r^2$, and $g_{zz} = 1$
    (c) $g_{rr} = 1$, $g_{\phi\phi} = r$, and $g_{zz} = 1$
    (d) $g_{rr} = dr^2$, $g_{\phi\phi} = r^2 \, d\phi^2$, and $g_{zz} = dz^2$

4.  The signature of

$$g_{ab} = \begin{pmatrix} 1 & 0 & 0 & 0 \\ 0 & -1 & 0 & 0 \\ 0 & 0 & -1 & 0 \\ 0 & 0 & 0 & -1 \end{pmatrix}$$

    is
    (a) $-2$
    (b) $2$
    (c) $0$
    (d) $1$

5.  In spherical coordinates, a vector has the following components: $X^a = \left(r, \frac{1}{r\sin\theta}, \frac{1}{\cos^2\theta}\right)$. The component $X_\phi$ is given by
    (a) $1/\cos^2\theta$
    (b) $\cos^2\theta$
    (c) $r^2\tan^2\theta$
    (d) $r^2/\cos^2\theta$

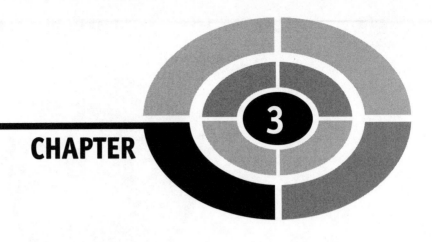

# More on Tensors

In this chapter we continue to lay down the mathematical framework of relativity. We begin with a discussion of manifolds. We are going to loosely define only what a manifold is so that the readers will have a general idea of what this concept means in the context of relativity. Next we will review and add to our knowledge of vectors and one forms, and then learn some new tensor properties and operations.

## Manifolds

To describe curved spacetime mathematically, we will use a mathematical concept known as a *manifold*. Basically speaking, a manifold is nothing more than a continuous space of points that may be curved (and complicated in other ways) globally, but locally it looks like plain old flat space. So in a small enough neighborhood Euclidean geometry applies. Think of the surface of a sphere or the surface of the earth as an example. Globally, of course, the earth is a curved surface. Imagine drawing a triangle with sides that went from the equator to

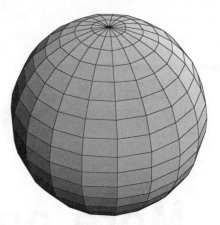

**Fig. 3-1.** The surface of a sphere is an example of a manifold. Pick a small enough patch, and space in the patch is Euclidean (flat).

the North Pole. For that kind of triangle, the familiar formulas of Euclidean geometry are not going to apply. But locally it is flat, and good old Euclidean geometry applies.

Other examples of manifolds include a torus (Fig. 3-2), or even a more abstract example like the set of rotations in cartesian coordinates.

A *differentiable manifold* is a space that is continuous and differentiable. It is intuitively obvious that we must describe spacetime by a differentiable manifold, because to do physics we need to be able to do calculus.

Generally speaking, a manifold cannot be completely covered by a single coordinate system. But we can "cover" the manifold with a set of open sets $U_i$ called *coordinate patches*. Each coordinate patch can be mapped to flat Euclidean space.

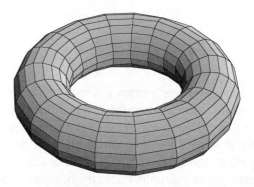

**Fig. 3-2.** A torus is an example of a manifold.

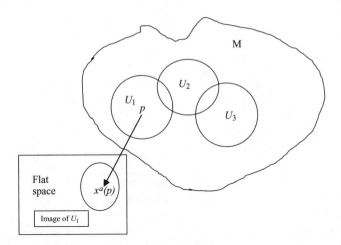

**Fig. 3-3.** A crude abstract illustration of a manifold. The manifold cannot be covered by a single coordinate system, but we cover it with a set of coordinate patches that we designate by $U_i$. These patches may overlap. The points in each patch can be mapped to flat Euclidean space. Here we illustrate a mapping for a coordinate patch we've called $U_{\bar{I}}$. A point $p$ that belongs to the manifold $M$ (say in $U_{\bar{I}}$) is mapped to a coordinate that we designate $x^a(p)$. (Houghston and Todd, 1992)

# Parameterized Curves

The notion of a curve as the plot of some function is a familiar one. In the context of relativity, however, it is more useful to think of curves in terms of *parameterization*. In this view, a point that moves through space traces out the curve.

The parameter of the curve, which we denote by $\lambda$, is a real number. We describe a curve by a set of parametric equations that give the coordinates along the curve for a given value of $\lambda$:

$$x^a = x^a(\lambda) \tag{3.1}$$

In $n$ dimensions, since there are $n$ coordinates $x^a$, there will be $n$ such equations.

**EXAMPLE 3-1**
Consider the curve traced out by the unit circle in the plane. Describe a parametric equation for the curve.

**SOLUTION 3-1**
We call the parameter $\theta$. Since we are in two dimensions, we need two functions that will determine $(x, y)$ for a given value of $\theta$. Since the equation of a circle is given by $x^2 + y^2 = r^2$, in the case of the unit circle the equation that describes

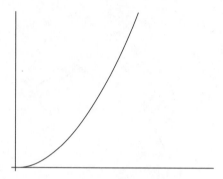

**Fig. 3-4.** A parabola with $x \geq 0$.

the curve is

$$x^2 + y^2 = 1$$

This is an easy equation to describe parametrically. Since we know that $\cos^2 \theta + \sin^2 \theta = 1$, we choose the parametric representation to be

$$x = \cos \theta \quad \text{and} \quad y = \sin \theta$$

**EXAMPLE 3-2**
Find a parameterization for the curve $y = x^2$ such that $x \geq 0$.

**SOLUTION 3-2**
The curve is shown in Fig. 3-4. To parameterize this curve with the restriction that $x \geq 0$, we can take $x$ to be some positive number $\lambda$. To ensure that the number is positive we square it; i.e.,

$$x = \lambda^2$$

Since $y = x^2$, it follows that $y = \lambda^4$ will work.

# Tangent Vectors and One Forms, Again

A curved space or spacetime is one that is going to change from place to place. As such, in a curved spacetime, one cannot speak of a vector that stretches from point to point. Instead, we must define all quantities like vectors and one forms locally. In the previous chapter, we basically tossed out the idea that a basis vector is defined as a partial derivative along some coordinate direction. Now let's explore that notion a bit more carefully. We can do so with parameterized curves.

Let $x^a(\lambda)$ be a parameterized curve. Then the components of the tangent vector to the curve are given by

$$\frac{dx^a}{d\lambda} \tag{3.2}$$

**EXAMPLE 3-3**
Describe the tangent vectors to the curves in Examples 3-1 and 3-2.

**SOLUTION 3-3**
In Example 3-1, we parameterized the curve traced out by the unit circle with $(\cos\theta,\ \sin\theta)$. Using (3.2), we see that the components of the tangent vector to the unit circle are given by $(-\sin\theta,\ \cos\theta)$. Calling the tangent vector $\vec{\tau}$ and writing the basis vectors in cartesian coordinates as $e_x$ and $e_y$, we have

$$\vec{\tau} = -\sin\theta\, e_x + \cos\theta\, e_y$$

Now we consider Example 3-2, where the given curve was $y = x^2$. We found that this curve could parameterize the curve with $(\lambda^2,\ \lambda^4)$. Using (3.2), we obtain the tangent vector $\vec{v} = (2\lambda,\ 4\lambda^3)$. Let's invert the relation used to describe the curve. We have

$$\lambda = \sqrt{x}$$

Therefore, we can write $\vec{v}$ as

$$\vec{v} = 2\sqrt{x}e_x + 4x^{3/2}e_y$$

Now that we've found out how to make a vector that is tangent to a parameterized curve, the next step is to find a basis. Let's expand this idea by considering some arbitrary continuous and differentiable function $f$. We can compute the derivative of $f$ in the direction of the curve by $\frac{df}{d\lambda}$. Therefore, $\frac{d}{d\lambda}$ is a vector that maps $f$ to a real number that is given by $\frac{df}{d\lambda}$. The chain rule allows us to write this as

$$\frac{df}{d\lambda} = \frac{df}{d\lambda}\frac{\partial x^a}{\partial x^a} = \frac{dx^a}{d\lambda}\frac{\partial f}{\partial x^a}$$

The function $f$ is arbitrary, so we can write

$$\frac{d}{d\lambda} = \frac{dx^a}{d\lambda} \frac{\partial}{\partial x^a}$$

In other words, we have expanded the vector $\frac{d}{d\lambda}$ in terms of a set of basis vectors. The components of the vector are $\frac{dx^a}{d\lambda}$ and the basis vectors are given by $\frac{\partial}{\partial x^a}$. This is the origin of the argument that the basis vectors $e_a$ are given by partial derivatives along the coordinate directions.

### EXAMPLE 3-4

The fact that the basis vectors are given in terms of partial derivatives with respect to the coordinates provides an explanation as to why we can write the transformation matrices as

$$\Lambda^{a'}_{\ b} = \frac{\partial x^a}{\partial x^{b'}}$$

### SOLUTION 3-4

This can be done by applying the chain rule to a basis vector. We have, in the primed coordinates,

$$e_{a'} = \frac{\partial}{\partial x^{a'}} = \frac{\partial}{\partial x^{a'}} \left( \frac{\partial x^b}{\partial x^b} \right) = \frac{\partial x^b}{\partial x^{a'}} \frac{\partial}{\partial x^b} = \Lambda^b_{\ a'} e_b$$

Similarly, going the other way, we have

$$e_b = \frac{\partial}{\partial x^b} = \frac{\partial}{\partial x^b} \left( \frac{\partial x^{a'}}{\partial x^{a'}} \right) = \frac{\partial x^{a'}}{\partial x^b} \frac{\partial}{\partial x^{a'}} = \Lambda^{a'}_{\ b} e_{a'}$$

Now let's explore some new notation that is frequently seen in books and the literature. We can write the dot product using a bracket-type notation $\langle\,,\,\rangle$. In the left slot, we place a one form and in the right side we place a vector. And so we write the dot product $p \cdot v$ as

$$p \cdot v = \langle \tilde{p}, \vec{v} \rangle = p_a v^a \tag{3.3}$$

Here $\tilde{p}$ is a one form. Using this notation, we can write the dot product between the basis one forms and basis vectors as

$$\langle \omega^a, e_b \rangle = \delta^a_b$$

This can be seen easily writing the bases in terms of partial derivatives:

$$\langle \omega^a, e_b \rangle = \left\langle \mathrm{d}x^a, \frac{\partial}{\partial x^b} \right\rangle = \frac{\partial x^a}{\partial x^b} = \delta_b^a$$

This type of notation makes it easy to find the components of vectors and one forms. We consider the dot product between an arbitrary vector $V$ and a basis one form:

$$\langle \omega^a, V \rangle = \langle \omega^a, V^b e_b \rangle = V^b \langle \omega^a, e_b \rangle = V^b \delta_b^a = V^a$$

Since the components of a vector are just numbers, we are free to pull them outside of the bracket $\langle\ ,\ \rangle$. We can use the same method to find the components of a one form:

$$\langle \sigma, e_b \rangle = \langle \sigma_a \omega^a, e_b \rangle = \sigma_a \langle \omega^a, e_b \rangle = \sigma_a \delta_b^a = \sigma_b$$

Now we see how we can derive the inner product between an arbitrary one form and vector:

$$\langle \sigma, V \rangle = \langle \sigma_a \omega^a, V^b e_b \rangle = \sigma_a V^b \langle \omega^a, e_b \rangle = \sigma_a V^b \delta_b^a = \sigma_a V^a$$

These operations are linear. In particular,

$$\langle \sigma, aV + bW \rangle = a \langle \sigma, V \rangle + b \langle \sigma, W \rangle$$
$$\langle a\sigma + b\rho, V \rangle = a \langle \sigma, V \rangle + b \langle \rho, V \rangle$$

where $a, b$ are scalars, $\sigma, \rho$ are one forms, and $V, W$ are vectors.

# Tensors as Functions

A tensor is a function that maps vectors and one forms to the real numbers. The components of a tensor are found by passing basis one forms and basis vectors as arguments. For example, we consider a rank 2 tensor $T$. If we pass two basis one forms as argument, we get

$$T \left( \omega^a, \omega^b \right) = T^{ab}$$

A tensor with raised indices has contravariant components and is therefore expanded in terms of basis vectors; i.e.,

$$T = T^{ab} e_a \otimes e_b$$

We have already seen an example of a tensor with lowered indices, the metric tensor

$$g_{ab} \omega^a \otimes \omega^b = g_{ab} \, dx^a \otimes dx^b$$

(in a coordinate basis). A tensor is not fixed with raised or lowered indices; we recall from the last chapter that we can raise or lower indices using the metric. We can also have tensors with mixed indices. For each raised index we need a basis vector, and for each lowered index we need a basis one form when writing out the tensor as an expansion (the way you would write a vector expanding in a basis). For example

$$S = S^{ab}{}_c \, e_a \otimes e_b \otimes \omega^c$$

We get the components in the opposite way; that is, to get the upper index pass a one form, and to get the lower index pass a basis vector:

$$S^{ab}{}_c = S\left(\omega^a, \omega^b, e_c\right)$$

We can pass arbitrary vectors and one forms to a tensor. Remember, the components of vectors and one forms are just numbers. So we can write

$$S(\sigma, \rho, V) = S\left(\sigma_a \omega^a, \rho_b \omega^b, V^c e_c\right) = \sigma_a \rho_b V^c S\left(\omega^a, \omega^b, e_c\right) = \sigma_a \rho_b V^c S^{ab}{}_c$$

The quantity $\sigma_a \rho_b V^c S^{ab}{}_c$ is a *number*, which is consistent with the notion that a tensor maps vectors and one forms to numbers. Note that the summation convention is used.

# Tensor Operations

We now summarize a few basic algebraic operations that can be carried out with tensors to produce new tensors. These operations basically mirror the types of things you can carry out with vectors. For example, we can add two tensors of the same type to get a new tensor:

$$R^{ab}{}_c = S^{ab}{}_c + T^{ab}{}_c$$

It follows that we can subtract two tensors of the same type to get a new tensor of the same type:

$$Q^a{}_b = S^a{}_b - \frac{T^a}{a}b$$

We can also multiply a tensor by a scalar $a$ to get a new tensor

$$S_{ab} = a T_{ab}$$

Note that in these examples, the placement of indices and number of indices are arbitrary. We are simply providing specific examples. The only requirement is that all of the tensors in these types of operations have to be of the same type.

We can use addition, subtraction, and scalar multiplication to derive the symmetric and antisymmetric parts of a tensor. A tensor is symmetric if $B_{ab} = B_{ba}$ and antisymmetric if $T_{ab} = -T_{ba}$. The symmetric part of a tensor is given by

$$T_{(ab)} = \frac{1}{2}(T_{ab} + T_{ba}) \tag{3.4}$$

and the antisymmetric part of a tensor is

$$T_{[ab]} = \frac{1}{2}(T_{ab} - T_{ba}) \tag{3.5}$$

We can extend this to more indices, but we won't worry about that for the time being. Often, the notation is extended to include multiple tensors. For instance,

$$V_{(a}W_{b)} = \frac{1}{2}(V_a W_b + W_b V_a)$$

Tensors of different types can be multiplied together. If we multiply a tensor of type $(m, n)$ by a tensor of type $(p, q)$, the result is a tensor of type $(m + p, n + q)$. For example

$$R^{ab}S^c{}_{de} = T^{abc}{}_{de}$$

*Contraction* can be used to turn an $(m, n)$ tensor into an $(m - 1, n - 1)$ tensor. This is done by setting a raised and lowered index equal:

$$R_{ab} = R^c{}_{acb} \tag{3.6}$$

Remember, repeated indices indicate a sum.

The Kronecker delta can be used to manipulate tensor expressions. Use the following rule: When a raised index in a tensor matches the lowered index in the Kronecker delta, change it to the value of the raised index of the Kronecker delta. This sounds confusing, so we demonstrate it with an example

$$\delta_b^a T^{bc}{}_d = T^{ac}{}_d$$

Now consider the opposite. When a lowered index in a tensor matches a raised index in the Kronecker delta, set that index to the value of the lowered index of the Kronkecker delta:

$$\delta_d^c T^{ab}{}_c = T^{ab}{}_d$$

**EXAMPLE 3-5**
Show that if a tensor is symmetric then it is independent of basis.

**SOLUTION 3-5**
We can work this out easily using the tensor transformation properties. Considering $B_{ab} = B_{ba}$, we work out the left side:

$$B_{ab} = \Lambda^{c'}{}_a \Lambda^{d'}{}_b B_{c'd'} = \frac{\partial x^{c'}}{\partial x^a} \frac{\partial x^{d'}}{\partial x^b} B_{c'd'}$$

For the other side, since we can move the derivatives around, we find

$$B_{ba} = \Lambda^{d'}{}_b \Lambda^{c'}{}_a B_{d'c'} = \frac{\partial x^{d'}}{\partial x^b} \frac{\partial x^{c'}}{\partial x^a} B_{d'c'} = \frac{\partial x^{c'}}{\partial x^a} \frac{\partial x^{d'}}{\partial x^b} B_{d'c'}$$

Equating both terms, it immediately follows that $B_{c'd'} = B_{d'c'}$.
It is also true that if $B_{ab} = B_{ba}$, then $B^{cd} = B^{dc}$. Working this out,

$$B^{cd} = g^{ca} B_a{}^d = g^{ca} g^{db} B_{ab} = g^{ca} g^{db} B_{ba} = g^{ca} B^d{}_a = B^{dc}$$

**EXAMPLE 3-6**
Let $T^{ab}$ be antisymmetric. Show that

$$S_{[a} T_{bc]} = \frac{1}{3} \left( S_a T_{bc} - S_b T_{ac} + S_c T_{ab} \right)$$

**SOLUTION 3-6**

Since $T^{ab}$ is antisymmetric, we know that $T_{ab} = -T_{ba}$. The expression $A_{[abc]}$ is given by

$$A_{[abc]} = \frac{1}{6} \left( A_{abc} + A_{bca} + A_{cab} - A_{bac} - A_{acb} - A_{cba} \right)$$

Therefore, we find

$$S_{[a} T_{bc]} = \frac{1}{6} \left( S_a T_{bc} + S_b T_{ca} + S_c T_{ab} - S_b T_{ac} - S_a T_{cb} - S_c T_{ba} \right)$$

Now we use the antisymmetry of $T$, $T_{ab} = -T_{ba}$, to write

$$S_{[a} T_{bc]} = \frac{1}{6} \left( S_a T_{bc} - S_b T_{ac} + S_c T_{ab} - S_b T_{ac} + S_a T_{bc} + S_c T_{ab} \right)$$

$$= \frac{1}{6} \left( 2 S_a T_{bc} - 2 S_b T_{ac} + 2 S_c T_{ab} \right)$$

$$= \frac{1}{3} \left( S_a T_{bc} - S_b T_{ac} + S_c T_{ab} \right)$$

**EXAMPLE 3-7**

Let $Q^{ab} = Q^{ba}$ be a symmetric tensor and $R^{ab} = -R^{ba}$ be an antisymmetric tensor. Show that

$$Q^{ab} R_{ab} = 0$$

**SOLUTION 3-7**

Since $R^{ab} = -R^{ba}$, we can write

$$R_{ab} = \frac{1}{2} \left( R_{ab} + R_{ab} \right) = \frac{1}{2} \left( R_{ab} - R_{ba} \right)$$

Therefore, we have

$$Q^{ab} R_{ab} = \frac{1}{2} Q^{ab} \left( R_{ab} - R_{ba} \right) = \frac{1}{2} \left( Q^{ab} R_{ab} - Q^{ab} R_{ba} \right)$$

Note that the indices $a, b$ are repeated in both terms. This means they are dummy indices and we are free to change them. In the second term, we make the switch

$a \leftrightarrow b$, which gives

$$Q^{ab} R_{ab} = \frac{1}{2} \left( Q^{ab} R_{ab} - Q^{ab} R_{ba} \right) = \frac{1}{2} \left( Q^{ab} R_{ab} - Q^{ba} R_{ab} \right)$$

Now use the symmetry of $Q^{ab}$ to change the second term, giving the desired result:

$$\frac{1}{2} \left( Q^{ab} R_{ab} - Q^{ba} R_{ab} \right) = \frac{1}{2} \left( Q^{ab} R_{ab} - Q^{ab} R_{ab} \right) = \frac{1}{2} Q^{ab} \left( R_{ab} - R_{ab} \right) = 0$$

**EXAMPLE 3-8**
Show that if $Q^{ab} = Q^{ba}$ is a symmetric tensor and $T_{ab}$ is arbitrary, then

$$T_{ab} Q^{ab} = \frac{1}{2} Q^{ab} \left( T_{ab} + T_{ba} \right)$$

**SOLUTION 3-8**
Using the symmetry of $Q$, we have

$$T_{ab} Q^{ab} = T_{ab} \frac{1}{2} \left( Q^{ab} + Q^{ab} \right) = T_{ab} \frac{1}{2} \left( Q^{ab} + Q^{ba} \right)$$

Now multiply it by $T$, which gives

$$T_{ab} \frac{1}{2} \left( Q^{ab} + Q^{ba} \right) = \frac{1}{2} \left( T_{ab} Q^{ab} + T_{ab} Q^{ba} \right)$$

Again, the indices $a, b$ are repeated in both expressions. Therefore, they are dummy indices that can be changed. We swap them in the second term $a \leftrightarrow b$, which gives

$$\frac{1}{2} \left( T_{ab} Q^{ab} + T_{ab} Q^{ba} \right) = \frac{1}{2} \left( T_{ab} Q^{ab} + T_{ba} Q^{ab} \right) = \frac{1}{2} Q^{ab} \left( T_{ab} + T_{ba} \right)$$

# The Levi-Cevita Tensor

No book on relativity can really give you a good enough headache without mention of our friend, the Levi-Cevita tensor. This is

$$\varepsilon_{abcd} = \begin{cases} +1 & \text{for an even permutation of 0123} \\ -1 & \text{for an odd permutation of 0123} \\ \phantom{+}0 & \text{otherwise} \end{cases} \qquad (3.7)$$

We will see more of this in future chapters.

# Quiz

1.  If $T^{ab} = -T^{ba}$ then
    (a) $Q^{ab}T_{ab} = \frac{1}{2}\left(Q^{ab} + Q^{ba}\right)T_{ab}$
    (b) $Q^{ab}T_{ab} = \frac{1}{2}\left(Q^{ab} - Q^{ba}\right)T_{ab}$
    (c) $Q^{ab}T_{ab} = \frac{1}{2}\left(Q_{ab} - Q^{ba}\right)T_{ab}$
    (d) $Q^{ab}T_{ab} = \frac{1}{2}\left(Q_{ab} - Q_{ba}\right)T_{ab}$

2.  Let $x^a(\lambda)$ be a parameterized curve. The components of the tangent vector to this curve are best described by
    (a) $d\lambda/dx$
    (b) $f = x(\lambda)$
    (c) $dx^a/d\lambda$
    (d) $x = f(\lambda)$

3.  A tensor with components $T_a{}^b{}_c$ has an expansion in terms of basis vectors and one forms, given by
    (a) $T = T_a{}^b{}_c\,\omega^a \otimes \omega^b \otimes \omega^c$
    (b) $T = T_a{}^b{}_c\,\omega^a \otimes e_b \otimes \omega^c$
    (c) $T = T_a{}^b{}_c\,e_a \otimes \omega^b \otimes e_c$

4.  The symmetric part of $V_a W_b$ is best written as
    (a) $V_{(a}W_{b)} = \frac{1}{2}\left(V_a W_b - W_b V_a\right)$
    (b) $V_{(a}W_{b)} = \frac{1}{2}\left(V^a W_b + W^b V_a\right)$
    (c) $V_{(a}W_{b)} = \frac{1}{2}\left(V_a W_b + W_b V_a\right)$

5.  The Kronecker delta acts as
    (a) $\delta^a_b T^{bc}{}_d = -T^{ac}{}_d$
    (b) $\delta^a_b T^{bc}{}_d = T_{acd}$
    (c) $\delta^a_b T^{bc}{}_d = T_a{}^c{}_d$
    (d) $\delta^a_b T^{bc}{}_d = T^{ac}{}_d$

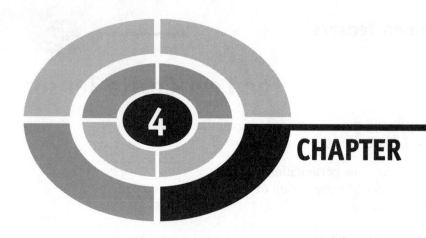

## CHAPTER 4

# Tensor Calculus

In this chapter we turn to the problem of finding the derivative of a tensor. In a curved space or spacetime, this is a bit of a thorny issue. As we will see, properly finding the derivative of a tensor, which should give us back a new tensor, is going to require some additional mathematical formalism. We will show how this works and then describe the metric tensor, which plays a central role in the study of gravity. Next we will introduce some quantities that are important in Einstein's equation.

## Testing Tensor Character

As we will see below, it is sometimes necessary to determine whether a given object is a tensor or not. The most straightforward way to determine whether an object is a given type of tensor is to check how it transforms. There are, however, a few useful tips that can serve as a guide as to whether or not a given quantity is a tensor.

The first test relies on the inner product. If the inner product

$$\sigma_a V^a = \phi$$

where $\phi$ is a scalar and is invariant for all vectors $V^a$, then $\sigma_a$ is a one form. We can carry this process further. If

$$T_{ab}\, V^a = U_b$$

such that $U_b$ is a one form, then $T_{ab}$ is a (0, 2) tensor. Another test on $T_{ab}$ is

$$T_{ab}\, V^a W^b = \phi$$

where $V^a$ and $W^b$ are vectors and $\phi$ is an invariant scalar, then $T_{ab}$ is a (0, 2) tensor.

# The Importance of Tensor Equations

In physics we seek invariance, i.e., we seek laws of physics written in an invariant form that is true for all observers. Tensors are a key ingredient in this recipe, because if a tensor equation is true in one coordinate system then it is true in all coordinate systems. This can greatly simplify analysis because we can often transform to a coordinate system where the mathematics will be easier. There we can find the form of a result we need and then express it in another coordinate system if desired.

A simple example of this is provided by the vacuum field equations. We will see that these can be expressed in the form of a (0, 2) tensor called the *Ricci tensor*, where

$$R_{ab} = 0$$

It is immediately obvious that this equation is true in any coordinate system. Let's transform to a different coordinate system using the [Lambda] matrix discussed

$$R_{ab} = \Lambda^{c'}{}_{a} \Lambda^{d'}{}_{b} R_{c'd'} = \frac{\partial x^{c'}}{\partial x^a} \frac{\partial x^{d'}}{\partial x^b} R_{c'd'}$$

On the right-hand side, the transformation of 0 is of course just 0, so we have

$$\frac{\partial x^{c'}}{\partial x^a} \frac{\partial x^{d'}}{\partial x^b} R_{c'd'} = 0$$

We can divide both sides by $\frac{\partial x^{c'}}{\partial x^d}\frac{\partial x^{d'}}{\partial x^b}$, which gives

$$R_{c'd'} = 0$$

# The Covariant Derivative

Consider the problem of taking the derivative of a vector. In ordinary cartesian coordinates, this is not really that complicated of an issue because the basis vectors are constant. So we can get the derivative of a vector by differentiating its components. However, in general, things are not so straightforward. In curved spaces the basis vectors themselves may vary from point to point. This means that when we take a derivative of a vector, we will have to differentiate the basis vectors as well.

Consider some vector $\vec{A}$. To compute the derivative, we expand the vector in a basis and then apply the Leibniz rule $(fg)' = f'g + g'f$ to obtain

$$\frac{\partial \vec{A}}{\partial x^a} = \frac{\partial}{\partial x^a}\left(A^b e_b\right) = \frac{\partial A^b}{\partial x^a}e_b + A^b \frac{\partial}{\partial x^a}\left(e_b\right)$$

In cartesian coordinates, we could just throw away the second term. But this isn't true in general. To see how this works in practice, we turn to a specific example.

We start with something familiar—the spherical polar coordinates. Cartesian coordinates are related to spherical coordinates in the following way:

$$x = r \sin\theta \cos\phi$$
$$y = r \sin\theta \sin\phi \qquad\qquad (4.1)$$
$$z = r \cos\theta$$

The first step in this exercise will be to work out the basis vectors in spherical coordinates in terms of the cartesian basis vectors. We'll do this using the procedures outlined in Chapter 2. This means we will need the transformation matrices that allow us to move back and forth between the two coordinate systems. Again, we denote the transformation matrix by the symbol $\Lambda^a{}_{b'}$, where an element of this matrix is given by

$$\frac{\partial x^a}{\partial x^{b'}}$$

We will consider the unprimed coordinates to be $(x, y, z)$ and the primed coordinates to be $(r, \theta, \phi)$. In that case, the transformation matrix assumes the form

$$\Lambda^a{}_{b'} = \begin{pmatrix} \frac{\partial x}{\partial r} & \frac{\partial y}{\partial r} & \frac{\partial z}{\partial r} \\ \frac{\partial x}{\partial \theta} & \frac{\partial y}{\partial \theta} & \frac{\partial z}{\partial \theta} \\ \frac{\partial x}{\partial \phi} & \frac{\partial y}{\partial \phi} & \frac{\partial z}{\partial \phi} \end{pmatrix} \tag{4.2}$$

Using the relationship among the coordinates described by (4.1), we obtain

$$\Lambda^a{}_{b'} = \begin{pmatrix} \sin\theta\cos\phi & \sin\theta\sin\phi & \cos\theta \\ r\cos\theta\cos\phi & r\cos\theta\sin\phi & -r\sin\theta \\ -r\sin\theta\sin\phi & r\sin\theta\cos\phi & 0 \end{pmatrix} \tag{4.3}$$

Now we have the machinery we need to work out the form of the basis vectors. We have already seen many times that the basis vectors transform as $e_{b'} = \Lambda^a{}_{b'} e_a$. So let's write down the basis vectors for spherical coordinates as expansions in terms of the cartesian basis using (4.3). We get

$$\begin{aligned}
e_r &= \Lambda^b{}_r e_b = \Lambda^x{}_r e_x + \Lambda^y{}_r e_y + \Lambda^z{}_r e_z \\
&= \sin\theta\cos\phi\, e_x + \sin\theta\sin\phi\, e_y + \cos\theta\, e_z \\
e_\theta &= \Lambda^b{}_\theta e_b = \Lambda^x{}_\theta e_x + \Lambda^y{}_\theta e_y + \Lambda^z{}_\theta e_z \\
&= r\cos\theta\cos\phi\, e_x + r\cos\theta\sin\phi\, e_y - r\sin\theta\, e_z \\
e_\phi &= \Lambda^b{}_\phi e_b = \Lambda^x{}_\phi e_x + \Lambda^y{}_\phi e_y + \Lambda^z{}_\phi e_z \\
&= -r\sin\theta\sin\phi\, e_x + r\sin\theta\cos\phi\, e_y
\end{aligned} \tag{4.4}$$

Now let's see what happens when we differentiate these basis vectors. As an example, we compute the derivatives of $e_r$ with respect to each of the spherical coordinates. Remember, the cartesian basis vectors $\{e_x, e_y, e_z\}$ are constant, so we don't have to worry about them. Proceeding, we have

$$\frac{\partial e_r}{\partial r} = \frac{\partial}{\partial r}\left(\sin\theta\cos\phi\, e_x + \sin\theta\sin\phi\, e_y + \cos\theta\, e_z\right) = 0$$

Our first attempt hasn't yielded anything suspicious. But let's compute the $\theta$ derivative. This one gives

$$\frac{\partial e_r}{\partial \theta} = \frac{\partial}{\partial \theta} \left( \sin\theta \cos\phi \, e_x + \sin\theta \sin\phi \, e_y + \cos\theta \, e_z \right)$$

$$= \cos\theta \cos\phi \, e_x + \cos\theta \sin\phi \, e_y - \sin\theta \, e_z$$

$$= \frac{1}{r} e_\theta$$

Now that's a bit more interesting. Instead of computing the derivative and getting zero, we find another basis vector, scaled by $1/r$. Let's go further and proceed by computing the derivative with respect to $\phi$. In this case, we get

$$\frac{\partial e_r}{\partial \phi} = \frac{\partial}{\partial \phi} \left( \sin\theta \cos\phi \, e_x + \sin\theta \sin\phi \, e_y + \cos\theta \, e_z \right)$$

$$= - \sin\theta \sin\phi \, e_x + \sin\theta \cos\phi \, e_y$$

$$= \frac{1}{r} e_\phi$$

Again, we've arrived at another basis vector.

It turns out that when differentiating basis vectors, there is a general relationship that gives the derivative of a basis vector in terms of a weighted sum. The sum is just an expansion in terms of the basis vectors with weighting coefficients denoted by $\Gamma^a_{\;bc}$.

$$\frac{\partial e_a}{\partial x^b} = \Gamma^c_{\;ab} \, e_c \tag{4.5}$$

The $\Gamma^a_{\;bc}$ are functions of the coordinates, as we saw in the examples we've calculated so far. Looking at the results we got above, we can identify the results in the following way:

$$\frac{\partial e_r}{\partial \theta} = \frac{1}{r} e_\theta \;\Rightarrow\; \Gamma^\theta_{\;r\theta} = \frac{1}{r}$$

$$\frac{\partial e_r}{\partial \phi} = \frac{1}{r} e_\phi \;\Rightarrow\; \Gamma^\phi_{\;r\phi} = \frac{1}{r}$$

Consider another derivative

$$\frac{\partial e_\phi}{\partial \theta} = \frac{\partial}{\partial \theta} \left(-r \sin\theta \sin\phi\, e_x + r \sin\theta \cos\phi\, e_y\right)$$

$$= -r \cos\theta \sin\phi\, e_x + r \cos\theta \cos\phi\, e_y$$

$$= \cos\theta \left(-r \sin\phi\, e_x + r \cos\phi\, e_y\right)$$

$$= \frac{\cos\theta}{\sin\theta} \left(-r \sin\theta \sin\phi\, e_x + r \sin\theta \cos\phi\, e_y\right)$$

$$= \cot\theta\, e_\phi$$

Comparison with (4.5) leads us to conclude that

$$\Gamma^\phi{}_{\phi\theta} = \cot\theta$$

The coefficient functions we have derived here are known as *Christoffel symbols* or an *affine connection.*

Basically, these quantities represent correction terms. A derivative operator needs to differentiate a tensor and give a result that is another tensor. In particular, the derivative of a tensor field that has valence $\binom{m}{n}$ should give a tensor field of valence $\binom{m}{n+1}$. We have seen one reason why we need the correction terms: outside of ordinary cartesian coordinates, the derivative of a vector is going to involve derivatives of the basis vectors as well. Another reason for introducing the Christoffel symbols is that the partial derivative of a tensor is not a tensor. First let's remind ourselves how the components of a vector transform:

$$X^{a'} = \frac{\partial x^{a'}}{\partial x^b} X^b$$

Keeping this in mind, we have

$$\partial_{c'} X^{a'} = \frac{\partial}{\partial x^{c'}} \left(\frac{\partial x^{a'}}{\partial x^b} X^b\right)$$

$$= \frac{\partial}{\partial x^{c'}} \left(\frac{\partial x^d}{\partial x^d}\right) \left(\frac{\partial x^{a'}}{\partial x^b} X^b\right)$$

$$= \frac{\partial x^d}{\partial x^{c'}} \frac{\partial}{\partial x^d} \left( \frac{\partial x^{a'}}{\partial x^b} X^b \right)$$

$$= \frac{\partial x^d}{\partial x^{c'}} \frac{\partial^2 x^{a'}}{\partial x^d \partial x^b} X^b + \frac{\partial x^d}{\partial x^{c'}} \frac{\partial x^{a'}}{\partial x^b} \frac{\partial X^b}{\partial x^d}$$

Now how does a $(1, 1)$ tensor transform? It does so like this:

$$T^{a'}{}_{b'} = \frac{\partial x^{a'}}{\partial x^c} \frac{\partial x^d}{\partial x^{b'}} T^c{}_d$$

That's the kind of transformation we got from the partial derivative above, in the second term on the last line:

$$\partial_{c'} X^{a'} = \frac{\partial x^d}{\partial x^{c'}} \frac{\partial^2 x^{a'}}{\partial x^d \partial x^b} X^b + \frac{\partial x^d}{\partial x^{c'}} \frac{\partial x^{a'}}{\partial x^b} \frac{\partial X^b}{\partial x^d}$$

But the first term leaves us out of luck as far as getting another tensor—thus the need for a correction term. Let's go back and look at the formula we had for the derivative of a vector $A$:

$$\frac{\partial A}{\partial x^a} = \frac{\partial}{\partial x^a} \left( A^b \right) e_b + A^b \frac{\partial}{\partial x^a} (e_b)$$

Now let's use (4.5) to rewrite the second term. This gives

$$A^b \frac{\partial}{\partial x^a} (e_b) = A^b \, \Gamma^c{}_{ba} e_c$$

Putting this into the formula that gives the derivative of a vector, we have

$$\frac{\partial A}{\partial x^a} = \frac{\partial}{\partial x^a} \left( A^b \right) e_b + \Gamma^c{}_{ba} A^b e_c$$

We rearranged the order of quantities in the second term for later convenience. Now, recall that any repeated indices that are both *up* and *down* in an expression are dummy indices, and so can be relabeled. In the second term, we swap $b \leftrightarrow c$ to change $\Gamma^c{}_{ba} A^b e_c \rightarrow \Gamma^b{}_{ca} A^c e_b$, and the expression for the derivative of a

vector becomes

$$\frac{\partial A}{\partial x^a} = \frac{\partial}{\partial x^a}\left(A^b\right)e_b + \Gamma^b{}_{ca}A^c e_b = \left(\frac{\partial A^b}{\partial x^a} + \Gamma^b{}_{ca}A^c\right)e_b$$

The expression in parentheses is the covariant derivative of a vector $A$. We denote the covariant derivative by $\nabla_b A^a$:

$$\nabla_b A^a = \frac{\partial A^a}{\partial x^b} + \Gamma^b{}_{ca}A^c \tag{4.6}$$

It can be verified that this object is a $(1, 1)$ tensor by checking how it transforms. To see how the Christoffel symbols transform, we look at (4.5). Writing this in primed coordinates, we have

$$\Gamma^{c'}{}_{a'b'}e_{c'} = \frac{\partial e_{a'}}{\partial x^{b'}} \tag{4.7}$$

Now, the basis vectors transform according to $e_{c'} = \Lambda^d{}_{c'}e_d = \frac{\partial x^d}{\partial x^{c'}}e_d$. The left-hand side then becomes

$$\Gamma^{c'}{}_{a'b'}e_{c'} = \Gamma^{c'}{}_{a'b'}\Lambda^d{}_{c'}e_d = \Gamma^{c'}{}_{a'b'}\frac{\partial x^d}{\partial x^{c'}}e_d \tag{4.8}$$

Now let's tackle the right-hand side of (4.7). We obtain

$$\frac{\partial e_{a'}}{\partial x^{b'}} = \frac{\partial}{\partial x^{b'}}\left(\Lambda^m{}_{a'}e_m\right) = \frac{\partial}{\partial x^{b'}}\left(\frac{\partial x^m}{\partial x^{a'}}e_m\right)$$

$$= \frac{\partial x^n}{\partial x^{b'}}\frac{\partial}{\partial x^n}\left(\frac{\partial x^m}{\partial x^{a'}}e_m\right)$$

$$= \frac{\partial x^n}{\partial x^{b'}}\left(\frac{\partial^2 x^m}{\partial x^n \partial x^{a'}}e_m + \frac{\partial x^m}{\partial x^{a'}}\frac{\partial e_m}{\partial x^n}\right)$$

Now we know that $\frac{\partial e_m}{\partial x^n}$ gives another Christoffel symbol, so we write this as

$$\frac{\partial e_{a'}}{\partial x^{b'}} = \frac{\partial x^n}{\partial x^{b'}}\left(\frac{\partial^2 x^m}{\partial x^n \partial x^{a'}}e_m + \frac{\partial x^m}{\partial x^{a'}}\Gamma^l{}_{mn}e_l\right)$$

In the first term inside parentheses, we are going to swap the dummy indices. We change $m \to d$ and now write the expression as

$$\frac{\partial e_{a'}}{\partial x^{b'}} = \frac{\partial x^n}{\partial x^{b'}} \left( \frac{\partial^2 x^d}{\partial x^n \partial x^{a'}} e_d + \frac{\partial x^m}{\partial x^{a'}} \Gamma^l_{mn} e_l \right)$$

In the second term, there is a dummy index $l$ that is associated with the basis vector. We would like to factor out the basis vector and so we change $l \to d$, which gives

$$\frac{\partial e_{a'}}{\partial x^{b'}} = \frac{\partial x^n}{\partial x^{b'}} \left( \frac{\partial^2 x^d}{\partial x^n \partial x^{a'}} e_d + \frac{\partial x^m}{\partial x^{a'}} \Gamma^d_{mn} e_d \right)$$

$$= \frac{\partial x^n}{\partial x^{b'}} \left( \frac{\partial^2 x^d}{\partial x^n \partial x^{a'}} + \frac{\partial x^m}{\partial x^{a'}} \Gamma^d_{mn} \right) e_d \qquad (4.9)$$

$$= \left( \frac{\partial x^n}{\partial x^{b'}} \frac{\partial^2 x^d}{\partial x^n \partial x^{a'}} + \frac{\partial x^n}{\partial x^{b'}} \frac{\partial x^m}{\partial x^{a'}} \Gamma^d_{mn} \right) e_d$$

We started out with (4.7). Setting the result for the left side of (4.7) given in (4.8) equal to the result found in (4.9) gives us the following:

$$\Gamma^{c'}_{a'b'} \frac{\partial x^d}{\partial x^{c'}} e_d = \left( \frac{\partial x^n}{\partial x^{b'}} \frac{\partial^2 x^d}{\partial x^n \partial x^{a'}} + \frac{\partial x^n}{\partial x^{b'}} \frac{\partial x^m}{\partial x^{a'}} \Gamma^d_{mn} \right) e_d$$

The basis vector $e_d$ appears on both sides, so we drop it and write

$$\Gamma^{c'}_{a'b'} \frac{\partial x^d}{\partial x^{c'}} = \frac{\partial x^n}{\partial x^{b'}} \frac{\partial^2 x^d}{\partial x^n \partial x^{a'}} + \frac{\partial x^n}{\partial x^{b'}} \frac{\partial x^m}{\partial x^{a'}} \Gamma^d_{mn}$$

Finally, we obtain the transformation law for the Christoffel symbols by dividing both sides through by $\frac{\partial x^d}{\partial x^{c'}}$. This gives

$$\Gamma^{c'}_{a'b'} = \frac{\partial x^{c'}}{\partial x^d} \frac{\partial x^n}{\partial x^{b'}} \frac{\partial^2 x^d}{\partial x^n \partial x^{a'}} + \frac{\partial x^{c'}}{\partial x^d} \frac{\partial x^n}{\partial x^{b'}} \frac{\partial x^m}{\partial x^{a'}} \Gamma^d_{mn} \qquad (4.10)$$

Now we see why the Christoffel symbols act as correction terms. The first piece of this can be used to cancel out the extra piece found in the transformation of a partial derivative of a tensor. That way, the covariant derivative transforms as it should, as a (1, 1) tensor.

Of course we are going to differentiate other objects besides vectors. The covariant derivative of a one form is given by

$$\nabla_b \sigma_a = \partial_b \sigma_a - \Gamma^c{}_{ab} \sigma_c \qquad (4.11)$$

This suggests the procedure to be used for the covariant derivative of an arbitrary tensor: First take the partial derivative of the tensor, and then add a $\Gamma^c{}_{ab}$ term for each contravariant index and subtract a $\Gamma^c{}_{ab}$ term for each covariant index. For example,

$$\nabla_c T^a{}_b = \partial_c T^a{}_b + \Gamma^a{}_{cd} T^d{}_b - \Gamma^d{}_{bc} T^a{}_d$$

$$\nabla_c T_{ab} = \partial_c T_{ab} - \Gamma^d{}_{ac} T_{db} - \Gamma^d{}_{cb} T_{ad}$$

$$\nabla_c T^{ab} = \partial_c T^{ab} + \Gamma^a{}_{cd} T^{db} + \Gamma^b{}_{dc} T^{ad}$$

The covariant derivative of a scalar function is just the partial derivative:

$$\nabla_a \phi = \partial_a \phi \qquad (4.12)$$

**EXAMPLE 4-1**
Consider polar coordinates. Find the covariant derivative $\nabla_a V^a$ of $\vec{V} = r^2 \cos\theta \, e_r - \sin\theta \, e_\theta$.

**SOLUTION 4-1**
The summation convention is in effect. Writing it out explicitly, we have

$$\nabla_a V^a = \nabla_r V^r + \nabla_\theta V^\theta$$

Using (4.6), we have

$$\nabla_r V^r = \frac{\partial V^r}{\partial r} + \Gamma^r{}_{cr} V^c = \frac{\partial V^r}{\partial r} + \Gamma^r{}_{rr} V^r + \Gamma^r{}_{\theta r} V^\theta$$

$$\nabla_\theta V^\theta = \frac{\partial V^\theta}{\partial \theta} + \Gamma^\theta{}_{c\theta} V^c = \frac{\partial V^\theta}{\partial \theta} + \Gamma^\theta{}_{r\theta} V^r + \Gamma^\theta{}_{\theta\theta} V^\theta$$

One can show that the Christoffel symbols for polar coordinates are (Exercise)

$$\Gamma^\theta{}_{r\theta} = \Gamma^\theta{}_{\theta r} = \frac{1}{r} \quad \text{and} \quad \Gamma^r{}_{\theta\theta} = -r$$

while all other components are zero, and so we obtain

$$\nabla_r V^r = \frac{\partial V^r}{\partial r} + \Gamma^r{}_{rr} V^r + \Gamma^r{}_{\theta r} V^\theta = \frac{\partial V^r}{\partial r}$$

$$\nabla_\theta V^\theta = \frac{\partial V^\theta}{\partial \theta} + \Gamma^\theta{}_{r\theta} V^r + \Gamma^\theta{}_{\theta\theta} V^\theta = \frac{\partial V^\theta}{\partial \theta} + \frac{1}{r} V^r$$

The sum then becomes

$$\nabla_a V^a = \nabla_r V^r + \nabla_\theta V^\theta = \frac{\partial V^r}{\partial r} + \frac{1}{r} V^r + \frac{\partial V^\theta}{\partial \theta}$$

For $\vec{V} = r^2 \cos\theta\, e_r - \sin\theta\, e_\theta$, this results in

$$\nabla_a V^a = \frac{\partial V^r}{\partial r} + \frac{1}{r} V^r + \frac{\partial V^\theta}{\partial \theta} = 2r \cos\theta + r \cos\theta - \cos\theta = 3r \cos\theta - \cos\theta$$

$$= \cos\theta\,(3r - 1)$$

**EXAMPLE 4-2**
Suppose that for some vector field $U^a$, $U^a U_a$ is a constant and $\nabla_a U_b - \nabla_b U_a = 0$. Show that $U^a \nabla_a U^b = 0$.

**SOLUTION 4-2**
Since $U^a U_a$ is a constant, the derivative must vanish. We compute the derivative of this product as

$$\nabla_b (U^a U_a) = U^a \nabla_b U_a + U_a \nabla_b U^a$$

Now we use $\nabla_a U_b - \nabla_b U_a = 0$ to rewrite the first term on the right-hand side, and then use index raising and lowering with the metric to write $U_b = g_{bc} U^c$, which gives

$$U^a \nabla_b U_a + U_a \nabla_b U^a = U^a \nabla_a U_b + U_a \nabla_b U^a$$

$$= U^a \nabla_a g_{bc} U^c + U_a \nabla_b U^a$$

$$= U^a U^c \nabla_a g_{bc} + g_{bc} U^a \nabla_a U^c + U_a \nabla_b U^a$$

$$= g_{bc} U^a \nabla_a U^c + U_a \nabla_b U^a$$

To obtain the last line, we used $\nabla_a g_{bc} = 0$. We now concentrate on the second term. To use $\nabla_a U_b - \nabla_b U_a = 0$ to change the indices, first we need to lower the index on $U^a$ inside the derivative. Remember, we are free to pull $g_{ab}$ outside the derivative since $\nabla_a g_{bc} = 0$. So we get

$$g_{bc} U^a \nabla_a U^c + U_a \nabla_b U^a = g_{bc} U^a \nabla_a U^c + U_a \nabla_b g^{ac} U_c$$

$$= g_{bc} U^a \nabla_a U^c + g^{ac} U_a \nabla_b U_c$$

$$= g_{bc} U^a \nabla_a U^c + g^{ac} U_a \nabla_c U_b$$

The first term is already in the form we need to prove the result. To get the second term in that form, we are going to have to raise the indices on the $U$'s. We do this and recall that $g^{ab} g_{bc} = \delta^a_c$, which gives

$$g_{bc} U^a \nabla_a U^c + g^{ac} U_a \nabla_c U_b = g_{bc} U^a \nabla_a U^c + g^{ac} g_{ae} U^e \nabla_c g_{bd} U^d$$

$$= g_{bc} U^a \nabla_a U^c + g^{ac} g_{ae} g_{bd} U^e \nabla_c U^d$$

$$= g_{bc} U^a \nabla_a U^c + \delta^c_e g_{bd} U^e \nabla_c U^d$$

$$= g_{bc} U^a \nabla_a U^c + g_{bd} U^c \nabla_c U^d$$

To obtain the last line, we used the Kronecker delta to set $e \to c$. Now note that $c$ is a dummy index. We change it to $a$ to match the first term

$$g_{bc} U^a \nabla_a U^c + g_{bd} U^c \nabla_c U^d = g_{bc} U^a \nabla_a U^c + g_{bd} U^a \nabla_a U^d$$

Now we focus on the other dummy index $d$. We're free to change it so we set it equal to $c$ and have

$$g_{bc} U^a \nabla_a U^c + g_{bd} U^a \nabla_a U^d = g_{bc} U^a \nabla_a U^c + g_{bc} U^a \nabla_a U^c$$

$$= g_{bc} \left( U^a \nabla_a U^c + U^a \nabla_a U^c \right)$$

$$= 2 g_{bc} U^a \nabla_a U^c$$

We started by taking the derivative of a constant, which is zero, $\nabla_b \left( U^a U_a \right) = 0$, and so this result must vanish. We divide the above expression by 2 and get

$$g_{bc} U^a \nabla_a U^c = 0$$

There are two possibilities: If $g_{bc} = 0$, then the equation is trivially zero. If it is not zero, we can divide both sides by $g_{bc}$ and get the desired result

$$U^a \nabla_a U^c = 0$$

# The Torsion Tensor

The torsion tensor is defined in terms of the connection as

$$T^a_{\ bc} = \Gamma^a_{\ bc} - \Gamma^a_{\ cb} = 2\Gamma^a_{\ [bc]} \tag{4.13}$$

It is easy to show that even though the connection is not a tensor, the difference between two connections is a tensor. In general relativity, the torsion tensor is taken to vanish, which means that the connection is symmetric; i.e.,

$$\Gamma^a_{\ bc} = \Gamma^a_{\ cb} \tag{4.14}$$

In some theories of gravity (known as Einstein-Cartan theories), the torsion tensor does not vanish. We will not study such cases in this book.

# The Metric and Christoffel Symbols

In $n$ dimensions there are $n^3$ functions called *Christoffel symbols of the first kind*, $\Gamma_{abc}$. In a coordinate basis (see Chapter 5) these functions can be derived from the metric tensor using the relationship

$$\Gamma_{abc} = \frac{1}{2}\left(\frac{\partial g_{bc}}{\partial x^a} + \frac{\partial g_{ca}}{\partial x^b} - \frac{\partial g_{ab}}{\partial x^c}\right) \tag{4.15}$$

More generally,

$$\Gamma_{abc} = \frac{1}{2}\left(\frac{\partial g_{bc}}{\partial x^a} + \frac{\partial g_{ca}}{\partial x^b} - \frac{\partial g_{ab}}{\partial x^c} + C_{abc} + C_{acb} - C_{bca}\right)$$

where the $C_{abc}$ are called *commutation coefficients* (see Chapter 5).

**EXAMPLE 4-3**
Show that

$$\frac{\partial g_{ab}}{\partial x^c} = \Gamma_{abc} + \Gamma_{bca}$$

**SOLUTION 4-3**
This is a simple problem to solve. We simply use (4.15) and permute indices, which gives

$$\Gamma_{abc} = \frac{1}{2}\left(\frac{\partial g_{bc}}{\partial x^a} + \frac{\partial g_{ca}}{\partial x^b} - \frac{\partial g_{ab}}{\partial x^c}\right)$$

$$\Gamma_{bca} = \frac{1}{2}\left(\frac{\partial g_{ca}}{\partial x^b} + \frac{\partial g_{ab}}{\partial x^c} - \frac{\partial g_{bc}}{\partial x^a}\right)$$

Adding, we find

$$\Gamma_{abc} + \Gamma_{bca} = \frac{1}{2}\left(\frac{\partial g_{bc}}{\partial x^a} + \frac{\partial g_{ca}}{\partial x^b} - \frac{\partial g_{ab}}{\partial x^c}\right) + \frac{1}{2}\left(\frac{\partial g_{ca}}{\partial x^b} + \frac{\partial g_{ab}}{\partial x^c} - \frac{\partial g_{bc}}{\partial x^a}\right)$$

$$= \frac{1}{2}\left(\frac{\partial g_{bc}}{\partial x^a} - \frac{\partial g_{bc}}{\partial x^a} + \frac{\partial g_{ca}}{\partial x^b} + \frac{\partial g_{ca}}{\partial x^b} - \frac{\partial g_{ab}}{\partial x^c} + \frac{\partial g_{ab}}{\partial x^c}\right)$$

$$= \frac{1}{2}\left(2\frac{\partial g_{ca}}{\partial x^b}\right) = \frac{\partial g_{ca}}{\partial x^b}$$

We can obtain the Christoffel symbols of the second kind (which are usually referred to simply as the Christoffel symbols) by raising an index with the metric

$$\Gamma^a{}_{bc} = g^{ad}\Gamma_{dbc}$$

Using this we can write (4.15) in a more popular form:

$$\Gamma^a{}_{bc} = \frac{1}{2}g^{ad}\left(\frac{\partial g_{b\bar{d}}}{\partial x^a} + \frac{\partial g_{c\bar{d}}}{\partial x^b} - \frac{\partial g_{ab}}{\partial x^{\bar{d}}}\right) \tag{4.16}$$

**EXAMPLE 4-4**
Find the Christoffel symbols for the 2-sphere of radius $a$

$$ds^2 = a^2 \, d\theta^2 + a^2 \sin^2 \theta \, d\phi^2$$

**SOLUTION 4-4**
The metric is given by

$$g_{ab} = \begin{pmatrix} a^2 & 0 \\ 0 & a^2 \sin^2 \theta \end{pmatrix}$$

and so $g_{\theta\theta} = a^2$ and $g_{\phi\phi} = a^2 \sin^2 \theta$. The inverse metric is

$$g^{ab} = \begin{pmatrix} \frac{1}{a^2} & 0 \\ 0 & \frac{1}{a^2 \sin^2 \theta} \end{pmatrix}$$

Therefore, $g^{\theta\theta} = \frac{1}{a^2}$ and $g^{\phi\phi} = \frac{1}{a^2 \sin^2 \theta}$. The only nonzero derivative is

$$\frac{\partial g_{\phi\phi}}{\partial \theta} = \frac{\partial}{\partial \theta} \left( a^2 \sin^2 \theta \right) = 2a^2 \sin \theta \cos \theta \qquad (4.17)$$

We find the Christoffel symbols using (4.16). Considering the first nonzero term of $g^{ab}$, we set $a = d = \theta$ in (4.16) and obtain

$$\Gamma^\theta{}_{bc} = \frac{1}{2} g^{\theta\theta} \left( \frac{\partial g_{b\theta}}{\partial x^c} + \frac{\partial g_{c\theta}}{\partial x^b} - \frac{\partial g_{bc}}{\partial \theta} \right) = -\frac{1}{2} g^{\theta\theta} \left( \frac{\partial g_{bc}}{\partial \theta} \right)$$

We set $\frac{\partial g_{c\theta}}{\partial x^b} = \frac{\partial g_{\theta b}}{\partial x^c} = 0$ because $g_{\theta\phi} = g_{\phi\theta} = 0$, $g_{\theta\theta} = a^2$, where $a$ is a constant, and so all derivatives of these terms vanish. The only possibility is

$$\Gamma^\theta{}_{\phi\phi} = -\frac{1}{2} g^{\theta\theta} \left( \frac{\partial g_{\phi\phi}}{\partial \theta} \right) = -\frac{1}{2} \left( \frac{1}{a^2} \right) (2a^2 \sin \theta \cos \theta) = -\sin \theta \cos \theta$$

The only other nonzero possibility for this metric is the term involving $g^{\phi\phi}$, and so we set $a = d = \phi$ in (4.16). This gives

$$\Gamma^\phi{}_{bc} = \frac{1}{2} g^{\phi\phi} \left( \frac{\partial g_{b\phi}}{\partial x^c} + \frac{\partial g_{c\phi}}{\partial x^b} - \frac{\partial g_{bc}}{\partial \phi} \right)$$

Only a derivative with respect to $\theta$ is going to be nonzero. Therefore we drop the first term and consider

$$\Gamma^\phi{}_{bc} = \frac{1}{2} g^{\phi\phi} \left( \frac{\partial g_{c\phi}}{\partial x^b} - \frac{\partial g_{bc}}{\partial \phi} \right)$$

First we take $b = \theta$, $c = \phi$, which gives

$$\Gamma^\phi{}_{\phi\theta} = \frac{1}{2} g^{\phi\phi} \left( \frac{\partial \phi}{\partial g_{\theta\phi}} - \frac{\partial \theta}{\partial g_{\phi\phi}} \right) = \frac{1}{2} g^{\phi\phi} \frac{\partial g_{\phi\phi}}{\partial \theta}$$

$$= \frac{1}{2} \left( \frac{1}{a^2 \sin^2 \theta} \right) (2a^2 \sin \theta \cos \theta) = \frac{\cos \theta}{\sin \theta} = \cot \theta$$

A similar procedure with $b = \theta$, $c = \phi$ gives $\Gamma^\phi{}_{\theta\phi} = \cot \theta$. All other Christoffel symbols are zero.

### EXAMPLE 4-5
A metric that is used in the study of colliding gravitational waves is the *Kahn-Penrose* metric (see Fig. 4-1). The coordinates used are $(u, v, x, y)$. With $u \geq 0$ and $v < 0$, this metric assumes the form

$$ds^2 = 2 \, du \, dv - (1 - u)^2 \, dx^2 - (1 + u)^2 \, dy^2$$

Find the Christoffel symbols of the first kind (using (4.15) for this metric.

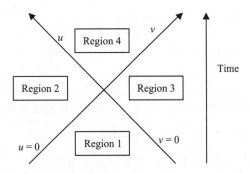

**Fig. 4-1.** The division of spacetime for the Kahn-Penrose solution used in the study of gravitational waves. We consider the metric in region 2.

## SOLUTION 4-5

In this problem, we use the shorthand notation $\partial_a = \frac{\partial}{\partial x^a}$ and so write (4.15) as

$$\Gamma_{abc} = \frac{1}{2}(\partial_a g_{bc} + \partial_b g_{ca} - \partial_c g_{ab})$$

We can arrange the components of the metric in a matrix as follows, defining $(x^0, x^1, x^2, x^3) \to (u, v, x, y)$

$$[g_{ab}] = \begin{pmatrix} 0 & 1 & 0 & 0 \\ 1 & 0 & 0 & 0 \\ 0 & 0 & -(1-u)^2 & 0 \\ 0 & 0 & 0 & -(1+u)^2 \end{pmatrix}$$

and so we have

$$g_{uv} = g_{vu} = 1, \qquad g_{xx} = -(1-u)^2, \qquad g_{yy} = -(1+u)^2$$

One approach is to simply write down all of the possible combinations of co-ordinates in (4.15), but we can use the form of the metric to quickly whittle down the possibilities. First we note that

$$\partial_a g_{uv} = \partial_a g_{vu} = 0$$

The only nonzero derivatives are

$$\partial_u g_{xx} = \partial_u[-(1-u)^2] = 2(1-u)$$
$$\partial_u g_{yy} = \partial_u[-(1+u)^2] = -2(1+u)$$

Consider terms involving $x$ first. Looking at $\Gamma_{abc} = \frac{1}{2}(\partial_a g_{bc} + \partial_b g_{ca} - \partial_c g_{ab})$, the possible combinations of $a,b,c$ that will give a term involving $\partial_u g_{xx}$ are

$$a = b = x, \quad c = u$$
$$a = u, \quad b = c = x$$
$$a = c = x, \quad b = u$$

Taking the first of these possibilities, we have

$$\Gamma_{xxu} = \frac{1}{2}(\partial_x g_{xu} + \partial_x g_{ux} - \partial_u g_{xx}) = -\frac{1}{2}\partial_u g_{xx} = -\frac{1}{2}[2(1-u)] = -1 + u$$

In the second case, we obtain

$$\Gamma_{uxx} = \frac{1}{2}(\partial_u g_{xx} + \partial_x g_{xu} - \partial_x g_{ux}) = \frac{1}{2}\partial_u g_{xx} = \frac{1}{2}[2(1-u)] = 1 - u$$

You can write down the third permutation and see that it gives the same result as this one. The next step is to consider the terms involving $\partial_u g_{yy}$. A similar exercise shows that

$$\Gamma_{uyy} = -1 - u$$
$$\Gamma_{yyu} = 1 + u$$

It is easy to see that if the components of the metric are constant, all of the Christoffel symbols vanish. Note that the Christoffel symbols of the first kind are symmetric in their first two indices:

$$\Gamma_{abc} = \Gamma_{bac}$$

**EXAMPLE 4-6**
Show that if the metric is diagonal, then

$$\Gamma^a{}_{ab} = \frac{\partial}{\partial x^b}\left(\frac{1}{2}\ln g_{aa}\right)$$

**SOLUTION 4-6**
If the metric is diagonal then the inverse components of the metric are easily found to be $g^{aa} = \frac{1}{g_{aa}}$. Using (4.15), we obtain

$$\Gamma^a{}_{ab} = g^{ac}\Gamma_{abc} = g^{aa}\Gamma_{aba} = \frac{1}{g_{aa}}\Gamma_{aba} = \frac{1}{g_{aa}}\left(\frac{1}{2}\frac{\partial g_{aa}}{\partial x^b}\right) = \frac{\partial}{\partial x^b}\left(\frac{1}{2}\ln g_{aa}\right)$$

We say a connection $\Gamma^a{}_{bc}$ is *metric-compatible* if the covariant derivative of the metric vanishes:

$$\nabla_c g_{ab} = 0 \qquad\qquad (4.18)$$

**EXAMPLE 4-7**

Show that $\nabla_c g_{ab} = 0$ using (4.16).

**SOLUTION 4-7**

The covariant derivative of the metric tensor is

$$\nabla_c g_{ab} = \partial_c g_{ab} - g_{db} \Gamma^d{}_{ac} - g_{ad} \Gamma^d{}_{bc}$$

Now we apply (4.16) to rewrite the Christoffel symbols in this equation in terms of the metric. We have

$$\Gamma^d{}_{ac} = \frac{1}{2} g^{de} \left( \partial_a g_{ec} + \partial_c g_{ea} - \partial_e g_{ac} \right)$$

$$\Gamma^d{}_{bc} = \frac{1}{2} g^{de} \left( \partial_b g_{ec} + \partial_c g_{eb} - \partial_e g_{bc} \right)$$

We now insert these terms in the covariant derivative of the metric:

$$\nabla_c g_{ab} = \partial_c g_{ab} - g_{db} \left[ \frac{1}{2} g^{de} \left( \partial_a g_{ec} + \partial_c g_{ea} - \partial_e g_{ac} \right) \right]$$

$$- g_{ad} \left[ \frac{1}{2} g^{de} \left( \partial_b g_{ec} + \partial_c g_{eb} - \partial_e g_{bc} \right) \right]$$

$$= \partial_c g_{ab} - \frac{1}{2} g_{db} g^{de} \partial_a g_{ec} - \frac{1}{2} g_{db} g^{de} \partial_c g_{ea} + \frac{1}{2} g_{db} g^{de} \partial_e g_{ac}$$

$$- \frac{1}{2} g_{ad} g^{de} \partial_b g_{ec} - \frac{1}{2} g_{ad} g^{de} \partial_c g_{eb} + \frac{1}{2} g_{ad} g^{de} \partial_e g_{bc}$$

Now we use the fact that $g_{ab} g^{bc} = \delta_a^c$ to make the following changes

$$g_{db} g^{de} = \delta_b^e \Rightarrow e \to b \quad \text{where } g_{db} g^{de} \text{ appears}$$

$$g_{ad} g^{de} = \delta_a^e \Rightarrow e \to a \quad \text{where } g_{ad} g^{de} \text{ appears}$$

The above expression then reduces to

$$\nabla_c g_{ab} = \partial_c g_{ab} - \frac{1}{2}\delta_b^e \partial_a g_{ec} - \frac{1}{2}\delta_b^e \partial_c g_{ea} + \frac{1}{2}\delta_b^e \partial_e g_{ac} - \frac{1}{2}\delta_a^e \partial_b g_{ec} - \frac{1}{2}\delta_a^e \partial_c g_{eb}$$

$$+ \frac{1}{2}\delta_a^e \partial_e g_{bc}$$

$$= \partial_c g_{ab} - \frac{1}{2}\partial_a g_{bc} - \frac{1}{2}\partial_c g_{ba} + \frac{1}{2}\partial_b g_{ac} - \frac{1}{2}\partial_b g_{ac} - \frac{1}{2}\partial_c g_{ab} + \frac{1}{2}\partial_a g_{bc}$$

$$= \partial_c g_{ab} + \frac{1}{2}\left[(\partial_a g_{bc} - \partial_a g_{bc}) + (\partial_b g_{ac} - \partial_b g_{ac})\right] - \frac{1}{2}\partial_c g_{ba} - \frac{1}{2}\partial_c g_{ab}$$

$$= \partial_c g_{ab} - \frac{1}{2}\partial_c g_{ba} - \frac{1}{2}\partial_c g_{ab}$$

Now we use the symmetry of the metric to write $g_{ab} = g_{ba}$ and we see that the result vanishes:

$$\nabla_c g_{ab} = \partial_c g_{ab} - \frac{1}{2}\partial_c g_{ba} - \frac{1}{2}\partial_c g_{ab} = \partial_c g_{ab} - \frac{1}{2}\partial_c g_{ab} - \frac{1}{2}\partial_c g_{ab}$$

$$= \partial_c g_{ab} - \partial_c g_{ab} = 0$$

# The Exterior Derivative

So far we have talked a bit about one forms, i.e., (0, 1) tensors. It is possible to have $p$ forms, i.e., a (0, $p$) tensor. One way to produce higher forms is by using the *wedge product*. The wedge product of two one forms $\alpha$, $\beta$ is the two form that we denote by

$$\alpha \wedge \beta = \alpha \otimes \beta - \beta \otimes \alpha \tag{4.19}$$

One sees immediately from the definition that

$$\alpha \wedge \beta = -\beta \wedge \alpha \tag{4.20}$$

This implies that

$$\alpha \wedge \alpha = 0 \tag{4.21}$$

We can obtain yet "higher order" forms by taking more wedge products. For example we can construct a three form using the one forms $\alpha$, $\beta$, $\gamma$ by writing $\alpha \wedge \beta \wedge \gamma$. An arbitrary $p$ form can be written in terms of wedge products of basis one forms $\omega^a$ as follows:

$$\alpha = \frac{1}{p!}\alpha_{a_1 a_2 \cdots a_p}\omega^{a_1} \wedge \omega^{a_2} \wedge \cdots \wedge \omega^{a_p} \tag{4.22}$$

The wedge product is linear and associative, and so $(a\alpha \wedge b\beta) \wedge \gamma = a\alpha \wedge \gamma + b\beta \wedge \gamma$.

One use of the wedge product is in calculation of *exterior derivatives*, something we are going to be doing extensively in the next chapter. The exterior derivative is denoted by the symbol $d$ and it maps a $p$ form into a $p + 1$ form. As the symbol indicates, the way we calculate with it is simply by taking differentials of objects. Since it maps a $p$ form into a $p + 1$ form, one might suspect that the result will involve wedge products.

Luckily in relativity we won't have to move too far up the ladder (as far as $p$ goes). So we can consider just a few cases. The first is the exterior derivative of an ordinary function, which we take to be a "0-form." The exterior derivative of a function $f$ is a one form given by

$$\mathrm{d}f = \frac{\partial f}{\partial x^a}\mathrm{d}x^a \tag{4.23}$$

Note that the summation convention is in effect. For general forms, let $\alpha$ be a $p$ form and $\beta$ be a $q$ form. Then we have

$$\mathrm{d}(\alpha \wedge \beta) = \mathrm{d}\alpha \wedge \beta + (-1)^p \alpha \wedge \mathrm{d}\beta \tag{4.24}$$

If these are both one forms, then $\mathrm{d}(\alpha \wedge \beta) = \mathrm{d}\alpha \wedge \beta - \alpha \wedge \mathrm{d}\beta$. A nice result is that $\mathrm{d}^2 = 0$, and so

$$\mathrm{d}(\mathrm{d}\alpha) = 0 \tag{4.25}$$

One application that will be seen frequently is the exterior derivative of a one form that we loosely view as a function multiplied by a differential

$$\mathrm{d}(f_a \mathrm{d}x^a) = \mathrm{d}f_a \wedge \mathrm{d}x^a \tag{4.26}$$

A *closed* form $\alpha$ is one for which $\mathrm{d}\alpha = 0$, while an exact form $\alpha$ is one for which $\alpha = \mathrm{d}\beta$ (where $\alpha$ is a $p$ form and $\beta$ is a $p - 1$ form).

**EXAMPLE 4-8**
Find the exterior derivative of the following one forms:

$$\sigma = e^{f(r)}\, dt \quad \text{and} \quad \rho = e^{g(r)} \cos(\theta) \sin(\phi)\, dr$$

**SOLUTION 4-8**
In the first case, we obtain

$$d\sigma = d\left(e^{f(r)}\, dt\right) = \frac{\partial}{\partial r}\left(e^{f(r)}\right)\, dr \wedge dt = f'(r)\, e^{f(r)}\, dr \wedge dt$$

For the second one form, we get

$$
\begin{aligned}
d\rho &= d\left(e^{g(r)} \cos(\theta)\sin(\phi)\, dr\right) \\
&= g'(r)e^{g(r)} \cos(\theta)\sin(\phi)\, dr \wedge dr - e^{g(r)}\sin(\theta)\sin(\phi)\, d\theta \wedge dr \\
&\quad + e^{g(r)} \cos(\theta)\cos(\phi)\, d\phi \wedge dr \\
&= -e^{g(r)}\sin(\theta)\sin(\phi)\, d\theta \wedge dr + e^{g(r)}\cos(\theta)\cos(\phi)\, d\phi \wedge dr \\
&= e^{g(r)}\sin(\theta)\sin(\phi)\, dr \wedge d\theta + e^{g(r)}\cos(\theta)\cos(\phi)\, d\phi \wedge dr
\end{aligned}
$$

Note that we used $dr \wedge dr = 0$, and on the last line we used (4.20) to eliminate the minus sign.

# The Lie Derivative

The Lie derivative is defined by

$$L_V W = V^b \nabla_b W^a - W^b \nabla_b V^a \tag{4.27}$$

The Lie derivative of a (0, 2) tensor is

$$L_V T_{ab} = V^c \nabla_c S_{ab} + S_{cb}\nabla_a V^c + S_{ac}\nabla_b V^c \tag{4.28}$$

In Chapter 8, we will study *Killing vectors*, which satisfy

$$L_K g_{ab} = 0 \tag{4.29}$$

# The Absolute Derivative and Geodesics

Consider a curve parameterized by $\lambda$. The *absolute derivative* of a vector $V$ is given by

$$\frac{DV^a}{D\lambda} = \frac{dV^a}{d\lambda} + \Gamma^a{}_{bc} u^b V^c \tag{4.30}$$

where $u$ is a tangent vector to the curve. The absolute derivative is of interest in relativity because it allows us to find *geodesics*. We will see later that relativity theory tells us that free-falling particles (i.e., particles subject only to the gravitational field) follow paths defined by geodesics.

A geodesic can be loosely thought of as the "shortest distance between two points." If a curve is a geodesic, then a tangent vector $u$ satisfies

$$\frac{Du^a}{D\lambda} = \alpha u^a \tag{4.31}$$

where the tangent vector is defined by $u^a = \frac{dx^a}{d\lambda}$ and $\alpha$ is a scalar function of $\lambda$. Using $u^a = \frac{dx^a}{d\lambda}$ with (4.30), we obtain

$$\frac{d^2 x^a}{d\lambda^2} + \Gamma^a{}_{bc} \frac{dx^b}{d\lambda} \frac{dx^c}{d\lambda} = \alpha \frac{dx^a}{d\lambda} \tag{4.32}$$

We can reparameterize the curve and can do so in such a way that $\alpha = 0$. When we use a parameter for which this is the case we call it an *affine* parameter. If we call the affine parameter $s$, then (4.32) becomes

$$\frac{d^2 x^a}{ds^2} + \Gamma^a{}_{bc} \frac{dx^b}{ds} \frac{dx^c}{ds} = 0 \tag{4.33}$$

Geometrically, what this means is that the vector is "transported into itself." (see Figure 4-2). That is, given a vector field $\vec{u}$, vectors at nearby points are parallel and have the same length. We can write (4.33) in the more convenient form:

$$\frac{d^2 x^a}{ds^2} + \Gamma^a{}_{bc} \frac{dx^b}{ds} \frac{dx^c}{ds} = 0 \tag{4.34}$$

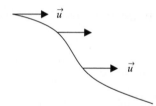

**Fig. 4-2.** A vector $u$ is parallel transported along a curve if the transported vector is parallel to the original vector and is of the same length.

The solutions to this equation, which give the coordinates in terms of the affine parameter, $x^a = x^a(s)$, are the geodesics or straightest possible curves for the geometry.

**EXAMPLE 4-9**
Find the geodesic equations for cylindrical coordinates.

**SOLUTION 4-9**
The line element for cylindrical coordinates is

$$ds^2 = dr^2 + r^2\,d\phi^2 + dz^2$$

It is easy to show that the only nonzero Christoffel symbols are

$$\Gamma^r{}_{\phi\phi} = -r$$

$$\Gamma^\phi{}_{r\phi} = \Gamma^\phi{}_{\phi r} = \frac{1}{r}$$

Using (4.34), we find the geodesic equations. First we set the indices $a = r$, $b = \phi$, $c = \phi$ to get

$$\frac{d^2r}{ds^2} - r\left(\frac{d\phi}{ds}\right)^2 = 0$$

Now taking $a = \phi$, $b = r$, $c = \phi$, we find

$$\frac{d^2\phi}{ds^2} + \frac{1}{r}\frac{dr}{ds}\frac{d\phi}{ds} = 0$$

Finally, since $\Gamma^z{}_{ab} = 0$ for all possible indices, we obtain

$$\frac{d^2 z}{ds^2} = 0$$

The geodesics are obtained by solving for the functions $r = r(s)$, $\phi = \phi(s)$, and $z = z(s)$. Since we are in flat space we might guess these are straight lines. This is easy to see from the last equation. Integrating once, we get

$$\frac{dz}{ds} = \alpha$$

where $\alpha$ is a constant. Integrating again we find

$$z(s) = \alpha s + \beta$$

where $\beta$ is another constant. Recalling $y = mx + b$, we see this is nothing other than your plain old straight line.

The geodesic equation provides a nice shortcut that can be used to obtain the Christoffel symbols. Following D'Inverno (1992), we make the following definition:

$$K = \frac{1}{2} g_{ab} \dot{x}^a \dot{x}^b \tag{4.35}$$

where we are using the "dot" notation to refer to derivatives with respect to the affine parameter $s$. We then compute

$$\frac{\partial K}{\partial x^a} = \frac{d}{ds} \left( \frac{\partial K}{\partial \dot{x}^a} \right) \tag{4.36}$$

for each coordinate and compare the result to the geodesic equation. The Christoffel symbols can then simply be read off. We demonstrate the technique with an example.

**EXAMPLE 4-10**
The metric for Rindler space (see Fig. 4-3) can be written as

$$ds^2 = \xi^2 \, d\tau^2 - d\xi^2$$

Use the geodesic equation and (4.35) to find the Christoffel symbols.

**SOLUTION 4-10**

For the Rindler metric, using (4.35) we find

$$K = \frac{1}{2}\left(\xi^2 \dot{\tau}^2 - \dot{\xi}^2\right) \tag{4.37}$$

Now we wish to find (4.36), which for the coordinate $\tau$ gives

$$\frac{\partial K}{\partial \tau} = 0$$

and so we have

$$\frac{d}{ds}\left(\frac{\partial K}{\partial \dot{\tau}}\right) = 0 \tag{4.38}$$

Using this result together with (4.37), we obtain

$$\frac{d}{ds}\left(\frac{\partial K}{\partial \dot{\tau}}\right) = 2\xi \dot{\xi} \dot{\tau} + \xi^2 \ddot{\tau} \tag{4.39}$$

Setting this equal to zero (due to (4.38)) and comparing with the geodesic equation, we find

$$\ddot{\tau} + \frac{2}{\xi}\dot{\xi}\dot{\tau} = 0$$

$$\Rightarrow \Gamma^\tau{}_{\xi\tau} = \frac{2}{\xi} = \Gamma^\tau{}_{\tau\xi}$$

A similar procedure applied to $K$ using the $\xi$ coordinate shows that

$$\Gamma^\xi{}_{\tau\tau} = \xi \tag{4.40}$$

# The Riemann Tensor

The final piece of the tensor calculus puzzle that we need to find the curvature for a given metric is the *Riemann tensor* (sometimes called simply the curvature tensor).

In terms of the metric connection (Christoffel symbols) it is given by

$$R^a{}_{bcd} = \partial_c \Gamma^a{}_{bd} - \partial_d \Gamma^a{}_{bc} + \Gamma^e{}_{bd}\Gamma^a{}_{ec} - \Gamma^e{}_{bc}\Gamma^a{}_{ed} \tag{4.41}$$

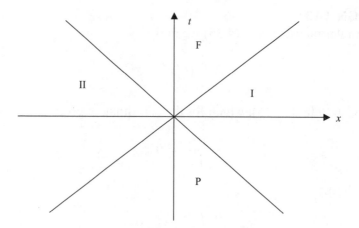

**Fig. 4-3.** Rindler coordinates describe the motion of observers moving in Minkowski spacetime that have constant acceleration. In the solved example, we consider an observer undergoing constant acceleration in the $+x$ direction. The metric shown is the metric for region 1.

We can lower the first index with the metric, which gives

$$R_{abcd} = g_{ae} R^e_{bcd}$$

This allows us to write (4.41) as

$$R_{abcd} = \partial_c \Gamma_{bda} - \partial_d \Gamma_{bca} + \Gamma_{ade} \Gamma^e_{bc} - \Gamma_{ace} \Gamma^e_{bd} \tag{4.42}$$

Using (4.15), it follows that we can write the Riemann tensor in terms of the metric as

$$R_{abcd} = \frac{1}{2} \left( \frac{\partial^2 g_{ad}}{\partial x^b \partial x^c} + \frac{\partial^2 g_{bc}}{\partial x^a \partial x^d} - \frac{\partial^2 g_{ac}}{\partial x^b \partial x^d} - \frac{\partial^2 g_{bd}}{\partial x^a \partial x^c} \right)$$
$$+ \Gamma_{ade} \Gamma^e_{bc} - \Gamma_{ace} \Gamma^e_{bd} \tag{4.43}$$

The Riemann tensor has several important symmetries. We state them without proof (they are easy, but tedious to derive using the definitions)

$$R_{abcd} = R_{cdab} = -R_{abdc} = -R_{bacd}$$
$$R_{abcd} + R_{acdb} + R_{adbc} = 0 \tag{4.44}$$

The Riemann tensor also satisfies the Bianchi identities

$$\nabla_a R_{debc} + \nabla_c R_{deab} + \nabla_b R_{deca} = 0 \qquad (4.45)$$

All together, in $n$ dimensions, there are $n^2 \left(n^2 - 1\right)/12$ independent nonzero components of the Riemann tensor. This fact together with the symmetries in (4.44) tells us that in two dimensions, the possible nonzero components of the Riemann tensor are

$$R_{1212} = R_{2121} = -R_{1221} = -R_{2112}$$

while in three dimensions the possible nonzero components of the Riemann tensor are

$$R_{1212} = R_{1313} = R_{2323} = R_{1213} = R_{1232} = R_{2123}, \qquad R_{1323} = R_{3132}$$

Computation of these quantities using a coordinate basis is extremely tedious, especially when we begin dealing with real spacetime metrics. In the next chapter we will introduce noncoordinate bases and a preferred way of calculating these components by hand. Calculations involving (4.41) are very laborious and should be left for the computer (using GR tensor for example). Nonetheless, it is good to go through the process at least once, so we shall demonstrate with a simple example.

**EXAMPLE 4-11**
Compute the components of the Riemann tensor for a unit 2-sphere, where

$$ds^2 = d\theta^2 + \sin^2 \theta \, d\phi^2$$

**SOLUTION 4-11**
The nonzero Christoffel symbols are

$$\Gamma^\theta_{\ \phi\phi} = -\sin\theta \cos\theta$$
$$\Gamma^\phi_{\ \phi\theta} = \Gamma^\phi_{\ \theta\phi} = \cot\theta$$

Using the fact that the nonzero components of the Riemann tensor in two diz-mensions are given by $R_{1212} = R_{2121} = -R_{1221} = -R_{2112}$, using (4.41) we calculate

$$R^{\theta}{}_{\phi\theta\phi} = \partial_{\theta}\Gamma^{\theta}{}_{\phi\phi} - \partial_{\phi}\Gamma^{\theta}{}_{\phi\theta} + \Gamma^{a}{}_{\phi\phi}\Gamma^{\theta}{}_{a\theta} - \Gamma^{a}{}_{\phi\theta}\Gamma^{\theta}{}_{a\phi}$$

$$= \partial_{\theta}\Gamma^{\theta}{}_{\phi\phi} + \Gamma^{\theta}{}_{\phi\phi}\Gamma^{\theta}{}_{\theta\theta} + \Gamma^{\phi}{}_{\phi\phi}\Gamma^{\theta}{}_{\phi\theta} - \Gamma^{\theta}{}_{\phi\theta}\Gamma^{\theta}{}_{\theta\phi} - \Gamma^{\phi}{}_{\phi\theta}\Gamma^{\theta}{}_{\phi\phi}$$

$$= \partial_{\theta}\Gamma^{\theta}{}_{\phi\phi} - \Gamma^{\theta}{}_{\phi\theta}\Gamma^{\theta}{}_{\theta\phi} - \Gamma^{\phi}{}_{\phi\theta}\Gamma^{\theta}{}_{\phi\phi}$$

Since $\Gamma^{\theta}{}_{\theta\phi} = 0$, this simplifies to

$$R^{\theta}{}_{\phi\theta\phi} = \partial_{\theta}\Gamma^{\theta}{}_{\phi\phi} - \Gamma^{\phi}{}_{\phi\theta}\Gamma^{\theta}{}_{\phi\phi}$$

$$= \partial_{\theta}(-\sin\theta\cos\theta) - (\cot\theta)(-\sin\theta\cos\theta)$$

$$= \sin^2\theta - \cos^2\theta + \left(\frac{\cos\theta}{\sin\theta}\right)(\sin\theta\cos\theta)$$

$$= \sin^2\theta - \cos^2\theta + \cos^2\theta$$

$$= \sin^2\theta$$

The other nonzero components can be found using the symmetry $R_{1212} = R_{2121} = -R_{1221} = -R_{2112}$.

# The Ricci Tensor and Ricci Scalar

The Riemann tensor can be used to derive two more quantities that are used to define the Einstein tensor. The first of these is the *Ricci tensor*, which is calculated from the Riemann tensor by contraction on the first and third indices:

$$R_{ab} = R^{c}{}_{acb} \tag{4.46}$$

The Ricci tensor is symmetric, so $R_{ab} = R_{ba}$. Using contraction on the Ricci tensor, we obtain the *Ricci scalar*

$$R = g^{ab}R_{ab} = R^{a}{}_{a} \tag{4.47}$$

Finally, the *Einstein tensor* is given by

$$G_{ab} = R_{ab} - \frac{1}{2} R g_{ab} \qquad (4.48)$$

**EXAMPLE 4-12**
Show that the Ricci scalar $R = 2$ for the unit 2-sphere.

**SOLUTION 4-12**
In the previous example, we found that $R^{\theta}{}_{\phi\theta\phi} = \sin^2 \theta$. The symmetry conditions of the Riemann tensor tell us that

$$R_{\theta\phi\theta\phi} = R_{\phi\theta\phi\theta} = -R_{\theta\phi\phi\theta} = -R_{\phi\theta\theta\phi}$$

The components of the metric tensor are

$$g_{\theta\theta} = g^{\theta\theta} = 1, \qquad g_{\phi\phi} = \sin^2 \theta, \qquad g^{\phi\phi} = \frac{1}{\sin^2 \theta}$$

Applying the symmetry conditions, we find

$$R_{\phi\theta\phi\theta} = \sin^2 \theta$$
$$R_{\theta\phi\phi\theta} = -\sin^2 \theta$$
$$R_{\phi\theta\theta\phi} = -\sin^2 \theta$$

Now we need to raise indices with the metric. This gives

$$R^{\phi}{}_{\theta\phi\theta} = g^{\phi\phi} R_{\phi\theta\phi\theta} = \left( \frac{1}{\sin^2 \theta} \right) \sin^2 \theta = 1$$

$$R^{\theta}{}_{\phi\phi\theta} = g^{\theta\theta} R_{\theta\phi\phi\theta} = -\sin^2 \theta$$

$$R^{\phi}{}_{\theta\theta\phi} = g^{\phi\phi} R_{\phi\theta\theta\phi} = \left( \frac{1}{\sin^2 \theta} \right) (-\sin^2 \theta) = -1$$

The components of the Ricci tensor are given by

$$R_{\theta\theta} = R^c{}_{\theta c\theta} = R^{\theta}{}_{\theta\theta\theta} + R^{\phi}{}_{\theta\phi\theta} = 1$$

$$R_{\phi\phi} = R^c{}_{\phi c\phi} = R^{\theta}{}_{\phi\theta\phi} + R^{\phi}{}_{\phi\phi\phi} = \sin^2\theta$$

$$R_{\theta\phi} = R_{\phi\theta} = R^c{}_{\theta c\phi} = R^{\theta}{}_{\theta\theta\phi} + R^{\phi}{}_{\theta\phi\phi} = 0$$

Now we contract indices to get the Ricci scalar

$$R = g^{ab} R_{ab} = g^{\theta\theta} R_{\theta\theta} + g^{\phi\phi} R_{\phi\phi} = 1 + \left(\frac{1}{\sin^2\theta}\right)\sin^2\theta = 1 + 1 = 2$$

# The Weyl Tensor and Conformal Metrics

We briefly mention one more quantity that will turn out to be useful in later studies. This is the *Weyl tensor* that can be calculated using the formula (in four dimensions)

$$C_{abcd} = R_{abcd} + \frac{1}{2}\left(g_{ad} R_{cb} + g_{bc} R_{da} - g_{ac} R_{db} - g_{bd} R_{ca}\right)$$

$$+ \frac{1}{6}\left(g_{ac} g_{db} - g_{ad} g_{cb}\right) R \tag{4.49}$$

This tensor is sometimes known as the *conformal tensor*. Two metrics are conformally related if

$$\bar{g}_{ab} = \omega^2(x)\, g_{ab} \tag{4.50}$$

for some differentiable function $\omega(x)$. A metric is *conformally flat* if we can find a function $\omega(x)$ such that the metric is conformally related to the Minkowski metric

$$g_{ab} = \omega^2(x)\, \eta_{ab} \tag{4.51}$$

A nice property of the Weyl tensor is that $C^a{}_{bcd}$ is the same for a given metric and any metric that is conformally related to it. This is the origin of the term conformal tensor.

**Quiz**

For Questions 1–6, consider the following line element:

$$ds^2 = dr^2 + r^2\, d\theta^2 + r^2\, \sin^2\theta\, d\phi^2$$

1.  The components of the metric tensor are
    (a) $g_{rr} = r$, $g_{\theta\theta} = r\,\sin\theta$, $g_{\phi\phi} = r^2\,\sin^2\theta$
    (b) $g_{rr} = r$, $g_{\theta\theta} = r^2$, $g_{\phi\phi} = r^2\,\sin^2\theta$
    (c) $g_{rr} = 1$, $g_{\theta\theta} = r^2$, $g_{\phi\phi} = r^2\,\sin^2\theta$
    (d) $g_{rr} = r$, $g_{\theta\theta} = r^2$, $g_{\phi\phi} = r^2\,\sin^2\theta$

2.  Compute the Christoffel symbols of the first kind. $\Gamma_{\theta\phi\phi}$ is
    (a) $r^2\,\sin\theta\,\cos\theta$
    (b) $r\,\sin\theta\,\cos\theta$
    (c) $r^2\,\sin^2\theta$
    (d) $\sin\theta\,\cos\theta$

3.  Now calculate the Christoffel symbols of the second kind. $\Gamma^{\phi}{}_{\phi\theta}$ is
    (a) $\frac{1}{r}$
    (b) $\frac{1}{r}\frac{\cos\theta}{\sin\theta}$
    (c) $\cot\theta$
    (d) $-\frac{1}{r^2}$

4.  Calculate the Riemann tensor. $R_{r\theta\theta\phi}$ is
    (a) $\sin\theta$
    (b) $r^3\,\sin\theta$
    (c) $\frac{\cos\theta}{r\,\sin\theta}$
    (d) $0$

5.  The determinant of the metric, $g$, is given by
    (a) $r^2\,\sin^4\theta$
    (b) $r^4\,\sin^2\theta$
    (c) $r^4\,\sin^4\theta$
    (d) $0$

    Now let $w^a = (r,\ \sin\theta,\ \sin\theta\,\cos\phi)$ and $v^a = (r,\ r^2\,\cos\theta,\ \sin\phi)$.

6.  The Lie derivative $u = L_v w$ has $u^{\phi}$ given by
    (a) $r^2\,\cos^2\theta\,\cos\phi - \sin\theta$
    (b) $r\,\cos^2\theta\,\cos\phi - \sin\theta$

    (c) $r^2 \cos\theta \, \cos^2\phi - \sin\theta$

    (d) $r^2 \cos^2\theta \, \cos\phi - \sin^2\theta$

For Questions 7 and 8, let $ds^2 = 2 \, du \, dv - (1-u)^2 \, dx^2 - (1+u)^2 \, dy^2$.

7.   $\Gamma^v{}_{xx}$ is
    (a) $1+u$
    (b) $-1+u$
    (c) $1-v$
    (d) $1-u$

8.   The Ricci scalar is
    (a) 1
    (b) 0
    (c) $u^2$
    (d) $v^2$

For Questions 9 and 10, let $ds^2 = y^2 \, \sin(x) \, dx^2 + x^2 \, \tan(y) \, dy^2$.

9.   $\Gamma^y{}_{yy}$ is given by
    (a) 0
    (b) $\cos^2 x$
    (c) $\dfrac{1+\tan^2 y}{2 \tan y}$
    (d) $\dfrac{1+\tan^2 x}{2 \tan y}$

10.   The Ricci scalar is
    (a) 0
    (b) $\dfrac{y \, \sin^2 x + y \, \sin^2 x \, \tan^2 y + x \, \cos x \, \tan^2 y}{x^2 y^2 \, \sin^2 x \, \tan^2 y}$
    (c) $\dfrac{x^2 \, \sin^2 x + y \, \sin^2 x \, \tan^2 y + x \, \cos x \, \tan^2 y}{y^2 \, \sin^2 x \, \tan^2 y}$
    (d) $y \, \sin^2 x + y \, \sin^2 x \, \cot^2 y + x \, \cos x \, \tan^2 y$

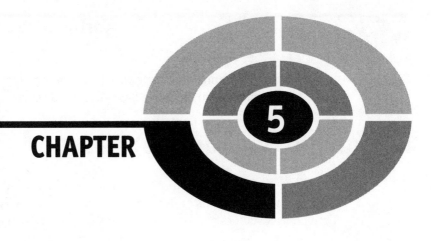

**CHAPTER**

**5**

# Cartan's Structure Equations

## Introduction

A coordinate basis is one for which the basis vectors are given by $e_a = \partial/\partial x^a$. With the exception of cartesian coordinates, a coordinate basis is not orthonormal. While it is possible to calculate using a coordinate basis, it is often not the best or easiest way to approach a problem. An alternative exists and this is to construct an orthonormal or "nonholonomic" basis. Physically, this means working in an observer's local frame. To express results in the global coordinates, we use a transformation that can be constructed easily by looking at the metric.

In this chapter we will begin by introducing this concept and then show how to transform between the two. Once we have this concept in place, we will develop a new set of equations that can be used to find the Riemann tensor for a given metric. This procedure appears a bit daunting at first but is actually much nicer to use than the methods we have discussed so far.

# Holonomic (Coordinate) Bases

The most natural choice of a basis that comes to mind is to define the basis vectors directly in terms of derivatives of coordinates. Given coordinates $x^a$, we define basis vectors and basis one forms in the following way:

$$e_a = \partial_a = \frac{\partial}{\partial x^a} \quad \text{and} \quad \omega^a = dx^a \tag{5.1}$$

When a basis set is defined exclusively in terms of derivatives with respect to the coordinates, we call it a *holonomic basis* or a *coordinate basis*. A given vector $V$ can be expanded in this basis as

$$V = V^a e_a$$

As an example, consider spherical polar coordinates

$$ds^2 = dr^2 + r^2 d\theta^2 + r^2 \sin^2\theta \, d\phi^2$$

The coordinate basis vectors are

$$e_r = \partial_r = \frac{\partial}{\partial r}, \qquad e_\theta = \partial_\theta = \frac{\partial}{\partial \theta}, \qquad e_\phi = \partial_\phi = \frac{\partial}{\partial \phi}$$

The important thing to notice here is that not all of these basis vectors are of unit length. In addition, they do not have the same dimensions. These considerations will lead us to a different basis that we will examine below.

In a coordinate basis, the basis vectors satisfy the following relationship:

$$e_a \cdot e_b = g_{ab} \tag{5.2}$$

Furthermore, we can write

$$v \cdot w = g_{ab} v^a w^b \tag{5.3}$$

Above, we briefly mentioned a problem with a coordinate basis. If we choose a coordinate basis, it may not be orthonormal. We can see this in the case of spherical polar coordinates using (5.2). Looking at the line element

$$ds^2 = dr^2 + r^2 d\theta^2 + r^2 \sin^2\theta\, d\phi^2$$

we see that the components of the metric are $g_{rr} = 1$, $g_{\theta\theta} = r^2$, and $g_{\phi\phi} = r^2 \sin^2\theta$. Now let's compute the lengths of the basis vectors. Using (5.2), we obtain

$$e_r \cdot e_r = 1 \;\Rightarrow\; |e_r| = \sqrt{g_{rr}} = 1$$

$$e_\theta \cdot e_\theta = r^2 \;\Rightarrow\; |e_\theta| = \sqrt{g_{\theta\theta}} = r$$

$$e_\phi \cdot e_\phi = r^2 \sin^2\theta \;\Rightarrow\; |e_\phi| = \sqrt{g_{\phi\phi}} = r \sin\theta$$

Since two of these basis vectors do not have unit length, this set which has been defined in terms of derivatives with respect to the coordinates is not orthonormal. To choose a basis that is orthonormal, we construct it such that the inner product of the basis vectors satisfies

$$g(e_a, e_b) = \eta_{ab}$$

This is actually easy to do, and as we will see, it makes the entire machinery of relativity much easier to deal with. A basis defined in this way is known as a *nonholonomic* or *noncoordinate basis*.

# Nonholonomic Bases

A *nonholonomic* basis is one such that the basis vectors are orthonormal with respect to the chosen metric. Another name for this type of basis is a *noncoordinate basis* and you will often hear the term *orthonormal tetrad* (more below). This type of basis is based on the fundamental ideas you are used to from freshman physics. A set of orthogonal vectors, each of unit length, are chosen for the basis. We indicate that we are working with an orthonormal basis by placing a "hat" or carat over the indices; i.e., basis vectors and basis one forms are written as

$$e_{\hat{a}}, \omega^{\hat{a}}$$

An orthonormal basis is of interest physically and has use beyond mere mathematics. Such a basis is used by a physical observer and represents a basis with respect to the local Lorentz frame, while the coordinate basis represents the global spacetime. As we move ahead in this chapter, we will learn how to transform between the two representations. We can expand any vector $V$ in terms of a coordinate or a noncoordinate basis. Just like any expansion, in terms of basis vectors, these are just different representations of the same vector

$$V = V^a e_a = V^{\hat{a}} e_{\hat{a}}$$

Since this basis represents the frame of the local Lorentz observer, we can use the flat space metric to raise and lower indices in that frame. As usual, the signs of the components can be read off the metric. For example, with a metric with the general form $ds^2 = dt^2 - d\vec{x}^2$, we have $\eta_{\hat{a}\hat{b}} = \text{diag}(1, -1, -1, -1)$. The basis vectors of a nonholonomic basis satisfy

$$e_{\hat{a}} \cdot e_{\hat{b}} = \eta_{\hat{a}\hat{b}} \tag{5.4}$$

In a nutshell, the basic idea of creating a nonholonomic basis is to scale it by the coefficient multiplying each differential in the line element. Let's illustrate this with an example. In the case of spherical polar coordinates, a noncoordinate basis is given by the following:

$$e_{\hat{r}} = \partial_r, \qquad e_{\hat{\theta}} = \frac{1}{r}\partial_\theta, \qquad e_{\hat{\phi}} = \frac{1}{r\sin\theta}\partial_\phi$$

An easy way to determine whether or not a given basis is holonomic is to calculate the *commutation coefficients* for the basis. We do this for the case of spherical polar coordinates in the next section.

# Commutation Coefficients

The *commutator* is defined to be

$$[A, B] = AB - BA$$

From calculus, we know that partial derivatives commute. Consider a function $f(x, y)$. It is true that

$$\frac{\partial^2 f}{\partial x \partial y} = \frac{\partial^2 f}{\partial y \partial x} \Rightarrow \frac{\partial^2 f}{\partial x \partial y} - \frac{\partial^2 f}{\partial y \partial x} = 0$$

Let's rewrite this in terms of the commutator of the derivatives acting on the function $f(x, y)$:

$$\frac{\partial^2 f}{\partial x \partial y} - \frac{\partial^2 f}{\partial y \partial x} = \left( \frac{\partial}{\partial x} \frac{\partial}{\partial y} - \frac{\partial}{\partial y} \frac{\partial}{\partial x} \right) f = \left[ \frac{\partial}{\partial x}, \frac{\partial}{\partial y} \right] f$$

So we can write

$$\left[ \frac{\partial}{\partial x}, \frac{\partial}{\partial y} \right] = 0$$

Looking at the holonomic basis for spherical polar coordinates, we had

$$e_r = \partial_r = \frac{\partial}{\partial r}, \qquad e_\theta = \partial_\theta = \frac{\partial}{\partial \theta}, \qquad e_\phi = \partial_\phi = \frac{\partial}{\partial \phi}$$

From the above arguments, it is clear that

$$[e_r, e_\theta] = [e_r, e_\phi] = [e_\theta, e_\phi] = 0$$

Now let's consider the nonholonomic basis we found for spherical polar coordinates. Let's compute the commutator $\left[ e_{\hat{r}}, e_{\hat{\theta}} \right]$. Since we're new to this process we carry along a test function as a crutch.

$$\left[ e_{\hat{r}}, e_{\hat{\theta}} \right] f = \left[ \partial_r, \frac{1}{r} \partial_\theta \right] f = \partial_r \left( \frac{1}{r} \partial_\theta f \right) - \frac{1}{r} \partial_\theta (\partial_r f)$$

$$= -\frac{1}{r^2} \partial_\theta f + \frac{1}{r} \partial_\theta (\partial_r f) - \frac{1}{r} \partial_\theta (\partial_r f)$$

$$= -\frac{1}{r^2} \partial_\theta f$$

Using the definitions given for the nonholonomic basis vectors, the end result is

$$\left[ e_{\hat{r}}, e_{\hat{\theta}} \right] f = -\frac{1}{r^2} \partial_\theta f$$

$$= -\frac{1}{r} e_{\hat{\theta}} f$$

The test function is just being carried along for the ride. Therefore, we can drop the test function and write

$$\left[e_{\hat{r}}, e_{\hat{\theta}}\right] = -\frac{1}{r} e_{\hat{\theta}}$$

This example demonstrates that the commutators of a nonholonomic basis do not always vanish. We can formalize this in the following way:

$$\left[e_i, e_j\right] = C_{ij}{}^k e_k \qquad (5.5)$$

The $C_{ij}{}^k$ are called *commutation coefficients*. The commutation coefficients are antisymmetric in the first two indices; i.e.

$$C_{ij}{}^k = -C_{ji}{}^k$$

If the following condition is met, then the basis set is holonomic.

$$C_{ij}{}^k = 0 \quad \forall i, j, k$$

We can also compute the commutation coefficients using the basis one forms, as we describe in the next section.

# Commutation Coefficients and Basis One Forms

It is also possible to determine whether or not a basis is holonomic by examining the basis one forms. A one form $\sigma$ can be expanded in terms of a set of coordinate basis one forms $\omega^a$ as

$$\sigma = \sigma_a \omega^a = \sigma_a \mathrm{d}x^a$$

In the same way that we can expand a vector in a different basis, we can also expand a one form in terms of a nonholonomic basis. Again using "hats" to denote the fact that we are working with a nonholonomic basis, we can write

$$\sigma = \sigma_{\hat{a}} \omega^{\hat{a}}$$

Both expansions represent the same one form. Given a particular set of basis one forms, it may be desirable to determine if it is holonomic. Once again we do this by calculating the commutation coefficients, but by a different method.

Given a set of basis one forms $\omega^a$, the commutation coefficients can be found by calculating $d\omega^a$. This quantity is related to commutation coefficients in the following way:

$$d\omega^a = -\frac{1}{2}C_{bc}{}^a\omega^b \wedge \omega^c \tag{5.6}$$

Now recall that for a coordinate basis, the basis one forms are given by

$$\omega^a = dx^a$$

In the previous chapter, we learned that the antisymmetry of the wedge product leads to the following result for an arbitrary $p$ form $\alpha$:

$$d(d\alpha) = 0$$

This means that for a coordinate basis, $d\omega^a = 0$. For spherical polar coordinates, if we choose the nonholonomic basis, the basis one forms are given by

$$\omega^{\hat{r}} = dr, \qquad \omega^{\hat{\theta}} = rd\theta, \qquad \omega^{\hat{\phi}} = r\sin\theta d\phi \tag{5.7}$$

Using (5.6), we can compute the commutation coefficients for this basis. For example,

$$d\omega^{\hat{\phi}} = d(r\sin\theta d\phi) = \sin\theta dr \wedge d\phi + r\cos\theta d\theta \wedge d\phi$$

Using the definitions given in (5.7), we can rewrite this expression in terms of the basis one forms. First, notice that

$$\sin\theta dr \wedge d\phi = dr \wedge \sin\theta d\phi = \frac{1}{r}dr \wedge r\sin\theta d\phi = \frac{1}{r}\omega^{\hat{r}} \wedge \omega^{\hat{\phi}}$$

For the second term, we find

$$r \cos \theta \, d\theta \wedge d\phi = \frac{\cos \theta}{r} (r \, d\theta) \wedge r \, d\phi = \frac{\cos \theta}{r \sin \theta} (r \, d\theta) \wedge r \sin \theta \, d\phi$$

$$= \frac{\cot \theta}{r} \omega^{\hat\theta} \wedge \omega^{\hat\phi}$$

Putting these results together, we obtain

$$d\omega^{\hat\phi} = \frac{1}{r} \omega^{\hat{r}} \wedge \omega^{\hat\phi} + \frac{\cot \theta}{r} \omega^{\hat\theta} \wedge \omega^{\hat\phi}$$

The antisymmetry of the wedge product means we can write this expression as

$$d\omega^{\hat\phi} = \frac{1}{r} \omega^{\hat{r}} \wedge \omega^{\hat\phi} + \frac{\cot \theta}{r} \omega^{\hat\theta} \wedge \omega^{\hat\phi}$$

$$= \frac{1}{2} \left( \frac{1}{r} \omega^{\hat{r}} \wedge \omega^{\hat\phi} - \frac{1}{r} \omega^{\hat\phi} \wedge \omega^{\hat{r}} \right) + \frac{1}{2} \left( \frac{\cot \theta}{r} \omega^{\hat\theta} \wedge \omega^{\hat\phi} - \frac{\cot \theta}{r} \omega^{\hat\phi} \wedge \omega^{\hat\theta} \right)$$

Now we compare with (5.6) to read off the commutation coefficients. We find

$$C_{\hat{r}\hat\phi}{}^{\hat\phi} = -C_{\hat\phi\hat{r}}{}^{\hat\phi} = -\frac{1}{r} \quad \text{and} \quad C_{\hat\theta\hat\phi}{}^{\hat\phi} = -C_{\hat\phi\hat\theta}{}^{\hat\phi} = -\frac{\cot \theta}{r}$$

Remember, if the commutation coefficients vanish, then the basis is holo-nomic. As we mentioned at the beginning of the chapter, it is often convenient to do calculations using the orthonormal basis but we may need to express re-sults in the coordinate basis. We now explore the techniques used to transform between the two.

# Transforming between Bases

We can work out a transformation law between the coordinate and noncoordinate basis vectors by using the coordinate components of the noncoordinate basis vectors. These components are denoted by $(e_{\hat{a}})^b$ and known as the *tetrad*. The meaning of these components is the same as we would find for any vector. In other words, we use them to expand a noncoordinate basis vector in terms of the basis vectors of a coordinate basis:

$$e_{\hat{a}} = (e_{\hat{a}})^b e_b \tag{5.8}$$

For example, we can expand the noncoordinate basis for spherical polar coordinates in terms of the coordinate basis vectors as follows:

$$e_{\hat{r}} = (e_{\hat{r}})^b \; e_b = (e_{\hat{r}})^r \, e_r + (e_{\hat{r}})^\theta \, e_\theta + (e_{\hat{r}})^\phi \, e_\phi$$

$$e_{\hat{\theta}} = \left(e_{\hat{\theta}}\right)^b \, e_b = \left(e_{\hat{\theta}}\right)^r \, e_r + \left(e_{\hat{\theta}}\right)^\theta \, e_\theta + \left(e_{\hat{\theta}}\right)^\phi \, e_\phi$$

$$e_{\hat{\phi}} = \left(e_{\hat{\phi}}\right)^b \, e_b = \left(e_{\hat{\phi}}\right)^r \, e_r + \left(e_{\hat{\phi}}\right)^\theta \, e_\theta + \left(e_{\hat{\phi}}\right)^\phi \, e_\phi$$

Earlier we stated that the noncoordinate basis vectors were

$$e_{\hat{r}} = \partial_r, \qquad e_{\hat{\theta}} = \frac{1}{r}\partial_\theta, \qquad e_{\hat{\phi}} = \frac{1}{r \sin\theta}\partial_\phi$$

Comparison with the above indicates that

$$(e_{\hat{r}})^r = 1$$

$$\left(e_{\hat{\theta}}\right)^\theta = \frac{1}{r}$$

$$\left(e_{\hat{\phi}}\right)^\phi = \frac{1}{r \sin\theta}$$

All other components are zero. The components $(e_{\hat{a}})^b$ can be used to construct a transformation matrix that we label $\Lambda_{\hat{a}}{}^b$ as this matrix represents a transformation between the global coordinates and the local Lorentz frame of an observer. In the case of spherical polar coordinates, we have

$$\Lambda_{\hat{a}}{}^b = \begin{pmatrix} 1 & 0 & 0 \\ 0 & \frac{1}{r} & 0 \\ 0 & 0 & \frac{1}{r \sin\theta} \end{pmatrix}$$

Expressing the transformation relation in terms of the matrix, we have

$$e_{\hat{a}} = \Lambda_{\hat{a}}{}^b e_b$$

The matrix $\Lambda_{\hat{a}}{}^b$ is invertible. The components of the inverse matrix represent the reverse situation, which is expanding the coordinate basis vectors in terms of the noncoordinate basis. This expansion can be written as follows:

$$e_a = (e_a)^{\hat{b}} \, e_{\hat{b}}$$

We use the components $(e_a)^{\hat{b}}$ to construct the inverse matrix that we write as $\left(\Lambda^{-1}\right)_{\hat{b}}{}^{\hat{a}}$. In terms of the components, we have

$$(e_a)^{\hat{b}}\left(e_{\hat{b}}\right)^c = \delta_a^c \quad \text{and} \quad (e_{\hat{a}})^b\left(e_b\right)^{\hat{c}} = \delta_{\hat{a}}^{\hat{c}}$$

Moreover, we have

$$(e_a)^{\hat{c}}\left(e_b\right)_{\hat{c}} = \eta_{ab}$$

In the case of spherical polar coordinates, the inverse matrix is given by

$$\left(\Lambda^{-1}\right)_b{}^{\hat{a}} = \begin{pmatrix} 1 & 0 & 0 \\ 0 & r & 0 \\ 0 & 0 & r\sin\theta \end{pmatrix}$$

It is also possible to derive a transformation relationship for the basis one forms. Once again, we recall the form of the basis one forms when using a coordinate basis. In that case they are exact differentials:

$$\omega^a = \mathrm{d}x^a$$

The noncoordinate basis is related to the coordinate basis in the following way:

$$\omega^{\hat{a}} = \omega^{\hat{a}}{}_b\,\mathrm{d}x^b$$

In the case of the basis one forms, the components of the transformation matrix are given the label $\omega^{\hat{a}}{}_b$. To work this out for spherical coordinates, we consider a single term; i.e.,

$$\omega^{\hat{\phi}} = \omega^{\hat{\phi}}{}_b\,\mathrm{d}x^b = \omega^{\hat{\phi}}{}_r\,\mathrm{d}r + \omega^{\hat{\phi}}{}_\theta\,\mathrm{d}\theta + \omega^{\hat{\phi}}{}_\phi\,\mathrm{d}\phi = r\sin\theta\,\mathrm{d}\phi$$

We conclude that the only nonzero component is given by

$$\omega^{\hat{\phi}}{}_\phi = r\sin\theta$$

It is a simple matter to show that the transformation matrix, which this time is denoted by $\Lambda^{\hat{a}}{}_b$, is given by

$$\Lambda^{\hat{a}}{}_b = \begin{pmatrix} 1 & 0 & 0 \\ 0 & r & 0 \\ 0 & 0 & r\sin\theta \end{pmatrix}$$

This is just the inverse matrix we found when transforming the basis vectors. To express the coordinate basis one forms in terms of the noncoordinate basis one forms, we use the inverse of this matrix; i.e.,

$$dx^a = \left(\Lambda^{-1}\right)^a{}_{\hat{b}}\,\omega^{\hat{b}}$$

In the case of spherical polar coordinates, this matrix is given by

$$\left(\Lambda^{-1}\right)^a{}_{\hat{b}} = \begin{pmatrix} 1 & 0 & 0 \\ 0 & \frac{1}{r} & 0 \\ 0 & 0 & \frac{1}{r\sin\theta} \end{pmatrix}$$

These transformation matrices are related to those used with the basis vectors in the following way:

$$\Lambda^{\hat{a}}{}_b = \left(\Lambda^{-1}\right)_b{}^{\hat{a}} \quad \text{and} \quad \Lambda_{\hat{a}}{}^b = \left(\Lambda^{-1}\right)^b{}_{\hat{a}}$$

# A Note on Notation

Consider a set of coordinates $\left[x^0, x^1, x^2, x^3\right]$. Suppose that we are working in a coordinate basis, i.e., $e_a = \partial/\partial x^a$. In this case, the metric or line element is written as

$$g = ds^2 = g_{ab}\,dx^a \otimes dx^b$$

If we are working with an orthonormal tetrad, we write the metric in terms of the basis one forms. In other words, we write

$$ds^2 = g = \eta_{\hat{a}\hat{b}}\,\omega^{\hat{a}} \otimes \omega^{\hat{b}}$$

In many cases, the inner product (as represented by $\eta_{\hat{a}\hat{b}}$) is diagonal. If we have $\eta_{\hat{a}\hat{b}} = \text{diag}\,(1, -1, -1, -1)$, then we can write the metric in the following way:

$$ds^2 = g = \omega^{\hat{0}} \otimes \omega^{\hat{0}} - \omega^{\hat{1}} \otimes \omega^{\hat{1}} - \omega^{\hat{2}} \otimes \omega^{\hat{2}} - \omega^{\hat{3}} \otimes \omega^{\hat{3}}$$

$$\approx \left(\omega^{\hat{0}}\right)^2 - \left(\omega^{\hat{1}}\right)^2 - \left(\omega^{\hat{2}}\right)^2 - \left(\omega^{\hat{3}}\right)^2$$

We will be using this frequently throughout the book in specific examples. We now turn to the task of computing curvature using the orthonormal basis. This type of calculation is sometimes referred to by the name *tetrad methods*. The equations used to perform the calculations are *Cartan's structure equations*.

# Cartan's First Structure Equation and the Ricci Rotation Coefficients

The first step in computing curvature, using the methods we are going to outline in this chapter, is to find the *curvature one forms* and the *Ricci rotation coefficients*. The notation used for this method makes it look a bit more mathematically sophisticated than it really is. In fact you may find it quite a bit less tedious than the "straightforward" methods used to find the Christoffel symbols and Riemann tensor in the last chapter.

The main thrust of this technique is given a set of basis one forms $\omega^{\hat{a}}$, we wish to calculate the derivatives $d\omega^{\hat{a}}$. These quantities satisfy *Cartan's first structure equation*, which is

$$d\omega^{\hat{a}} = -\Gamma^{\hat{a}}_{\ \hat{b}} \wedge \omega^{\hat{b}} \tag{5.9}$$

Note that we are using hatted indices, which indicates we are working in the noncoordinate basis. The $\Gamma^{\hat{a}}_{\ \hat{b}}$ are called *curvature one forms*, and they can be written in terms of the basis one forms $\omega^{\hat{a}}$ as follows:

$$\Gamma^{\hat{a}}_{\ \hat{b}} = \Gamma^{\hat{a}}_{\ \hat{b}\hat{c}}\, \omega^{\hat{c}} \tag{5.10}$$

We have introduced a new quantity, $\Gamma^{\hat{a}}_{\ \hat{b}\hat{c}}$, which are the Ricci rotation coefficients. They are related to the Christoffel symbols but are, in fact, different. We use them to obtain the Christoffel symbols by applying a transformation to the coordinate basis. We will illustrate this below.

The curvature one forms satisfy certain symmetry relations that will be useful during calculation. In particular

$$\Gamma_{\hat{a}\hat{b}} = -\Gamma_{\hat{b}\hat{a}} \tag{5.11}$$

$$\Gamma^{\hat{0}}{}_{\hat{i}} = \Gamma^{\hat{i}}{}_{\hat{0}} \quad \text{and} \quad \Gamma^{\hat{i}}{}_{\hat{j}} = -\Gamma^{\hat{j}}{}_{\hat{i}} \tag{5.12}$$

Here *i and j* are spatial indices and $i \neq j$. To see how these relations work, we can raise and lower indices with the metric in the local frame. If $\eta_{\hat{a}\hat{b}} = \text{diag}\,(1, -1, -1, -1)$, then using (5.11) we have

$$\Gamma^{\hat{i}}{}_{\hat{j}} = \eta^{\hat{i}\hat{i}}\Gamma_{\hat{i}\hat{j}} = -\Gamma_{\hat{i}\hat{j}} = \Gamma_{\hat{j}\hat{i}} = \eta_{\hat{j}\hat{j}}\Gamma^{\hat{j}}{}_{\hat{i}} = -\Gamma^{\hat{j}}{}_{\hat{i}}$$

This works because the elements of $\eta_{\hat{a}\hat{b}}$ are only on the diagonal, so we must have equal indices. Moreover, since we are dealing only with spatial components in this case, each instance of $\eta_{ij}$ introduces a minus sign. Of course the signs used here are by convention. Try working this out with $\eta_{\hat{a}\hat{b}} = \text{diag}\,(-1, 1, 1, 1)$ to see if things work out differently.

When calculating $d\omega^{\hat{a}}$, it is helpful to recall the following. Let $\alpha$ and $\beta$ be two arbitrary forms. Then

$$\begin{aligned} d\,(d\alpha) &= 0 \\ \alpha \wedge \beta &= -\beta \wedge \alpha \end{aligned} \tag{5.13}$$

So we recall (selecting the coordinate $r$ only for concreteness) that $d\,(dr) = 0$ and $dr \wedge dr = 0$.

The Christoffel symbols can be obtained from the Ricci rotation coefficients using the transformation matrices that take us from the *orthonormal basis* to the *coordinate basis*. In particular, we have

$$\Gamma^{a}{}_{bc} = \left(\Lambda^{-1}\right)^{a}{}_{\hat{d}}\,\Gamma^{\hat{d}}{}_{\hat{e}\hat{f}}\,\Lambda^{\hat{e}}{}_{b}\,\Lambda^{\hat{f}}{}_{c} \tag{5.14}$$

where $\Gamma^{a}{}_{bc}$ are the Christoffel symbols. Remember, a plain index indicates that the coordinate basis is being used, while the hatted index indicates that an orthonormal basis is being used. In this book when we write $\Gamma^{\hat{d}}{}_{\hat{e}\hat{f}}$, we are referring to the Ricci rotation coefficients in the orthonormal basis.

The techniques used in this chapter are far less tedious than using the coordinate methods for calculating the connection. In the next section, we will carry the method forward and see how we can calculate the Riemann tensor. The procedure used here will be to calculate both sides of (5.9) and compare to make an educated guess as to what the curvature one forms are (this is sometimes called the *guess method*). Once we have done that, we can use the symmetry properties in (5.12) to see if we can find any more of the curvature one forms. Let's apply this method to an example.

**EXAMPLE 5-1**
Consider the Tolman-Bondi-de Sitter metric, given by

$$ds^2 = dt^2 - e^{-2\Psi(t,r)}dr^2 - R^2(t,r)\,d\theta^2 - R^2(t,r)\sin^2\theta d\phi^2 \qquad (5.15)$$

This metric arises, for example, in the study of spherical dust with a cosmological constant. Find the Ricci rotation coefficients for this metric.

**SOLUTION 5-1**
First we need to examine the metric (5.15) and write down the basis one forms. Since we are asked to find the Ricci rotation coefficients, we will work with the noncoordinate orthonormal basis. Writing (5.15) as

$$ds^2 = \left(\omega^{\hat{t}}\right)^2 - \left(\omega^{\hat{r}}\right)^2 - \left(\omega^{\hat{\theta}}\right)^2 - \left(\omega^{\hat{\phi}}\right)^2$$

we identify the noncoordinate basis one forms as

$$\omega^{\hat{t}} = dt, \quad \omega^{\hat{r}} = e^{-\Psi(t,r)}dr, \quad \omega^{\hat{\theta}} = R(t,r)\,d\theta, \quad \omega^{\hat{\phi}} = R(t,r)\sin\theta d\phi \qquad (5.16)$$

As we proceed, at times we will denote differentiation with respect to time with a "dot" (i.e., $df/dt = \dot{f}$) and differentiation with respect to $r$ with a prime (i.e., $df/dr = f'$).

Now we apply (5.9) to each basis one form. The first one does not give us much information, since

$$d\omega^{\hat{t}} = d\,(dt) = 0$$

Let's move on to the second one form. This one gives us

$$d\omega^{\hat{r}} = d\left(e^{-\Psi(t,r)}dr\right) = -\frac{\partial\Psi}{\partial t}e^{-\Psi(t,r)}dt \wedge dr - \frac{\partial\Psi}{\partial r}e^{-\Psi(t,r)}dr \wedge dr$$

Since $dr \wedge dr = 0$, this immediately simplifies to

$$d\omega^{\hat{r}} = -\frac{\partial \Psi}{\partial t}e^{-\Psi(t,r)}dt \wedge dr$$

The next step is to rewrite the expression we've derived in terms of the basis one forms. When doing these types of calculations, it is helpful to write the coordinate differentials in terms of the basis one forms. Looking at (5.16), we quickly see that

$$dt = \omega^{\hat{t}}, \quad dr = e^{\Psi(t,r)}\omega^{\hat{r}}, \quad d\theta = \frac{1}{R(t,r)}\omega^{\hat{\theta}}, \quad d\phi = \frac{1}{R(t,r)\sin\theta}\omega^{\hat{\phi}}$$

$$(5.17)$$

Therefore, we obtain

$$d\omega^{\hat{r}} = -\frac{\partial \Psi}{\partial t}e^{-\Psi(t,r)}dt \wedge dr = -\frac{\partial \Psi}{\partial t}e^{-\Psi(t,r)}\omega^{\hat{t}} \wedge dr$$

$$= -\frac{\partial \Psi}{\partial t}e^{-\Psi(t,r)}\omega^{\hat{t}} \wedge e^{\Psi(t,r)}\omega^{\hat{r}}$$

$$= -\frac{\partial \Psi}{\partial t}\omega^{\hat{t}} \wedge \omega^{\hat{r}}$$

Now we use (5.13) to write this as

$$d\omega^{\hat{r}} = \frac{\partial \Psi}{\partial t}\omega^{\hat{r}} \wedge \omega^{\hat{t}} \qquad (5.18)$$

Now let's use (5.9) to write out what $d\omega^{\hat{r}}$ should be. We find

$$d\omega^{\hat{r}} = -\Gamma^{\hat{r}}_{\hat{b}} \wedge \omega^{\hat{b}} = -\Gamma^{\hat{r}}_{\hat{t}} \wedge \omega^{\hat{t}} - \Gamma^{\hat{r}}_{\hat{r}} \wedge \omega^{\hat{r}} - \Gamma^{\hat{r}}_{\hat{\theta}} \wedge \omega^{\hat{\theta}} - \Gamma^{\hat{r}}_{\hat{\phi}} \wedge \omega^{\hat{\phi}}$$

Comparing this expression with (5.18), we guess that

$$\Gamma^{\hat{r}}_{\hat{t}} = -\frac{\partial \Psi}{\partial t}\omega^{\hat{r}} \qquad (5.19)$$

Moving to the next basis one form, we get

$$d\omega^{\hat{\theta}} = d\left(R(t,r)d\theta\right)$$

$$= \dot{R}dt \wedge d\theta + R'dr \wedge d\theta$$

$$= \frac{\dot{R}}{R}dt \wedge Rd\theta + \frac{R'}{R}e^{\Psi}\left(e^{-\Psi}dr\right) \wedge Rd\theta$$

$$= \frac{\dot{R}}{R}\omega^{\hat{t}} \wedge \omega^{\hat{\theta}} + \frac{R'}{R}e^{\Psi}\omega^{\hat{r}} \wedge \omega^{\hat{\theta}}$$

Let's reverse the order of the terms using (5.13). The result is

$$d\omega^{\hat{\theta}} = -\frac{\dot{R}}{R}\omega^{\hat{\theta}} \wedge \omega^{\hat{t}} - \frac{R'}{R}e^{\Psi}\omega^{\hat{\theta}} \wedge \omega^{\hat{r}}$$

Using (5.9), we have

$$d\omega^{\hat{\theta}} = -\Gamma^{\hat{\theta}}_{\hat{b}} \wedge \omega^{\hat{b}}$$

$$= -\Gamma^{\hat{\theta}}_{\hat{t}} \wedge \omega^{\hat{t}} - \Gamma^{\hat{\theta}}_{\hat{r}} \wedge \omega^{\hat{r}} - \Gamma^{\hat{\theta}}_{\hat{\theta}} \wedge \omega^{\hat{\theta}} - \Gamma^{\hat{\theta}}_{\hat{\phi}} \wedge \omega^{\hat{\phi}}$$

Comparing this with the result we found just above, we conclude that

$$\Gamma^{\hat{\theta}}_{\hat{t}} = \frac{\dot{R}}{R}\omega^{\hat{\theta}} \quad \text{and} \quad \Gamma^{\hat{\theta}}_{\hat{r}} = \frac{R'}{R}e^{\Psi(t,r)}\omega^{\hat{\theta}} \tag{5.20}$$

Finally, we tackle $\omega^{\hat{\phi}}$. This term gives

$$d\omega^{\hat{\phi}} = d\left(R\sin\theta d\phi\right)$$

$$= \dot{R}\sin\theta dt \wedge d\phi + R'\sin\theta dr \wedge d\phi + \cos\theta Rd\theta \wedge d\phi$$

$$= \frac{\dot{R}}{R}dt \wedge R\sin\theta d\phi + \frac{R'}{R}e^{\Psi(t,r)}\left(e^{-\Psi(t,r)}dr\right) \wedge R\sin\theta d\phi$$

$$+ \frac{\cos\theta}{R\sin\theta}R\,d\theta \wedge R\sin\theta d\phi$$

Writing the differentials in terms of the basis one forms, and then reversing the order of each term, we get

$$d\omega^{\hat{\phi}} = \frac{\dot{R}}{R}\omega^{\hat{t}} \wedge \omega^{\hat{\phi}} + \frac{R'}{R}e^{\Psi(t,r)}\omega^{\hat{r}} \wedge \omega^{\hat{\phi}} + \frac{\cot\theta}{R}\omega^{\hat{\theta}} \wedge \omega^{\hat{\phi}}$$

$$= -\frac{\dot{R}}{R}\omega^{\hat{\phi}} \wedge \omega^{\hat{t}} - \frac{R'}{R}e^{\Psi(t,r)}\omega^{\hat{\phi}} \wedge \omega^{\hat{r}} - \frac{\cot\theta}{R}\omega^{\hat{\phi}} \wedge \omega^{\hat{\theta}}$$

Using (5.9), we have

$$d\omega^{\hat{\phi}} = -\Gamma^{\hat{\phi}}{}_{\hat{t}} \wedge \omega^{\hat{t}} - \Gamma^{\hat{\phi}}{}_{\hat{r}} \wedge \omega^{\hat{r}} - \Gamma^{\hat{\phi}}{}_{\hat{\theta}} \wedge \omega^{\hat{\theta}} - \Gamma^{\hat{\phi}}{}_{\hat{\phi}} \wedge \omega^{\hat{\phi}}$$

Comparing this with the result we obtained above, we conclude that

$$\Gamma^{\hat{\phi}}{}_{\hat{t}} = \frac{\dot{R}}{R}\omega^{\hat{\phi}}, \qquad \Gamma^{\hat{\phi}}{}_{\hat{r}} = \frac{R'}{R}e^{\Psi(t,r)}\omega^{\hat{\phi}}, \qquad \Gamma^{\hat{\phi}}{}_{\hat{\theta}} = \frac{\cot\theta}{R}\omega^{\hat{\phi}} \qquad (5.21)$$

Earlier we noted that we could obtain no information by calculating $d\omega^{\hat{t}}$. For the other terms, we have basically gone as far as we can using the "guess" method. Now we can use the symmetry properties shown in (5.12) to find the other curvature one forms. Specifically, we have

$$\Gamma^{\hat{t}}{}_{\hat{r}} = \Gamma^{\hat{r}}{}_{\hat{t}} = -\frac{\partial\Psi}{\partial t}\omega^{\hat{r}}$$

$$\Gamma^{\hat{t}}{}_{\hat{\theta}} = \Gamma^{\hat{\theta}}{}_{\hat{t}} = \frac{\dot{R}}{R}\omega^{\hat{\theta}}$$

$$\Gamma^{\hat{t}}{}_{\hat{\phi}} = \Gamma^{\hat{\phi}}{}_{\hat{t}} = \frac{\dot{R}}{R}\omega^{\hat{\phi}}$$

Now paying attention only to spatial indices, we conclude that

$$\Gamma^{\hat{r}}{}_{\hat{\theta}} = -\Gamma^{\hat{\theta}}{}_{\hat{r}} = -\frac{R'}{R}e^{\Psi(t,r)}\omega^{\hat{\theta}}$$

$$\Gamma^{\hat{r}}{}_{\hat{\phi}} = -\Gamma^{\hat{\phi}}{}_{\hat{r}} = -\frac{R'}{R}e^{\Psi(t,r)}\omega^{\hat{\phi}}$$

$$\Gamma^{\hat{\theta}}{}_{\hat{\phi}} = -\Gamma^{\hat{\phi}}{}_{\hat{\theta}} = -\frac{\cot\theta}{R}\omega^{\hat{\phi}}$$

Now that we have calculated the curvature one forms, we can obtain the Ricci rotation coefficients using (5.10), which we restate here:

$$\Gamma^{\hat{a}}{}_{\hat{b}} = \Gamma^{\hat{a}}{}_{\hat{b}\hat{c}}\omega^{\hat{c}}$$

We start by considering $\Gamma^{\hat{t}}{}_{\hat{b}}$. It turns out that $\Gamma^{\hat{t}}{}_{\hat{t}} = 0$, and so we can skip this term. Moving on, we have

$$\Gamma^{\hat{t}}{}_{\hat{r}} = \Gamma^{\hat{t}}{}_{\hat{r}\hat{t}}\omega^{\hat{t}} + \Gamma^{\hat{t}}{}_{\hat{r}\hat{r}}\omega^{\hat{r}} + \Gamma^{\hat{t}}{}_{\hat{r}\hat{\theta}}\omega^{\hat{\theta}} + \Gamma^{\hat{t}}{}_{\hat{r}\hat{\phi}}\omega^{\hat{\phi}}$$

We noted above that $\Gamma^{\hat{t}}{}_{\hat{r}} = -\frac{\partial \Psi(t,r)}{\partial t}\omega^{\hat{r}}$. Comparing the two expressions, we conclude that

$$\Gamma^{\hat{t}}{}_{\hat{r}\hat{r}} = -\frac{\partial \Psi(t,r)}{\partial t} \tag{5.22}$$

In addition, we conclude that $\Gamma^{\hat{t}}{}_{\hat{r}\hat{t}} = \Gamma^{\hat{t}}{}_{\hat{r}\hat{\theta}} = \Gamma^{\hat{t}}{}_{\hat{r}\hat{\phi}} = 0$. Moving to the next coordinate, we have

$$\Gamma^{\hat{t}}{}_{\hat{\theta}} = \Gamma^{\hat{t}}{}_{\hat{\theta}\hat{c}}\omega^{\hat{c}} = \Gamma^{\hat{t}}{}_{\hat{\theta}\hat{t}}\omega^{\hat{t}} + \Gamma^{\hat{t}}{}_{\hat{\theta}\hat{r}}\omega^{\hat{r}} + \Gamma^{\hat{t}}{}_{\hat{\theta}\hat{\theta}}\omega^{\hat{\theta}} + \Gamma^{\hat{t}}{}_{\hat{\theta}\hat{\phi}}\omega^{\hat{\phi}}$$

Above we found that $\Gamma^{\hat{t}}{}_{\hat{\theta}} = \Gamma^{\hat{\theta}}{}_{\hat{t}} = \frac{\dot{R}}{R}\omega^{\hat{\theta}}$, and so the only nonzero term is that involving $\omega^{\hat{\theta}}$, and we conclude that

$$\Gamma^{\hat{t}}{}_{\hat{\theta}\hat{\theta}} = \frac{\dot{R}}{R} \tag{5.23}$$

Similarly, we find

$$\Gamma^{\hat{t}}{}_{\hat{\phi}\hat{\phi}} = \frac{\dot{R}}{R} \tag{5.24}$$

Now we move on to terms of the form $\Gamma^{\hat{r}}{}_{\hat{b}}$. First we have

$$\Gamma^{\hat{r}}{}_{\hat{t}} = \Gamma^{\hat{r}}{}_{\hat{t}\hat{t}}\omega^{\hat{t}} + \Gamma^{\hat{r}}{}_{\hat{t}\hat{r}}\omega^{\hat{r}} + \Gamma^{\hat{r}}{}_{\hat{t}\hat{\theta}}\omega^{\hat{\theta}} + \Gamma^{\hat{r}}{}_{\hat{t}\hat{\phi}}\omega^{\hat{\phi}}$$

Now, in (5.19), we found that $\Gamma^{\hat{r}}_{\hat{t}} = -\dot{\Psi}(t, r)\omega^{\hat{r}}$. Comparison of the two leads us to conclude that

$$\Gamma^{\hat{r}}_{\hat{t}\hat{r}} = -\frac{\partial \Psi}{\partial t}$$

Next, skipping the $\Gamma^{\hat{r}}_{\hat{r}}$ term since it vanishes by (5.12) (set $i = j$), we go to

$$\Gamma^{\hat{r}}_{\hat{\theta}} = \Gamma^{\hat{r}}_{\hat{\theta}\hat{c}}\omega^{\hat{c}} = \Gamma^{\hat{r}}_{\hat{\theta}\hat{t}}\omega^{\hat{t}} + \Gamma^{\hat{r}}_{\hat{\theta}\hat{r}}\omega^{\hat{r}} + \Gamma^{\hat{r}}_{\hat{\theta}\hat{\theta}}\omega^{\hat{\theta}} + \Gamma^{\hat{r}}_{\hat{\theta}\hat{\phi}}\omega^{\hat{\phi}}$$

Comparison with our earlier result, where we found that $\Gamma^{\hat{r}}_{\hat{\theta}} = -\Gamma^{\hat{\theta}}_{\hat{r}} = -\frac{R'}{R}e^{\Psi(t,r)}\omega^{\hat{\theta}}$, leads us to conclude that

$$\Gamma^{\hat{r}}_{\hat{\theta}\hat{\theta}} = -\frac{R'}{R}e^{\Psi(t,r)}$$

Finally, we have

$$\Gamma^{\hat{r}}_{\hat{\phi}} = \Gamma^{\hat{r}}_{\hat{\phi}\hat{c}}\omega^{\hat{c}} = \Gamma^{\hat{r}}_{\hat{\phi}\hat{t}}\omega^{\hat{t}} + \Gamma^{\hat{r}}_{\hat{\phi}\hat{r}}\omega^{\hat{r}} + \Gamma^{\hat{r}}_{\hat{\phi}\hat{\theta}}\omega^{\hat{\theta}} + \Gamma^{\hat{r}}_{\hat{\phi}\hat{\phi}}\omega^{\hat{\phi}}$$

Earlier we found that $\Gamma^{\hat{r}}_{\hat{\phi}} = -\frac{R'}{R}e^{\Psi(t,r)}\omega^{\hat{\phi}}$. Comparing this to the above, it must be the case that

$$\Gamma^{\hat{r}}_{\hat{\phi}\hat{\phi}} = -\frac{R'}{R}e^{\Psi(t,r)}$$

A similar procedure applied to terms of the form $\Gamma^{\hat{\theta}}_{\hat{b}}$ and $\Gamma^{\hat{\phi}}_{\hat{b}}$ gives the remaining Ricci rotation coefficients

$$\Gamma^{\hat{\theta}}_{\hat{t}\hat{\theta}} = \Gamma^{\hat{\phi}}_{\hat{t}\hat{\phi}} = \frac{\dot{R}}{R}, \quad \Gamma^{\hat{\theta}}_{\hat{r}\hat{\theta}} = \Gamma^{\hat{\phi}}_{\hat{r}\hat{\phi}} = \frac{R'}{R}e^{\Psi(t,r)}, \quad \Gamma^{\hat{\phi}}_{\hat{\theta}\hat{\phi}} = -\Gamma^{\hat{\theta}}_{\hat{\phi}\hat{\phi}} = -\frac{\cot\theta}{R}$$

We have finished calculating the Ricci rotation coefficients, which can be used to obtain the Christoffel symbols or as we will see later, to compute the curvature. At this point, we can demonstrate how to transform these quantities to the coordinate frame to give the Christoffel symbols. This is not strictly necessary but it may be desired. Let's consider a simple example; the procedure is the same when applied to all terms. We repeat the formula we need here, namely (5.14):

$$\Gamma^{a}_{bc} = \left(\Lambda^{-1}\right)^{a}_{\hat{d}}\Gamma^{\hat{d}}_{\hat{e}\hat{f}}\Lambda^{\hat{e}}_{b}\Lambda^{\hat{f}}_{c}$$

First, we need to construct the transformation matrix. This is easy enough to do; we just read off the coefficients of the metric. To keep you from having to flip back several pages we restate it here:

$$ds^2 = dt^2 - e^{-2\Psi(t,r)}dr^2 - R^2(t,r)\,d\theta^2 - R^2(t,r)\sin^2\theta\,d\phi^2$$

The diagonal elements of the matrix can just be read off the metric. Therefore, the transformation matrix is

$$\Lambda^{\hat{a}}{}_b = \begin{pmatrix} 1 & 0 & 0 & 0 \\ 0 & e^{-\Psi(t,r)} & 0 & 0 \\ 0 & 0 & R(t,r) & 0 \\ 0 & 0 & 0 & R(t,r)\sin\theta \end{pmatrix} \tag{5.25}$$

and the inverse is

$$\left(\Lambda^{-1}\right)^{a}{}_{\hat{b}} = \begin{pmatrix} 1 & 0 & 0 & 0 \\ 0 & e^{\Psi(t,r)} & 0 & 0 \\ 0 & 0 & \frac{1}{R} & 0 \\ 0 & 0 & 0 & \frac{1}{R\sin\theta} \end{pmatrix} \tag{5.26}$$

In this case, finding the Christoffel symbols is rather easy since the matrix is diagonal. As an example, let's compute $\Gamma^\phi{}_{\theta\phi}$. Using the formula, this is

$$\Gamma^\phi{}_{\theta\phi} = \left(\Lambda^{-1}\right)^\phi{}_{\hat{\phi}}\Gamma^{\hat{\phi}}{}_{\hat{\theta}\hat{\phi}}\Lambda^{\hat{\theta}}{}_\theta\Lambda^{\hat{\phi}}{}_\phi = \left(\frac{1}{R\sin\theta}\right)\left(-\frac{\cot\theta}{R}\right)(R)(R\sin\theta)$$
$$= -\cot\theta$$

This example has shown us how to carry out the first step needed to compute curvature, getting the Christoffel symbols. Usually, we will carry our calculations further in the local frame using the Ricci rotation coefficients. The procedure used to do so is explored in the next section.

# Computing Curvature

Working in a coordinate basis to find all the components of the affine connection and then calculating an endless series of derivatives to get the Riemann tensor is a hairy mess that one would like to avoid. Thankfully, a method was developed by Cartan that is a bit more sophisticated but saves a great deal of tedium.

To get the curvature and ultimately the Einstein tensor so that we can learn about gravitational fields, we need the Riemann tensor and the quantities that can be derived from it. Previously we learned that

$$R^a{}_{bcd} = \partial_c \Gamma^a{}_{bd} - \partial_d \Gamma^a{}_{bc} + \Gamma^e{}_{bd}\Gamma^a{}_{ec} - \Gamma^e{}_{bc}\Gamma^a{}_{ed}$$

This is a daunting formula that seems to ooze airs of tedious calculation. Only a person who was completely insane would enjoy calculating such a beast. Luckily, Cartan has saved us with a more compact equation that is a bit easier on the mind once you get the hang of it. The key insight is to notice that we obtain the Riemann tensor from the Christoffel symbols by differentiating them. In the previous section, we calculated a set of curvature one forms, which contain the Ricci rotation coefficients as their components in the local frame. So it makes perfect sense that we should differentiate these things to get the curvature tensor. We do so in a way applicable to forms and define a new set of quantities called the curvature two forms which we label by $\Omega^{\hat{a}}{}_{\hat{b}}$ in this book. They are given by

$$\Omega^{\hat{a}}{}_{\hat{b}} = \mathrm{d}\Gamma^{\hat{a}}{}_{\hat{b}} + \Gamma^{\hat{a}}{}_{\hat{c}} \wedge \Gamma^{\hat{c}}{}_{\hat{b}} \tag{5.27}$$

It turns out that they are related to the Riemann tensor in the following way:

$$\Omega^{\hat{a}}{}_{\hat{b}} = \frac{1}{2} R^{\hat{a}}{}_{\hat{b}\hat{c}\hat{d}}\omega^{\hat{c}} \wedge \omega^{\hat{d}} \tag{5.28}$$

Now, notice that the Riemann tensor in this equation is expressed in the orthonormal basis (there are "hats" on the indices). That means we need to transform to the coordinate basis if we need or want to work there. This is done by using the following handy transformation formula:

$$R^a{}_{bcd} = \left(\Lambda^{-1}\right)^a{}_{\hat{e}} R^{\hat{e}}{}_{\hat{f}\hat{g}\hat{h}} \Lambda^{\hat{f}}{}_b \Lambda^{\hat{g}}{}_c \Lambda^{\hat{h}}{}_d \tag{5.29}$$

Let's do a few calculations to see how to use these quantities. We start with a very simple case and then consider one that's a bit more complicated.

**EXAMPLE 5-2**
Perhaps the simplest metric for which we have nonzero curvature is the unit sphere, where

$$ds^2 = d\theta^2 + \sin^2\theta \, d\phi^2$$

Find the Ricci scalar using Cartan's structure equations.

**SOLUTION 5-2**

This is a very simple metric. The basis one forms (for the orthonormal basis) are given by

$$\omega^{\hat\theta} = \mathrm{d}\theta \quad\text{and}\quad \omega^{\hat\phi} = \sin\theta\,\mathrm{d}\phi$$

In the quiz, you will show that the nonzero Ricci rotation coefficient for this metric is given by

$$\Gamma^{\hat\phi}{}_{\hat\theta} = \cot\theta\,\omega^{\hat\phi}$$

$$\Rightarrow\ \Gamma^{\hat\phi}{}_{\hat\theta\hat\phi} = \cot\theta$$

In this case, (5.27) becomes

$$\Omega^{\hat\phi}{}_{\hat\theta} = \mathrm{d}\Gamma^{\hat\phi}{}_{\hat\theta} + \Gamma^{\hat\phi}{}_{\hat c} \wedge \Gamma^{\hat c}{}_{\hat\theta} = \mathrm{d}\Gamma^{\hat\phi}{}_{\hat\theta} + \Gamma^{\hat\phi}{}_{\hat\theta} \wedge \Gamma^{\hat\theta}{}_{\hat\theta} + \Gamma^{\hat\phi}{}_{\hat\phi} \wedge \Gamma^{\hat\phi}{}_{\hat\theta}$$

$$= \mathrm{d}\Gamma^{\hat\phi}{}_{\hat\theta} = \mathrm{d}\left(\cot\theta\,\omega^{\hat\phi}\right)$$

$$= \mathrm{d}\left(\frac{\cos\theta}{\sin\theta}\sin\theta\,\mathrm{d}\phi\right) = \mathrm{d}\left(\cos\theta\,\mathrm{d}\phi\right)$$

Moving from the first to the second line, we used $\Gamma^{\hat\theta}{}_{\hat\theta} = \Gamma^{\hat\phi}{}_{\hat\phi} = 0$. Moving on, we obtain

$$\Omega^{\hat\phi}{}_{\hat\theta} = \mathrm{d}\left(\cos\theta\,\mathrm{d}\phi\right) = -\sin\theta\,\mathrm{d}\theta \wedge \mathrm{d}\phi$$

The next step is to rewrite the differentials in terms of the nonholonomic basis one forms. Looking at our definitions $\omega^{\hat\theta} = \mathrm{d}\theta$ and $\omega^{\hat\phi} = \sin\theta\,\mathrm{d}\phi$, we see that we can write $\mathrm{d}\phi = \frac{1}{\sin\theta}\omega^{\hat\phi}$, and so we obtain

$$\Omega^{\hat\phi}{}_{\hat\theta} = -\sin\theta\,\mathrm{d}\theta \wedge \mathrm{d}\phi = -\omega^{\hat\theta} \wedge \omega^{\hat\phi} = \omega^{\hat\phi} \wedge \omega^{\hat\theta} \tag{5.30}$$

This result can be used to obtain the components of the Riemann tensor via (5.28). Since there are only two dimensions, this equation will be very simple, because the only nonzero terms are those that contain $\omega^{\hat a} \wedge \omega^{\hat b}$ where $\hat a \neq \hat b$. This gives

$$\Omega^{\hat\phi}{}_{\hat\theta} = \frac{1}{2}R^{\hat\phi}{}_{\hat\theta\hat c\hat d}\omega^{\hat c} \wedge \omega^{\hat d} = \frac{1}{2}R^{\hat\phi}{}_{\hat\theta\hat\theta\hat\phi}\omega^{\hat\theta} \wedge \omega^{\hat\phi} + \frac{1}{2}R^{\hat\phi}{}_{\hat\theta\hat\phi\hat\theta}\omega^{\hat\phi} \wedge \omega^{\hat\theta}$$

Now, we use $\alpha \wedge \beta = -\beta \wedge \alpha$ to rewrite this as

$$\Omega^{\hat{\phi}}{}_{\hat{\theta}} = \frac{1}{2}\left(R^{\hat{\phi}}{}_{\hat{\theta}\hat{\theta}\hat{\phi}} - R^{\hat{\phi}}{}_{\hat{\theta}\hat{\phi}\hat{\theta}}\right)\omega^{\hat{\theta}} \wedge \omega^{\hat{\phi}}$$

To get this into a form where we can read off the components of the Riemann tensor from (5.30), we need to use the symmetries of the Riemann tensor. First we need to lower some indices. Since the metric is $ds^2 = d\theta^2 + \sin^2\theta\, d\phi^2$, $\eta_{\hat{a}\hat{b}}$ is nothing but the identity matrix:

$$\eta_{\hat{a}\hat{b}} = \begin{pmatrix} 1 & 0 \\ 0 & 1 \end{pmatrix}$$

Now recall the symmetries of the Riemann tensor:

$$R_{abcd} = -R_{bacd} = -R_{abdc}$$

These symmetries still apply in the local frame where we are using the nonholonomic basis. Therefore, we can write

$$R^{\hat{\phi}}{}_{\hat{\theta}\hat{\phi}\hat{\theta}} = \eta^{\hat{\phi}\hat{\phi}}R_{\hat{\phi}\hat{\theta}\hat{\phi}\hat{\theta}} = R_{\hat{\phi}\hat{\theta}\hat{\phi}\hat{\theta}} = -R_{\hat{\phi}\hat{\theta}\hat{\theta}\hat{\phi}} = -\eta_{\hat{\phi}\hat{\phi}}R^{\hat{\phi}}{}_{\hat{\theta}\hat{\theta}\hat{\phi}} = -R^{\hat{\phi}}{}_{\hat{\theta}\hat{\theta}\hat{\phi}}$$

This allows us to rewrite the following:

$$\frac{1}{2}\left(R^{\hat{\phi}}{}_{\hat{\theta}\hat{\theta}\hat{\phi}} - R^{\hat{\phi}}{}_{\hat{\theta}\hat{\phi}\hat{\theta}}\right) = \frac{1}{2}\left(R^{\hat{\phi}}{}_{\hat{\theta}\hat{\theta}\hat{\phi}} + R^{\hat{\phi}}{}_{\hat{\theta}\hat{\theta}\hat{\phi}}\right) = R^{\hat{\phi}}{}_{\hat{\theta}\hat{\theta}\hat{\phi}}$$

With this change, we get

$$\Omega^{\hat{\phi}}{}_{\hat{\theta}} = R^{\hat{\phi}}{}_{\hat{\theta}\hat{\theta}\hat{\phi}}\omega^{\hat{\theta}} \wedge \omega^{\hat{\phi}}$$

Comparison with (5.30) gives

$$R^{\hat{\phi}}{}_{\hat{\theta}\hat{\theta}\hat{\phi}} = -1$$

Another application of the symmetries of the Riemann tensor shows us that

$$R^{\hat{\phi}}{}_{\hat{\theta}\hat{\phi}\hat{\theta}} = R^{\hat{\theta}}{}_{\hat{\phi}\hat{\theta}\hat{\phi}} = +1$$

$$R^{\hat{\theta}}{}_{\hat{\theta}\hat{\theta}\hat{\phi}} = R_{\hat{\theta}\hat{\theta}\hat{\theta}\hat{\phi}} = -R_{\hat{\theta}\hat{\theta}\hat{\theta}\hat{\phi}}, \qquad R^{\hat{\theta}}{}_{\hat{\theta}\hat{\theta}\hat{\phi}} = 0$$

The second line holds true because $R_{abcd} = -R_{bacd}$ and if $a = -a \Rightarrow a = 0$. This also holds for $R^{\hat{\phi}}{}_{\hat{\theta}\hat{\phi}\hat{\phi}}$.

At this point, we can calculate the Ricci tensor and the Ricci scalar. Since the latter quantity is a scalar, it is invariant and it is not necessary to transform to the global coordinates. From the previous argument, we find

$$R_{\hat{\theta}\hat{\phi}} = R^{\hat{a}}{}_{\hat{\theta}\hat{a}\hat{\phi}} = R^{\hat{\theta}}{}_{\hat{\theta}\hat{\theta}\hat{\phi}} + R^{\hat{\phi}}{}_{\hat{\theta}\hat{\phi}\hat{\phi}} = 0$$

The other terms are

$$
\begin{aligned}
R_{\hat{\theta}\hat{\theta}} &= R^{\hat{a}}{}_{\hat{\theta}\hat{a}\hat{\theta}} = R^{\hat{\theta}}{}_{\hat{\theta}\hat{\theta}\hat{\theta}} + R^{\hat{\phi}}{}_{\hat{\theta}\hat{\phi}\hat{\theta}} = 0 + 1 = 1 \\
R_{\hat{\phi}\hat{\phi}} &= R^{\hat{a}}{}_{\hat{\phi}\hat{a}\hat{\phi}} = R^{\hat{\theta}}{}_{\hat{\phi}\hat{\theta}\hat{\phi}} + R^{\hat{\phi}}{}_{\hat{\phi}\hat{\phi}\hat{\phi}} = 1 + 0 = 1
\end{aligned}
\tag{5.31}
$$

From this, it is a simple matter to calculate the Ricci scalar

$$R = \eta^{\hat{a}\hat{b}} R_{\hat{a}\hat{b}} = \eta^{\hat{\theta}\hat{\theta}} R_{\hat{\theta}\hat{\theta}} + \eta^{\hat{\phi}\hat{\phi}} R_{\hat{\phi}\hat{\phi}} = R_{\hat{\theta}\hat{\theta}} + R_{\hat{\phi}\hat{\phi}} = 1 + 1 = 2$$

The Ricci scalar can be used to give a basic characterization of the intrinsic curvature of a given geometry. The value of the Ricci scalar tells us what the geometry "looks like" locally. If the Ricci scalar is positive, as it is in this case, the surface looks like a sphere. If it is negative, then the surface looks like a saddle. We can think about this in terms of drawing a triangle. If R > 0, then the angles add upto more than 180 degrees, while if R < 0 they add up to less than 180 degrees. These observations lead to the designations of 'positive curvature' and 'negative curvature'. Now, if R = 0, then the geometry is flat and the angles of a triangle add upto the expected 180 degrees.

**EXAMPLE 5-3**
The Robertson-Walker metric

$$ds^2 = -dt^2 + \frac{a^2(t)}{1 - kr^2} dr^2 + a^2(t)r^2 d\theta^2 + a^2(t)r^2 \sin^2\theta d\phi^2$$

describes a homogeneous, isotropic, and expanding universe. The constant $k = -1, 0, 1$ depending on whether the universe is open, flat, or closed. Find the components of the Riemann tensor using tetrad methods.

## SOLUTION 5-3

Looking at the metric, we see that we can define the following orthonormal basis of one forms:

$$\omega^{\hat{t}} = dt, \quad \omega^{\hat{r}} = \frac{a(t)}{\sqrt{1-kr^2}} dr, \quad \omega^{\hat{\theta}} = ra(t)d\theta, \quad \omega^{\hat{\phi}} = ra(t)\sin\theta d\phi$$

$$(5.32)$$

From this point on to save a bit of writing we will write $a(t) = a$. Using (5.9) to calculate the curvature, one forms gives

$$d\omega^{\hat{t}} = d\,(dt) = 0$$

$$d\omega^{\hat{r}} = d\left(\frac{a}{\sqrt{1-kr^2}} dr\right) = \frac{\dot{a}dt \wedge dr}{\sqrt{1-kr^2}} = -\frac{\dot{a}}{a}\omega^{\hat{r}} \wedge \omega^{\hat{t}}$$

$$d\omega^{\hat{\theta}} = d\,(rad\theta) = r\dot{a}dt \wedge d\theta + adr \wedge d\theta$$

$$= \frac{\dot{a}}{a}\omega^{\hat{t}} \wedge \omega^{\hat{\theta}} + \frac{a\sqrt{1-kr^2}}{\sqrt{1-kr^2}}\frac{ra}{ra}dr \wedge d\theta$$

$$= -\frac{\dot{a}}{a}\omega^{\hat{\theta}} \wedge \omega^{\hat{t}} + \frac{\sqrt{1-kr^2}}{ra}\omega^{\hat{r}} \wedge \omega^{\hat{\theta}}$$

$$= -\frac{\dot{a}}{a}\omega^{\hat{\theta}} \wedge \omega^{\hat{t}} - \frac{\sqrt{1-kr^2}}{ra}\omega^{\hat{\theta}} \wedge \omega^{\hat{r}}$$

$$d\omega^{\hat{\phi}} = d\,(ra\sin\theta d\phi)$$

$$= r\dot{a}\sin\theta dt \wedge d\phi + a\sin\theta dr \wedge d\phi + ra\cos\theta d\theta \wedge d\phi$$

$$= -\frac{\dot{a}}{a}\omega^{\hat{\phi}} \wedge \omega^{\hat{t}} - \frac{\sqrt{1-kr^2}}{ra}\omega^{\hat{\phi}} \wedge \omega^{\hat{r}} - \frac{\cot\theta}{ra}\omega^{\hat{\phi}} \wedge \omega^{\hat{\theta}}$$

Using the following relations found by writing out the right-hand side of (5.9)

$$d\omega^{\hat{r}} = -\Gamma^{\hat{r}}{}_{\hat{t}} \wedge \omega^{\hat{t}} - \Gamma^{\hat{r}}{}_{\hat{r}} \wedge \omega^{\hat{r}} - \Gamma^{\hat{r}}{}_{\hat{\theta}} \wedge \omega^{\hat{\theta}} - \Gamma^{\hat{r}}{}_{\hat{\phi}} \wedge \omega^{\hat{\phi}}$$

$$d\omega^{\hat{\theta}} = -\Gamma^{\hat{\theta}}{}_{\hat{t}} \wedge \omega^{\hat{t}} - \Gamma^{\hat{\theta}}{}_{\hat{r}} \wedge \omega^{\hat{r}} - \Gamma^{\hat{\theta}}{}_{\hat{\theta}} \wedge \omega^{\hat{\theta}} - \Gamma^{\hat{\theta}}{}_{\hat{\phi}} \wedge \omega^{\hat{\phi}}$$

$$d\omega^{\hat{\phi}} = -\Gamma^{\hat{\phi}}{}_{\hat{t}} \wedge \omega^{\hat{t}} - \Gamma^{\hat{\phi}}{}_{\hat{r}} \wedge \omega^{\hat{r}} - \Gamma^{\hat{\phi}}{}_{\hat{\theta}} \wedge \omega^{\hat{\theta}} - \Gamma^{\hat{\phi}}{}_{\hat{\phi}} \wedge \omega^{\hat{\phi}}$$

we read off the following curvature one-forms:

$$\Gamma^{\hat{r}}{}_{\hat{t}} = \frac{\dot{a}}{a}\omega^{\hat{r}}, \qquad \Gamma^{\hat{\theta}}{}_{\hat{t}} = \frac{\dot{a}}{a}\omega^{\hat{\theta}}, \qquad \Gamma^{\hat{\theta}}{}_{\hat{r}} = \frac{\sqrt{1-kr^2}}{ra}\omega^{\hat{\theta}}$$

$$\Gamma^{\hat{\phi}}{}_{\hat{t}} = \frac{\dot{a}}{a}\omega^{\hat{\phi}}, \qquad \Gamma^{\hat{\phi}}{}_{\hat{r}} = \frac{\sqrt{1-kr^2}}{ra}\omega^{\hat{\phi}}, \qquad \Gamma^{\hat{\phi}}{}_{\hat{\theta}} = \frac{\cot\theta}{ra}\omega^{\hat{\phi}}$$

$$(5.33)$$

We can raise and lower indices using $\eta_{\hat{a}\hat{b}}$. Notice the form of the metric, which indicates that in this case we must set $\eta_{\hat{a}\hat{b}} = \text{diag}(-1, 1, 1, 1)$. Let's see how this works with the curvature one forms:

$$\Gamma^{\hat{r}}{}_{\hat{t}} = \eta^{\hat{r}\hat{r}}\Gamma_{\hat{r}\hat{t}} = \Gamma_{\hat{r}\hat{t}} = -\Gamma_{\hat{t}\hat{r}} = -\eta_{\hat{t}\hat{t}}\Gamma^{\hat{t}}{}_{\hat{r}} = \Gamma^{\hat{t}}{}_{\hat{r}}$$

$$\Gamma^{\hat{\theta}}{}_{\hat{r}} = \eta^{\hat{\theta}\hat{\theta}}\Gamma_{\hat{\theta}\hat{r}} = \Gamma_{\hat{\theta}\hat{r}} = -\Gamma_{\hat{r}\hat{\theta}} = -\eta_{\hat{r}\hat{r}}\Gamma^{\hat{r}}{}_{\hat{\theta}} = -\Gamma^{\hat{r}}{}_{\hat{\theta}}$$

Now we calculate the curvature two forms using (5.27). We explicitly calculate $\Omega^{\hat{\theta}}{}_{\hat{r}}$:

$$\Omega^{\hat{\theta}}{}_{\hat{r}} = d\Gamma^{\hat{\theta}}{}_{\hat{r}} + \Gamma^{\hat{\theta}}{}_{\hat{c}} \wedge \Gamma^{\hat{c}}{}_{\hat{r}}$$

$$= d\Gamma^{\hat{\theta}}{}_{\hat{r}} + \Gamma^{\hat{\theta}}{}_{\hat{t}} \wedge \Gamma^{\hat{t}}{}_{\hat{r}} + \Gamma^{\hat{\theta}}{}_{\hat{r}} \wedge \Gamma^{\hat{r}}{}_{\hat{r}} + \Gamma^{\hat{\theta}}{}_{\hat{\theta}} \wedge \Gamma^{\hat{\theta}}{}_{\hat{r}} + \Gamma^{\hat{\theta}}{}_{\hat{\phi}} \wedge \Gamma^{\hat{\phi}}{}_{\hat{r}}$$

$$= d\Gamma^{\hat{\theta}}{}_{\hat{r}} + \Gamma^{\hat{\theta}}{}_{\hat{t}} \wedge \Gamma^{\hat{t}}{}_{\hat{r}} + \Gamma^{\hat{\theta}}{}_{\hat{\phi}} \wedge \Gamma^{\hat{\phi}}{}_{\hat{r}}$$

Now, for the first term in this expression we have

$$\Gamma^{\hat{\theta}}{}_{\hat{r}} = \frac{\sqrt{1-kr^2}}{ra}\omega^{\hat{\theta}} = \frac{\sqrt{1-kr^2}}{ra}(ra\,d\theta) = \sqrt{1-kr^2}\,d\theta$$

Therefore

$$d\Gamma^{\hat{\theta}}{}_{\hat{r}} = d\left(\sqrt{1-kr^2}\,d\theta\right) = -\frac{kr}{\sqrt{1-kr^2}}dr \wedge d\theta$$

$$= -\frac{k}{a^2}\omega^{\hat{r}} \wedge \omega^{\hat{\theta}}$$

For the other two terms, we obtain

$$\Gamma^{\hat{\theta}}{}_{\hat{t}} \wedge \Gamma^{\hat{t}}{}_{\hat{r}} = \frac{\dot{a}}{a}\omega^{\hat{\theta}} \wedge \frac{\dot{a}}{a}\omega^{\hat{r}} = \frac{\dot{a}^2}{a^2}\omega^{\hat{\theta}} \wedge \omega^{\hat{r}}$$

$$\Gamma^{\hat\theta}{}_{\hat\phi} \wedge \Gamma^{\hat\phi}{}_{\hat r} = -\frac{\cot\theta}{ra}\omega^{\hat\phi} \wedge \frac{\sqrt{1-kr^2}}{ra}\omega^{\hat\phi} = 0$$

Therefore, the curvature two form is

$$\Omega^{\hat\theta}{}_{\hat r} = -\frac{k}{a^2}\omega^{\hat r} \wedge \omega^{\hat\theta} + \frac{\dot a^2}{a^2}\omega^{\hat\theta} \wedge \omega^{\hat r}$$

$$= \frac{k}{a^2}\omega^{\hat\theta} \wedge \omega^{\hat r} + \frac{\dot a^2}{a^2}\omega^{\hat\theta} \wedge \omega^{\hat r}$$

$$= \frac{\dot a^2 + k}{a^2}\omega^{\hat\theta} \wedge \omega^{\hat r} \qquad (5.34)$$

Using (5.28), we can obtain the components of the Riemann tensor. First, using the symmetries of the Riemann tensor together with $\eta_{\hat a \hat b} = \mathrm{diag}(-1, 1, 1, 1)$, note that

$$R^{\hat\theta}{}_{\hat r \hat r \hat\theta} = \eta^{\hat\theta\hat\theta}R_{\hat\theta\hat r\hat r\hat\theta} = R_{\hat\theta\hat r\hat r\hat\theta} = -R_{\hat\theta\hat r\hat\theta\hat r} = -\eta_{\hat\theta\hat\theta}R^{\hat\theta}{}_{\hat r\hat\theta\hat r} = -R^{\hat\theta}{}_{\hat r\hat\theta\hat r}$$

and so

$$\Omega^{\hat\theta}{}_{\hat r} = \frac{1}{2}R^{\hat\theta}{}_{\hat r\hat\theta\hat r}\omega^{\hat\theta} \wedge \omega^{\hat r} + \frac{1}{2}R^{\hat\theta}{}_{\hat r\hat r\hat\theta}\omega^{\hat r} \wedge \omega^{\hat\theta}$$

$$= \frac{1}{2}R^{\hat\theta}{}_{\hat r\hat\theta\hat r}\omega^{\hat\theta} \wedge \omega^{\hat r} - \frac{1}{2}R^{\hat\theta}{}_{\hat r\hat r\hat\theta}\omega^{\hat\theta} \wedge \omega^{\hat r}$$

$$= \frac{1}{2}\left(R^{\hat\theta}{}_{\hat r\hat\theta\hat r} - R^{\hat\theta}{}_{\hat r\hat r\hat\theta}\right)\omega^{\hat\theta} \wedge \omega^{\hat r} = \frac{1}{2}\left(R^{\hat\theta}{}_{\hat r\hat\theta\hat r} + R^{\hat\theta}{}_{\hat r\hat\theta\hat r}\right)\omega^{\hat\theta} \wedge \omega^{\hat r}$$

$$= R^{\hat\theta}{}_{\hat r\hat\theta\hat r}\omega^{\hat\theta} \wedge \omega^{\hat r}$$

Comparison with (5.34) leads us to conclude that

$$R^{\hat\theta}{}_{\hat r\hat\theta\hat r} = \frac{\dot a^2 + k}{a^2}$$

Similar calculations show that (check)

$$R^{\hat{t}}{}_{\hat{\theta}\hat{t}\hat{\theta}} = R^{\hat{t}}{}_{\hat{\phi}\hat{t}\hat{\phi}} = R^{\hat{t}}{}_{\hat{r}\hat{t}\hat{r}} = \frac{\ddot{a}}{a}$$

$$R^{\hat{r}}{}_{\hat{\theta}\hat{r}\hat{\theta}} = \frac{\dot{a}^2 + k}{a^2}$$

$$R^{\hat{\theta}}{}_{\hat{\phi}\hat{\theta}\hat{\phi}} = \frac{\dot{a}^2 + k}{a^2}, \qquad R^{\hat{\phi}}{}_{\hat{r}\hat{r}\hat{\phi}} = -\frac{\dot{a}^2}{a^2}$$

# Quiz

Consider spherical polar coordinates, where $ds^2 = dr^2 + r^2 d\theta^2 + r^2 \sin^2\theta \, d\phi^2$. Calculate the Ricci rotation coefficients.

1.  $\Gamma^{\hat{r}}{}_{\hat{\phi}\hat{\phi}}$ is
    (a) $-\frac{1}{r}$
    (b) $r \, \sin^2\theta$
    (c) $-r \, \sin\theta \, \cos\theta$
    (d) $-r \, \sin^2\phi$

2.  $\Gamma^{\hat{\theta}}{}_{\hat{\phi}\hat{\phi}}$ is given by
    (a) $\tan\theta$
    (b) $-\sin\theta \, \sin\phi$
    (c) $-\frac{\cot\theta}{r}$
    (d) $r^2 \sin\theta \, \cos\theta$

3.  Applying the appropriate transformation matrix to the Ricci rotation coefficients, one finds that $\Gamma^{r}{}_{\phi\phi}$ is
    (a) $r \, \sin^2\theta$
    (b) $-r \, \sin^2\phi$
    (c) $-\frac{\cot\theta}{r}$
    (d) $\frac{1}{r^2}$

4.  Consider the Rindler metric, $ds^2 = u^2 \, dv^2 - du^2$. One finds that nonzero Ricci rotation coefficients are
    (a) $\Gamma^{\hat{v}}{}_{\hat{v}\hat{v}} = \Gamma^{\hat{u}}{}_{\hat{u}\hat{v}} = -\frac{1}{u}$
    (b) The space is flat, so all the Ricci rotation coefficients vanish.
    (c) $\Gamma^{\hat{u}}{}_{\hat{v}\hat{v}} = \Gamma^{\hat{v}}{}_{\hat{u}\hat{v}} = -\frac{1}{u^2}$
    (d) $\Gamma^{\hat{u}}{}_{\hat{v}\hat{v}} = \Gamma^{\hat{v}}{}_{\hat{u}\hat{v}} = -\frac{1}{u}$

5. In spherical polar coordinates, the commutation coefficients $C_{r\theta}{}^\theta$, $C_{r\phi}{}^\phi$, and $C_{\theta\phi}{}^\phi$ are

   (a) $C_{r\theta}{}^\theta = C_{r\phi}{}^\phi = -\frac{1}{r}$, $C_{\theta\phi}{}^\phi = 0$

   (b) $C_{r\theta}{}^\theta = C_{r\phi}{}^\phi = -\frac{1}{r}$, $C_{\theta\phi}{}^\phi = -\tan\theta$

   (c) $C_{r\theta}{}^\theta = C_{r\phi}{}^\phi = -\frac{1}{r}$, $C_{\theta\phi}{}^\phi = -\frac{\cot\theta}{r}$

   (d) $C_{r\theta}{}^\theta = C_{r\phi}{}^\phi = -\frac{1}{r^2}$, $C_{\theta\phi}{}^\phi = -\frac{\cot\theta}{r}$

6. Consider the Tolman metric studied in Example 5-1. The Ricci rotation coefficient $\Gamma^{\hat\phi}{}_{\hat r\hat\phi}$ is given by

   (a) $-\frac{R'}{R}\,e^{\Psi(t,r)}$

   (b) $\frac{R'}{R}\,e^{\Psi(t,r)}$

   (c) $-e^{\Psi(t,r)}$

   (d) $-\frac{e^{\Psi(t,r)}}{R}$

7. The $G_{tt}$ component of the Einstein tensor for the Tolman metric is given by

   (a) $G_{tt} = 0$

   (b) $G_{tt} = \frac{1}{R}\left[-R\,e^{2\Psi}\left(2R'\Psi' + 2R'' + R^{-1}R'^2 - 2\dot R\dot\Psi R + 1 + \dot R^2\right)\right]$

   (c) $G_{tt} = \frac{1}{R^2}\left[R\,e^{2\Psi}\left(2R'\Psi' + 2R'' + R^{-1}R'^2 - 2\dot R\dot\Psi R + 1 + \dot R^2\right)\right]$

   (d) $G_{tt} = \frac{1}{R^2}\left[-R\,e^{2\Psi}\left(2R'\Psi' + 2R'' + R^{-1}R'^2 - 2\dot R\dot\Psi R + 1 + \dot R^2\right)\right]$

8. For the Robertson-Walker metric in Example 5-2, using $\eta_{\hat a\hat b} = \mathrm{diag}(-1,\ 1,\ 1,\ 1)$ to raise and lower indices, one finds that

   (a) $\Gamma^{\hat\phi}{}_{\hat\theta} = -\Gamma^{\hat\theta}{}_{\hat\phi}$

   (b) $\Gamma^{\hat\theta}{}_{\hat\theta} = -\Gamma^{\hat\theta}{}_{\hat\phi}$

   (c) $\Gamma^{\hat\phi}{}_{\hat\phi} = -\Gamma^{\hat\theta}{}_{\hat\phi}$

   (d) $\Gamma^{\hat\phi}{}_{\hat\theta} = \Gamma^{\hat\theta}{}_{\hat\phi}$

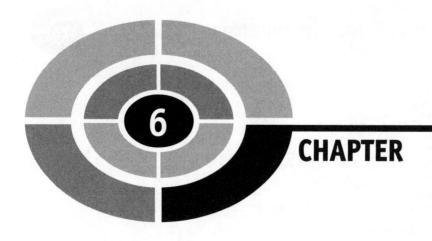

# The Einstein Field Equations

The physical principles that form the basis of Einstein's theory of gravitation have their roots in the famed Tower of Pisa experiments conducted by Galileo in the seventeenth century. Galileo did not actually drop balls from the famed leaning tower, but instead rolled them down the inclined planes. How the experiments were actually conducted is not of importance here; our concern is only with one fundamental fact they reveal—that all bodies in a gravitational field experience the same acceleration regardless of their mass or internal composition.

It is this fundamental result that allows us to arrive at our first *principle of equivalence*. In basic newtonian physics, a quantity called mass shows up in three basic equations—those that describe inertial forces, the force on a body due to a gravitational potential, and the force that a body produces when it is the source of a gravitational field. There is not really any a priori reason to assume that the mass that shows up in the equations describing each of these situations is one and the same. However, we will show that Galileo's results prove that this is the case.

# Equivalence of Mass in Newtonian Theory

In newtonian theory, there are three *types* of mass. The first two types describe the response of a body to inertial and gravitational forces, while the third type is used to describe the gravitational field that results when a given body acts as a source. More specifically,

- *Inertial mass*: The first appearance of mass in an elementary physics course is in the famous equation $F = ma$. Inertial mass is a measure of the ability of a body to resist changes in motion. In the following, we denote inertial mass by $m^I$.
- *Passive gravitational mass*: In newtonian theory, the force that a body feels due to a gravitational field described by a potential $\phi$ is given by $F = -m\nabla\phi$. The mass $m$ in this equation, which describes the reaction of a body to a given gravitational field, is called passive gravitational mass. We denote it by $m^p$.
- *Active gravitational mass*: This type of mass acts as the source of a gravitational field.

It is not obvious a priori that these types of masses should be equivalent. We now proceed to demonstrate that they are. We begin by considering the motion of two bodies in a gravitational field. Galileo showed that if we neglect air resistance, two bodies released simultaneously from a height $h$ will reach the ground at the same time. In other words, all bodies in a given gravitational field have the same acceleration. This is true regardless of their mass or internal compositions.

We consider the motion of two bodies in a gravitational field. The gravitational field exerts a force on a body and so we can use Newton's second law to write

$$F_1 = m_1^I a_1$$

$$F_2 = m_2^I a_2$$

Now the force on a body due to the gravitational field can be written in terms of the potential using $F = -m\nabla\phi$, where in this case $m$ is the passive gravitational mass of the body. And so we have

$$F_1 = m_1^I a_1 = -m_1^P \nabla\phi$$
$$F_2 = m_2^I a_2 = -m_2^P \nabla\phi$$

Using the second equation, we solve for the acceleration of mass 2:

$$m_2^I a_2 = -m_2^P \nabla\phi$$
$$\Rightarrow a_2 = -\frac{m_2^P}{m_2^I} \nabla\phi$$

However, the experimental results obtained by Galileo tell us that all bodies in a gravitational field fall with the same acceleration, which we denote by $g$. This means that $a_1 = a_2 = g$, and we have

$$a_2 = g = -\frac{m_2^P}{m_2^I} \nabla\phi$$

Since $a_1 = a_2 = g$, we can rewrite $F_1 = m_1^I a_1 = -m_1^P \nabla\phi$ in the following way:

$$F_1 = m_1^I a_1 = m_1^I g = -m_1^P \nabla\phi$$
$$g = -\frac{m_1^P}{m_1^I} \nabla\phi$$

Equating both expressions that we have obtained for $g$, we find that

$$\frac{m_1^P}{m_1^I} \nabla\phi = \frac{m_2^P}{m_2^I} \nabla\phi$$

Canceling $\nabla\phi$ from both sides, we get

$$\frac{m_1^P}{m_1^I} = \frac{m_2^P}{m_2^I}$$

Masses $m_1$ and $m_2$ used in this experiment are completely arbitrary, and we can substitute any body we like for mass $m_2$ and the result will be the same. Therefore, we conclude that the ratio of passive gravitational mass to inertial mass is a constant for any body. We can choose this constant to be unity and conclude that

$$m_1^{\mathrm{I}} = m_1^{\mathrm{p}}$$

That is, inertial mass and passive gravitational mass are *equivalent* (this result has been verified to high precision experimentally by the famous Eötvös experiment). We now show that active gravitational mass is equivalent to passive gravitational mass.

Consider two masses again labeled $m_1$ and $m_2$. We place mass $m_1$ at the origin and $m_2$ is initially located at some distance $r$ from $m_1$ along a radial line. The gravitational potential due to mass $m_1$ at a distance $r$ is given by

$$\phi_1 = -G \frac{m_1^{\mathrm{A}}}{r}$$

where $G$ is Newton's gravitational constant. The force on mass $m_2$ due to mass $m_1$ is given by

$$F_2 = -m_2^{\mathrm{p}} \nabla \phi_1$$

Since we are working with the radial coordinate only, the gradient can be written as

$$F_2 = -m_2^{\mathrm{p}} \nabla \phi_1 = -m_2^{\mathrm{p}} \left[ \frac{\partial}{\partial r} \left( -G \frac{m_1^{\mathrm{A}}}{r} \right) \hat{r} \right] = -\hat{r} G \frac{m_1^{\mathrm{A}} m_2^{\mathrm{p}}}{r^2}$$

Similarly, the force on mass $m_1$ due to the gravitational field produced by mass $m_2$ is

$$F_1 = \hat{r} G \frac{m_2^{\mathrm{A}} m_1^{\mathrm{p}}}{r^2}$$

To understand the difference in sign, note that in this case we have $-\hat{r}$ since the force points in the opposite direction. Now, Newton's third law tells us that $F_1 = -F_2$; therefore, we must have

$$G \frac{m_2^{\mathrm{A}} m_1^{\mathrm{p}}}{r^2} = G \frac{m_1^{\mathrm{A}} m_2^{\mathrm{p}}}{r^2}$$

We cancel the common terms $G$ and $r^2$, which give

$$m_2^A m_1^p = m_1^A m_2^p$$

This leads to

$$\frac{m_1^p}{m_1^A} = \frac{m_2^p}{m_2^A}$$

Again, we could choose any masses we like for this experiment. Therefore, this ratio must be a constant that we take to be unity, and we conclude that

$$m^A = m^p$$

that is, the active and passive gravitational mass for a body are equivalent. We have already found that the passive gravitational mass is equivalent to inertial mass, and so we have shown that

$$m = m^I = m^p = m^A$$

where we have used the single quantity $m$ to represent the mass of the body.

# Test Particles

Imagine that we are studying a region of spacetime where some distribution of matter and energy acts as a source of gravitational field that we call the *background field*. A *test particle* is one such that the gravitational field it produces is negligible as compared to the background field. In other words, the presence of the test particle will in no way change or alter the background field.

# The Einstein Lift Experiments

The Einstein lift experiments are a simple set of thought experiments that can be used to describe the equivalence principle. In Einstein's day, he used "lifts" or elevators to illustrate his points. We will do so in a more modern sense using spaceships. In all of these experiments, we consider scenarios with no rotation. We begin by considering a spaceship that is deep in interstellar space far from any source of gravitational fields. Furthermore, the spaceship is designed such that the astronaut inside has no way of communication with the outside universe;

in particular, there are no windows in the spacecraft so he cannot look outside to determine anything about his state of motion or location in the universe. When reading through these experiments, remember Newton's first law: A particle at rest or in uniform motion remains at rest or in uniform motion unless acted upon by an external force.

**Case 1.** In the first experiment, consider a spaceship that is not accelerating but is instead moving uniformly through space with respect to an inertial observer. The astronaut is holding a small ball, which he subsequently releases. What he will find is that, in accordance with Newton's first law, the ball simply remains at rest with respect to the astronaut where he released it. (see Fig. 6-1)

**Case 2.** We now consider an accelerating ship. The spaceship is still located deep in space far from any planets, stars, or other source of gravitational field.

**Fig. 6-1.** Art the astronaut in an unaccelerated spaceship in deep space, far from any gravitational fields. He releases a ball in front of him, and it remains there at rest with respect to the astronaut.

**Fig. 6-2.** An accelerating spaceship. This time when the astronaut releases the ball, it falls straight to the floor as he sees it.

But this time, the spaceship is accelerating with a constant acceleration $a$. For definiteness, we take the acceleration to be identical to the acceleration due to gravity at the surface of the earth; i.e., $a = g = 9.81$ m/s$^2$. If the astronaut releases the ball in this situation, he will find that from his perspective, it falls straight to the floor. (see Fig. 6-2)

**Case 3.** Turning to a third scenario, we now imagine a spaceship on earth that sits comfortably on the launchpad. The dimensions of the spacecraft are such that the tidal effects of gravity cannot be observed and that the rotational motion of the earth has no effect. We all know what happens when the astronaut releases the ball in this situation; it falls straight to the floor, just like it did in the previous situation. (see Fig. 6-3)

The situation inside the spacecraft on the launchpad is the same as the spacecraft with acceleration $g$ in deep space.

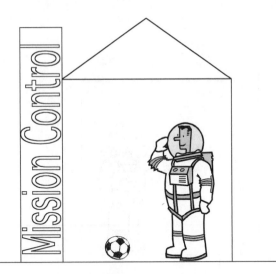

**Fig. 6-3.** Spaceship on the launchpad, resting on the surface of the earth. Drop a ball, and it falls straight to the floor.

**Case 4.** Finally, we consider one more situation: a spacecraft in free fall on earth. Let's say that the spaceship is in a mineshaft falling straight down. In this case, when the astronaut releases the ball, he will find a situation to that he encountered deep in space when not accelerating. When he releases the ball, it remains stationary where he released it. (see Fig. 6-4)

The point of these experiments is the following: In a region of spacetime that is small enough so that the tidal effects of a gravitational field cannot be observed, there are no experiments that can distinguish between a frame of reference that is in free fall in a gravitational field and one that is moving uniformly through space when no gravitational field is present.

More specifically, Cases 2 and 3 are indistinguishable to the astronaut. Assuming he cannot look outside, he can in no way differentiate whether or not he is accelerating deep in space with acceleration $g$ or if he is stationary on the surface of the earth. This implies that any frame that is accelerated in special relativity is indistinguishable from one in a gravitational field, provided that the region of spacetime used to make measurements is small enough; that is, tidal forces cannot be observed.

Cases 1 and 4 illustrate that there is no experiment that can distinguish uniform motion through space in the absence of a gravitational field from free fall *within* a gravitational field. Again, to the astronaut—provided that his environment is completely sealed—these situations seem identical. Cases 1 and 4 are illustrations of the *weak equivalence principle*.

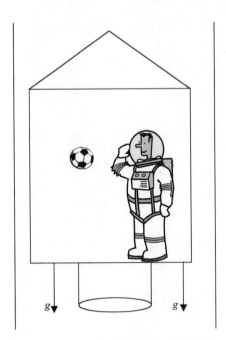

**Fig. 6-4.** A spaceship on earth, in free fall down a mineshaft. Art the astronaut releases a ball and finds to his astonishment that it remains at rest in front of him.

# The Weak Equivalence Principle

The weak equivalence principle is a statement about the universal nature of the gravitational field. Galileo found that all matter responds to the gravitational field in exactly the same way regardless of mass or internal composition. Moreover, special relativity teaches us about the equivalence of mass and energy. Combining these two lessons of physics leads us to the weak equivalence principle.

**Weak Equivalence Principle.**    The gravitational field couples in the same way to all mass and energy. The gravitational field is universal. This is a formal statement of the result—we obtained in the first part of the chapter—inertial and gravitational masses are equivalent.

# The Strong Equivalence Principle

Cases 2 and 4 in our thought experiments involving the spaceship illustrate the *strong equivalence principle*. This principle states that the laws of physics are the same in an accelerated reference frame and in a uniform and static gravitational field.

Note that an experiment that takes place over a large enough time interval or large enough region of space will reveal the tidal effects of gravity. Under these conditions, the equivalence principle would no longer apply. For example, consider two balls released from rest over the surface of the earth.

# The Principle of General Covariance

In Chapter 4 we noted that if an equation involving tensors is true in one coordinate system, it is true in all coordinate systems. This leads to the *principle of general covariance*, which simply states that the laws of physics, which should be invariant under a coordinate transformation, should be stated in tensorial form. Note that this *principle* is controversial. We merely state it here because it led Einstein in the development of his theory.

# Geodesic Deviation

In ordinary flat space, parallel lines always remain parallel. Now consider a more general space where the "straightest possible lines" are geodesics. What happens to geodesics that start off parallel in a curved space? You can get a hint by taking a look at the nearest map of the earth, you can find, showing the curved longitude lines that run from the North Pole to the South Pole. At the equator, these lines are parallel, but as you move North or South, neighboring lines begin to move together, or start off at the North Pole. Lines that emerge from the same point diverge as you move toward the equator. (see Fig. 6-5)

This behavior is typical of any curved geometry. In fact, in a curved space geodesics that start off parallel will eventually cross. Since gravity is just geometry, we expect to find this kind of behavior in the motion of particles on geodesics in spacetime.

In a gravitational field, the convergence of initially parallel geodesics is nothing more than an expression of gravitational tidal effects. Physically, this is exhibited in the shared acceleration between two particles in free fall in a gravitational field. Release two particles from some height $h$ above the earth. While the particles start off moving initially on parallel lines toward the ground, since they are on radial paths to the earth's center they will be seen to move toward each other if $h$ is large enough. This is a manifestation of the tidal effects of gravity. We study this phenomenon with equation of *geodesic deviation*. In your studies of gravity, you will often see the term *congruence*. A congruence is a

**Fig. 6-5.** In flat space, as seen on the left, lines that start off parallel remain parallel. In a curved space, however, this is not true. As shown on the right, lines that start off parallel end up converging on the sphere as you move from the equator to the North Pole. This type of "deviation" from being parallel is true in a general curved space or in a curved spacetime.

set of curves such that each point $p$ in the manifold lies on a single curve. To study geodesic deviation in spacetime, we consider a congruence of timelike geodesics. If we call the tangent vector to the curve $u^a$, then the congruence will represent a set of inertial worldlines if $u_a u^a = 1$.

We define the *connecting vector* as a vector that points from one geodesic to its neighbor. More specifically, it joins two points on neighboring curves at the same value of the affine parameter. This is illustrated in Fig. 6-6.

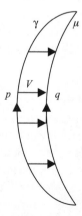

**Fig. 6-6.** Two curves $\gamma$ and $\mu$ in a manifold. Suppose each curve is parameterized by an affine parameter $\tau$. The deviation vector $V$ connects the curves at two points $p$ and $q$ such that $\gamma(\tau_1) = p$ and $\mu(\tau_1) = q$ for some value of the affine parameter $\tau = \tau_1$.

Now, consider a connecting vector $\eta^a$, which points from the geodesic of an inertial particle to the geodesic of another, infinitesimally close inertial particle. We say that $\eta^a$ is Lie propagated if its Lie derivative with respect to $u^a$ vanishes; i.e.,

$$L_u \eta^a = \boldsymbol{u}^b \nabla_b \eta^a - \eta^b \nabla_b u^a = 0$$

We will also use the following result.

**EXAMPLE 6-1**
Show that

$$\nabla_a \nabla_b V^c - \nabla_b \nabla_a V^c = R^c_{dab} V^d$$

**SOLUTION 6-1**
Recall from Chapter 4 that the covariant derivative of a vector is given by

$$\nabla_b V^a = \frac{\partial V^a}{\partial x^b} + \Gamma^a_{cb} V^c$$

Also note that

$$\nabla_c T^a_{\ b} = \partial_c T^a_{\ b} + \Gamma^a_{cd} T^d_{\ b} - \Gamma^d_{bc} T^a_{\ d} \qquad (6.1)$$

Proceeding, we have

$$\nabla_a \nabla_b V^c = \nabla_a \left( \partial_b V^c + \Gamma^c_{eb} V^e \right)$$

We can easily calculate the second derivative by treating $\partial_b V^c + \Gamma^c_{eb} V^e$ as a single tensor. Let's call it $S^c_b = \partial_b V^c + \Gamma^c_{eb} V^e$. Then using (6.1), we have

$$\nabla_a S^c_{\ b} = \partial_a S^c_{\ b} + \Gamma^c_{ad} S^d_b - \Gamma^d_{ba} S^c_{\ d}$$

Back substitution of $S^c_{\ b} = \partial_b V^c + \Gamma^c_{eb} V^e$ in this expression gives

$$\nabla_a \left( \partial_b V^c + \Gamma^c_{eb} V^e \right) = \partial_a \left( \partial_b V^c + \Gamma^c_{eb} V^e \right) + \Gamma^c_{ad} \left( \partial_b V^d + \Gamma^d_{eb} V^e \right)$$
$$- \Gamma^d_{ba} \left( \partial_d V^c + \Gamma^c_{ed} V^e \right)$$

A similar exercise shows that

$$\nabla_b \nabla_a V^c = \partial_b \left( \partial_b V^c + \Gamma^c_{ea} V^e \right) + \Gamma^c_{bd} \left( \partial_a V^d + \Gamma^d_{ea} V^e \right)$$
$$- \Gamma^d_{ab} (\partial_d V^c + \Gamma^c_{ed} V^e)$$

Let's calculate $\nabla_a \nabla_b V^c - \nabla_b \nabla_a V^c$ term by term. Since partial derivatives commute, and therefore $\partial_a \partial_b V^c - \partial_b \partial_a V^c = 0$, subtraction of the first terms gives

$$\partial_a \left( \partial_b V^c + \Gamma^c{}_{eb} V^e \right) - \partial_b \left( \partial_b V^c + \Gamma^c{}_{ea} V^e \right) = \partial_a \left( \Gamma^c{}_{eb} V^e \right) - \partial_b \left( \Gamma^c{}_{ea} V^e \right)$$

$$= V^e \left( \partial_a \Gamma^c{}_{eb} - \partial_b \Gamma^c{}_{ea} \right) + \Gamma^c{}_{eb} \partial_a V^e - \Gamma^c{}_{ea} \partial_b V^e \tag{6.2}$$

For the remaining terms, we use the fact that we are using a torsion-free connection, and therefore $\Gamma^a{}_{bc} = \Gamma^a{}_{cb}$, so we find

$$\Gamma^c{}_{ad} \left( \partial_b V^d + \Gamma^d{}_{eb} V^e \right) - \Gamma^d{}_{ba} \left( \partial_d V^c + \Gamma^c{}_{ed} V^e \right)$$

$$-\Gamma^c{}_{bd} \left( \partial_a V^d + \Gamma^d{}_{ea} V^e \right) + \Gamma^d{}_{ab} \left( \partial_d V^c + \Gamma^c{}_{ed} V^e \right)$$

$$= \Gamma^c{}_{ad} \left( \partial_b V^d + \Gamma^d{}_{eb} V^e \right) - \Gamma^d{}_{ab} \left( \partial_d V^c + \Gamma^c{}_{ed} V^e \right)$$

$$-\Gamma^c{}_{bd} \left( \partial_a V^d + \Gamma^d{}_{ea} V^e \right) + \Gamma^d{}_{ab} \left( \partial_d V^c + \Gamma^c{}_{ed} V^e \right)$$

$$= \Gamma^c{}_{ad} \left( \partial_b V^d + \Gamma^d{}_{eb} V^e \right) - \Gamma^c{}_{bd} \left( \partial_a V^d + \Gamma^d{}_{ea} V^e \right)$$

$$= \Gamma^c{}_{ad} \partial_b V^d - \Gamma^c{}_{bd} \partial_a V^d + \Gamma^c{}_{ad} \Gamma^d{}_{eb} V^e - \Gamma^c{}_{bd} \Gamma^d{}_{ea} V^e$$

We can go further by relabeling dummy indices and again using the fact that the Christoffel symbols are symmetric in the lower indices to rewrite this term as

$$\Gamma^c{}_{ad} \partial_b V^d - \Gamma^c{}_{bd} \partial_a V^d + \Gamma^c{}_{ad} \Gamma^d{}_{eb} V^e - \Gamma^c{}_{bd} \Gamma^d{}_{ea} V^e$$

$$= \Gamma^c{}_{ae} \partial_b V^e - \Gamma^c{}_{be} \partial_a V^e + \Gamma^c{}_{ad} \Gamma^d{}_{eb} V^e - \Gamma^c{}_{bd} \Gamma^d{}_{ea} V^e$$

$$= \Gamma^c{}_{ea} \partial_b V^e - \Gamma^c{}_{eb} \partial_a V^e + \Gamma^c{}_{ad} \Gamma^d{}_{eb} V^e - \Gamma^c{}_{bd} \Gamma^d{}_{ea} V^e$$

To get the final result, we add this expression to (6.2). However, notice that $\Gamma^c{}_{ea} \partial_b V^e - \Gamma^c{}_{eb} \partial_a V^e$ will cancel similar terms in (6.2), and so we are left with

$$\nabla_a \nabla_b V^c - \nabla_b \nabla_a V^c = V^e \left( \partial_a \Gamma^c{}_{eb} - \partial_b \Gamma^c{}_{ea} \right) + \Gamma^c{}_{ad} \Gamma^d{}_{eb} V^e - \Gamma^c{}_{bd} \Gamma^d{}_{ea} V^e$$

$$= \left( \partial_a \Gamma^c{}_{eb} - \partial_b \Gamma^c{}_{ea} + \Gamma^c{}_{ad} \Gamma^d{}_{eb} - \Gamma^c{}_{bd} \Gamma^d{}_{ea} \right) V^e$$

$$= \left( \partial_a \Gamma^c{}_{db} - \partial_b \Gamma^c{}_{da} + \Gamma^c{}_{ae} \Gamma^e{}_{db} - \Gamma^c{}_{be} \Gamma^e{}_{da} \right) V^d$$

In the last line, we swapped the dummy indices $d \leftrightarrow e$. From (4.41), we see that

$$R^a{}_{bcd} = \partial_c \Gamma^a{}_{bd} - \partial_d \Gamma^a{}_{bc} + \Gamma^e{}_{bd} \Gamma^a{}_{ec} - \Gamma^e{}_{bc} \Gamma^a{}_{ed}$$

and therefore we conclude that

$$\nabla_a \nabla_b V^c - \nabla_b \nabla_a V^c = R^c_{dab} V^d \qquad (6.3)$$

Furthermore, since the Lie derivative vanishes, we can write

$$u^b \nabla_b \eta^a = \eta^b \nabla_b u^a \qquad (6.4)$$

Now $\eta^c$ is a measure of the distance between two inertial particles. From elementary physics, you recall that velocity is the time derivative of position, i.e., $v = \mathrm{d}x/\mathrm{d}t$ and acceleration is $a = \mathrm{d}^2 x/\mathrm{d}t^2$. For inertial geodesics with tangent vector $u^a$ and parameter $\tau$, by analogy, we define the relative acceleration between two geodesics by

$$\begin{aligned}
\frac{\mathrm{D}^2 \eta^a}{\mathrm{D}\tau^2} &= u^b \nabla_b \left( u^c \nabla_c \eta^a \right) \\
&= u^b \nabla_b \left( \eta^c \nabla_c u^a \right) \\
&= u^b \left( \nabla_b \eta^c \nabla_c u^a + \eta^c \nabla_b \nabla_c u^a \right) \\
&= u^b \nabla_b \eta^c \nabla_c u^a + u^b \eta^c \nabla_b \nabla_c u^a
\end{aligned}$$

We can use (6.3) to write the last term as $\nabla_b \nabla_c u^a = \nabla_c \nabla_b u^a + R^a_{dbc} u^d$ to obtain

$$\begin{aligned}
\frac{\mathrm{D}^2 \eta^a}{\mathrm{D}\tau^2} &= u^b \nabla_b \eta^c \nabla_c u^a + u^b \eta^c \left( \nabla_c \nabla_b u^a + R^a_{dbc} u^d \right) \\
&= u^b \nabla_b \eta^c \nabla_c u^a + \eta^c u^b \nabla_c \nabla_b u^a + \eta^c u^b u^d R^a_{dbc} \\
&= \eta^b \nabla_b u^c \nabla_c u^a + \eta^c u^b \nabla_c \nabla_b u^a + \eta^c u^b u^d R^a_{dbc}
\end{aligned}$$

Relabeling dummy indices, we set $\eta^c u^b \nabla_c \nabla_b u^a = \eta^b u^c \nabla_b \nabla_c u^a$ and we have

$$\frac{\mathrm{D}^2 \eta^a}{\mathrm{D}\tau^2} = \eta^b \nabla_b u^c \nabla_c u^a + \eta^b u^c \nabla_b \nabla_c u^a + \eta^c u^b u^d R^a_{dbc}$$

Now the Leibniz rule $\nabla_b \left( u^c \nabla_c u^a \right) = \nabla_b u^c \nabla_c u^a + u^c \nabla_b \nabla_c u^a$ and so this becomes

$$\begin{aligned}
\frac{\mathrm{D}^2 \eta^a}{\mathrm{D}\tau^2} &= \eta^b \nabla_b u^c \nabla_c u^a + \eta^b u^c \nabla_b \nabla_c u^a + \eta^c u^b u^d R^a_{dbc} \\
&= \eta^b \left( \nabla_b u^c \nabla_c u^a + u^c \nabla_b \nabla_c u^a \right) + \eta^c u^b u^d R^a_{dbc} \\
&= \eta^b \left( \nabla_b \left( u^c \nabla_c u^a \right) \right) + \eta^c u^b u^d R^a_{dbc}
\end{aligned}$$

However, since is the tangent vector to a geodesic $u^c \nabla_c u^a = 0$. We can rearrange and then relabel dummy indices on the last term setting $\eta^c u^b u^d R^a{}_{dbc} = R^a{}_{dbc}$ $u^b u^d \eta^c = R^a{}_{bcd} u^b u^c \eta^d$, and so we obtain the equation of geodesic deviation

$$\frac{D^2 \eta^a}{D\tau^2} = R^a{}_{bcd} u^b u^c \eta^d$$

We can summarize this result in the following way: Gravity exhibits itself through tidal effects that cause inertial particles to undergo a mutual acceleration. Geometrically, this is manifest via spacetime curvature. We describe the relative acceleration between two geodesics using the equation of geodesic deviation.

# The Einstein Equations

In this section we introduce the Einstein equations and relate them to the equations used to describe gravity in the newtonian framework.

Newtonian gravity can be described by two equations. The first of these describes the path of a particle through space. If a particle is moving through a gravity field with potential $\phi$, then Newton's second law gives

$$F = ma = -m\nabla\phi$$

Canceling the mass term from both sides and writing the acceleration as the second derivative of position with respect to time, we have

$$\frac{d^2 x}{dt^2} = -\nabla\phi$$

This equation is analogous to the equation of geodesic deviation, for which we found

$$\frac{D^2 \eta^a}{D\tau^2} = R^a{}_{bcd} u^b u^c \eta^d$$

And so we have one piece of the puzzle: we know how to describe the behavior of matter in response to a gravitational field, which makes itself felt through the curvature. However, now consider the other equation used in newtonian gravity. This equation describes how mass acts as a source of gravitational field, i.e.,

Poisson's equation

$$\nabla^2 \phi = 4\pi G \rho$$

Einstein's equation will have a similar overall form. On the right-hand side, the source term in Newton's theory is the mass density in a given region of space. The lesson of special relativity is that mass and energy are equivalent. Therefore, we need to incorporate this idea into our new theory of gravity, and consider that all forms of mass-energy can be sources of gravitational fields. This is done by describing sources with the stress-energy tensor $T_{ab}$. This is a more general expression than mass density because it includes energy density as well. We will discuss it in more detail in the next and following chapters.

On the left side of Newton's equation, we see second derivatives of the potential. In relativity theory, the metric plays the role of gravitational potential. We have seen that through the relations

$$\Gamma^a_{\ bc} = \frac{1}{2} g^{ad} \left( \frac{\partial g_{bc}}{\partial x^d} + \frac{\partial g_{cd}}{\partial x^b} - \frac{\partial g_{db}}{\partial x^c} \right)$$

$$R^a_{\ bcd} = \partial_c \Gamma^a_{\ bd} - \partial_d \Gamma^a_{\ bc} + \Gamma^e_{\ bd} \Gamma^a_{\ ec} - \Gamma^e_{\ bc} \Gamma^a_{\ ed}$$

The curvature tensor encodes second derivatives of the metric. So if we are going to consider the metric to be analogous to gravitational potentials in Newton's theory, some term(s) involving the curvature tensor must appear on the left-hand side of the equations. The equation $\nabla^2 \phi = 4\pi G \rho$ actually relates the trace of $\nabla_i \nabla_j \phi$ to the mass density; therefore, we expect that the trace of the curvature tensor, which as we learned in Chapter 4 gives the Ricci tensor, will serve as the term on the left-hand side. So the equations will be something like

$$R_{ab} \propto T_{ab}$$

An important constraint on the form of Einstein's equations imposed by the appearance of the stress-energy tensor on the right side will be the conservation of momentum and energy, which as we will see in the next chapter is expressed by the relation

$$\nabla_b T^{ab} = 0$$

This constraint means that $R_{ab} \propto T_{ab}$ will not work because $\nabla_b R^{ab} \neq 0$. The contracted Bianchi identities (see problem 1) imply that $\nabla_a R^{ab} = \frac{1}{2} g^{ab} \nabla_a R$, where $R$ is the Ricci scalar. Therefore, if we instead use the Einstein tensor on the left-hand side, we will satisfy the laws of conservation of energy and momentum. That is, we set

$$G_{ab} = R_{ab} - \frac{1}{2}g_{ab}R$$

and then arrive at the field equations

$$G_{ab} = \kappa T_{ab}$$

where $\kappa$ is a constant that turns out to be $8\pi G$.

The *vacuum equations* are used to study the gravitational field in a region of spacetime outside of the source—i.e., where no matter and energy are present. For example, you can study the vacuum region of spacetime outside of a star. We can set $T_{ab} = 0$ and then the vacuum Einstein equations become

$$R_{ab} = 0 \tag{6.5}$$

# The Einstein Equations with Cosmological Constant

The cosmological constant was originally added to the equations by Einstein as a fudge factor. At the time, he and others believed that the universe was static. As we shall see Einstein's equations predict a dynamic universe, and so Einstein tinkered with the equations a bit to get them to fit his predispositions at the time. When the observations of Hubble proved beyond reasonable doubt that the universe was expanding, Einstein threw out the cosmological constant and described it as the biggest mistake of his life.

Recently, however, observation seems to indicate that some type of vacuum energy is at work in the universe, and so the cosmological constant is coming back in style. It is possible to include a small cosmological constant and still have a dynamic universe. If we define the vacuum energy of the universe to be

$$\rho_v = \frac{\Lambda}{8\pi G}$$

then including this term, Einstein's equations can be written as

$$R_{ab} - \frac{1}{2}g_{ab}R + g_{ab}\Lambda = 8\pi G T_{ab} \tag{6.6}$$

or

$$G_{ab} + g_{ab}\Lambda = 8\pi G T_{ab}$$

Therefore with this addition the Einstein tensor remains unchanged, and most of our work will involve calculating this beast. We demonstrate the solution of Einstein's equations with a cosmological constant term in the next example.

# An Example Solving Einstein's Equations in 2+1 Dimensions

We now consider an example that considers Einstein's equations in 2+1 dimensions. This means that we restrict ourselves to two spatial dimensions and time. Models based on 2+1 dimensions can be used to simplify the analysis while retaining important conceptual results. This is a technique that can be used in the study of quantum gravity for example. For a detailed discussion, see Carlip (1998).

In this example, we consider the gravitational collapse of an inhomogeneous, spherically symmetric dust cloud $T_{ab} = \rho u_a u_b$ with nonzero cosmological constant $\Lambda < 0$. This is a long calculation, so we divide it into three examples. This problem is based on a recently published paper (see References), so it will give you an idea of how relativity calculations are done in actual current research. The first example will help you review the techniques covered in Chapter 5.

**EXAMPLE 6-2**
Consider the metric

$$ds^2 = -dt^2 + e^{2b(t,r)}\,dr^2 + R(t,\ r)\,d\phi^2$$

and use Cartan's structure equations to find the components of the curvature tensor.

**SOLUTION 6-2**
With nonzero cosmological constant, Einstein's equation takes the form

$$G_{ab} + g_{ab}\Lambda = 8\pi\,T_{ab}$$

For the given metric, we define the following orthonormal basis one forms:

$$\omega^{\hat{t}} = dt, \qquad \omega^{\hat{r}} = e^{b(t,r)}\,dr, \qquad \omega^{\hat{\phi}} = R(t,\ r)\,d\phi \qquad (6.7)$$

These give us the following inverse relationships which will be useful in calculations:

$$dt = \omega^{\hat{t}}, \qquad dr = e^{-b(t,r)}\omega^{\hat{r}}, \qquad d\phi = \frac{1}{R(t,\ r)}\omega^{\hat{\phi}} \qquad (6.8)$$

With this basis defined, we have

$$\eta_{\hat{a}\hat{b}} = \begin{pmatrix} -1 & 0 & 0 \\ 0 & 1 & 0 \\ 0 & 0 & 1 \end{pmatrix}$$

which we can use to raise and lower indices.

We will find the components of the Einstein tensor using Cartan's methods. To begin, we calculate the Ricci rotation coefficients. Recall that Cartan's first structure equation is

$$d\omega^{\hat{a}} = -\Gamma^{\hat{a}}{}_{\hat{b}} \wedge \omega^{\hat{b}} \qquad (6.9)$$

Also recall that

$$\Gamma^{\hat{a}}{}_{\hat{b}} = \Gamma^{\hat{a}}{}_{\hat{b}\hat{c}}\omega^{\hat{c}} \qquad (6.10)$$

The first equation gives us no information, since we have

$$d\omega^{\hat{t}} = d(dt) = 0$$

Moving to $\omega^{\hat{r}}$, we find

$$d\omega^{\hat{r}} = d\left(e^{b(t,r)}\,dr\right) = \frac{\partial b}{\partial t}e^{b(t,r)}\,dt \wedge dr + \frac{\partial b}{\partial r}e^{b(t,r)}\,dr \wedge dr$$
$$= \frac{\partial b}{\partial t}e^{b(t,r)}\,dt \wedge dr \qquad (6.11)$$

Using (6.8), we rewrite this in terms of the basis one forms to get

$$d\omega^{\hat{r}} = \frac{\partial b}{\partial t}e^{b(t,r)}\,dt \wedge dr = \frac{\partial b}{\partial t}\omega^{\hat{t}} \wedge \omega^{\hat{r}} = -\frac{\partial b}{\partial t}\omega^{\hat{r}} \wedge \omega^{\hat{t}} \qquad (6.12)$$

Explicitly writing out Cartan's first structure equation for $d\omega^{\hat{r}}$, we find that

$$d\omega^{\hat{r}} = -\Gamma^{\hat{r}}{}_{\hat{b}} \wedge \omega^{\hat{b}}$$
$$= -\Gamma^{\hat{r}}{}_{\hat{t}} \wedge \omega^{\hat{t}} - \Gamma^{\hat{r}}{}_{\hat{r}} \wedge \omega^{\hat{r}} - \Gamma^{\hat{r}}{}_{\hat{\phi}} \wedge \omega^{\hat{\phi}} \qquad (6.13)$$

Comparing this with (6.12), which has only basis one forms $\omega^{\hat{t}}$ and $\omega^{\hat{r}}$ in the expression, we guess that the only nonzero term is given by $\Gamma^{\hat{r}}{}_{\hat{t}} \wedge \omega^{\hat{t}}$ and conclude that

$$\Gamma^{\hat{r}}{}_{\hat{t}} = \frac{\partial b}{\partial t} \omega^{\hat{r}} \tag{6.14}$$

Using (6.10), we expand the left-hand side to get

$$\Gamma^{\hat{r}}{}_{\hat{t}} = \Gamma^{\hat{r}}{}_{\hat{t}\hat{t}} \omega^{\hat{t}} + \Gamma^{\hat{r}}{}_{\hat{t}\hat{r}} \omega^{\hat{r}} + \Gamma^{\hat{r}}{}_{\hat{t}\hat{\phi}} \omega^{\hat{\phi}}$$

Comparing this with (6.14) shows that the only nonzero Ricci rotation coefficient in this expansion is

$$\Gamma^{\hat{r}}{}_{\hat{t}\hat{r}} = \frac{\partial b}{\partial t}$$

We now proceed to find other nonzero Ricci rotation coefficients that are related to this one via symmetries. Recall that

$$\Gamma_{\hat{a}\hat{b}} = -\Gamma_{\hat{b}\hat{a}}, \qquad \Gamma^{\hat{0}}{}_{\hat{i}} = \Gamma^{\hat{i}}{}_{\hat{0}}, \qquad \Gamma^{\hat{i}}{}_{\hat{j}} = -\Gamma^{\hat{j}}{}_{\hat{i}} \tag{6.15}$$

This means that $\Gamma^{\hat{r}}{}_{\hat{t}} = \Gamma^{\hat{t}}{}_{\hat{r}} = \frac{\partial b}{\partial t} \omega^{\hat{r}}$. Expanding $\Gamma^{\hat{t}}{}_{\hat{r}}$, we have

$$\Gamma^{\hat{t}}{}_{\hat{r}} = \Gamma^{\hat{t}}{}_{\hat{r}\hat{t}} \omega^{\hat{t}} + \Gamma^{\hat{t}}{}_{\hat{r}\hat{r}} \omega^{\hat{r}} + \Gamma^{\hat{t}}{}_{\hat{r}\hat{\phi}} \omega^{\hat{\phi}}$$

This tells us that $\Gamma^{\hat{t}}{}_{\hat{r}\hat{r}} = \frac{\partial b}{\partial t}$. Moving to the final basis one form, we have

$$d\omega^{\hat{\phi}} = d\left( R(t,\, r)\, d\phi \right) = \frac{\partial R}{\partial t}\, dt \wedge d\phi + \frac{\partial R}{\partial r}\, dr \wedge d\phi$$

$$= \frac{1}{R} \frac{\partial R}{\partial t} \omega^{\hat{t}} \wedge \omega^{\hat{\phi}} + \frac{1}{R} \frac{\partial R}{\partial r} e^{-b(t,r)} \omega^{\hat{r}} \wedge \omega^{\hat{\phi}} \tag{6.16}$$

$$= -\frac{1}{R} \frac{\partial R}{\partial t} \omega^{\hat{\phi}} \wedge \omega^{\hat{t}} + \frac{1}{R} \frac{\partial R}{\partial r} e^{-b(t,r)} \omega^{\hat{\phi}} \wedge \omega^{\hat{r}}$$

Using Cartan's structure equation, we can write

$$d\omega^{\hat{\phi}} = -\Gamma^{\hat{\phi}}{}_{\hat{t}} \wedge \omega^{\hat{t}} - \Gamma^{\hat{\phi}}{}_{\hat{r}} \wedge \omega^{\hat{r}} - \Gamma^{\hat{\phi}}{}_{\hat{\phi}} \wedge \omega^{\hat{\phi}} \tag{6.17}$$

Comparing this with (6.16), we conclude that

$$\Gamma^{\hat{\phi}}{}_{\hat{t}} = \frac{1}{R}\frac{\partial R}{\partial t}\omega^{\hat{\phi}} \quad \text{and} \quad \Gamma^{\hat{\phi}}{}_{\hat{r}} = \frac{1}{R}\frac{\partial R}{\partial r}e^{-b(t,r)}\omega^{\hat{\phi}} \tag{6.18}$$

Using (6.10) to expand each term, we find the following nonzero Ricci rotation coefficients:

$$\Gamma^{\hat{\phi}}{}_{\hat{t}\hat{\phi}} = \frac{1}{R}\frac{\partial R}{\partial t} \quad \text{and} \quad \Gamma^{\hat{\phi}}{}_{\hat{r}\hat{\phi}} = \frac{1}{R}\frac{\partial R}{\partial r}e^{-b(t,r)} \tag{6.19}$$

As an aside, note that $\Gamma_{\hat{c}\hat{a}\hat{b}} = -\Gamma_{\hat{a}\hat{c}\hat{b}}$. This means that any terms that match on the first two indices must vanish. For example,

$$\Gamma^{\hat{\phi}}{}_{\hat{\phi}\hat{r}} = \eta^{\hat{\phi}\hat{\phi}}\Gamma_{\hat{\phi}\hat{\phi}\hat{r}} = \Gamma_{\hat{\phi}\hat{\phi}\hat{r}} = -\Gamma_{\hat{\phi}\hat{\phi}\hat{r}}$$

$$\therefore \Gamma_{\hat{\phi}\hat{\phi}\hat{r}} = 0$$

$$\Rightarrow \Gamma^{\hat{\phi}}{}_{\hat{\phi}\hat{r}} = 0$$

Now let's apply the symmetries listed in (6.15) to find the other nonzero terms. For the first term, we get

$$\Gamma^{\hat{\phi}}{}_{\hat{t}} = \eta^{\hat{\phi}\hat{\phi}}\Gamma_{\hat{\phi}\hat{t}} = \Gamma_{\hat{\phi}\hat{t}} = -\Gamma_{\hat{t}\hat{\phi}} = -\eta_{\hat{t}\hat{t}}\Gamma^{\hat{t}}{}_{\hat{\phi}} = \Gamma^{\hat{t}}{}_{\hat{\phi}}$$

$$\Rightarrow \Gamma^{\hat{t}}{}_{\hat{\phi}} = \frac{1}{R}\frac{\partial R}{\partial t}\omega^{\hat{\phi}} \tag{6.20}$$

This leads us to the conclusion that

$$\Gamma^{\hat{t}}{}_{\hat{\phi}\hat{\phi}} = \frac{1}{R}\frac{\partial R}{\partial t}$$

Using (6.15), we see that $\Gamma^{\hat{\phi}}{}_{\hat{r}} = -\Gamma^{\hat{r}}{}_{\hat{\phi}}$. So we have

$$\Gamma^{\hat{r}}{}_{\hat{\phi}} = -\frac{1}{R}\frac{\partial R}{\partial r}\, e^{-b(t,r)}\omega^{\hat{\phi}}$$

$$\Rightarrow\ \Gamma^{\hat{r}}{}_{\hat{\phi}\hat{\phi}} = -\frac{1}{R}\frac{\partial R}{\partial r}\, e^{-b(t,r)}$$

We now proceed to use Cartan's second structure equation to find the components of the curvature tensor in the noncoordinate basis. Recall that the curvature two forms were defined via

$$\Omega^{\hat{a}}{}_{\hat{b}} = d\Gamma^{\hat{a}}{}_{\hat{b}} + \Gamma^{\hat{a}}{}_{\hat{c}} \wedge \Gamma^{\hat{c}}{}_{\hat{b}} = \frac{1}{2}R^{\hat{a}}{}_{\hat{b}\hat{c}\hat{d}}\omega^{\hat{c}} \wedge \omega^{\hat{d}} \tag{6.21}$$

We will solve two terms and leave the remaining terms as an exercise. Setting $\hat{a} = \hat{r}$ and $\hat{b} = \hat{t}$ in (6.21) gives

$$\Omega^{\hat{r}}{}_{\hat{t}} = d\Gamma^{\hat{r}}{}_{\hat{t}} + \Gamma^{\hat{r}}{}_{\hat{c}} \wedge \Gamma^{\hat{c}}{}_{\hat{t}}$$

$$= d\Gamma^{\hat{r}}{}_{\hat{t}} + \Gamma^{\hat{r}}{}_{\hat{t}} \wedge \Gamma^{\hat{t}}{}_{\hat{t}} + \Gamma^{\hat{r}}{}_{\hat{r}} \wedge \Gamma^{\hat{r}}{}_{\hat{t}} + \Gamma^{\hat{r}}{}_{\hat{\phi}} \wedge \Gamma^{\hat{\phi}}{}_{\hat{t}} \tag{6.22}$$

We begin by calculating $d\Gamma^{\hat{r}}{}_{\hat{t}}$ using (6.14) and recalling that $\omega^{\hat{r}} = e^{b(t,r)}\, dr$:

$$d\Gamma^{\hat{r}}{}_{\hat{t}} = d\left(\frac{\partial b}{\partial t}\omega^{\hat{r}}\right) = d\left(\frac{\partial b}{\partial t}\, e^{b(t,r)}dr\right)$$

$$= \frac{\partial^2 b}{\partial t^2}\, e^{b(t,r)}\, dt \wedge dr + \left(\frac{\partial b}{\partial t}\right)^2 e^{b(t,r)}\, dt \wedge dr + \frac{\partial b}{\partial t}\frac{\partial b}{\partial r}\, e^{b(t,r)}\, dr \wedge dr$$

Since $dr \wedge dr = 0$, this simplifies to

$$d\Gamma^{\hat{r}}{}_{\hat{t}} = \frac{\partial^2 b}{\partial t^2}\, e^{b(t,r)}\, dt \wedge dr + \left(\frac{\partial b}{\partial t}\right)^2 e^{b(t,r)}\, dt \wedge dr$$

$$= \left[\frac{\partial^2 b}{\partial t^2} + \left(\frac{\partial b}{\partial t}\right)^2\right]\omega^{\hat{t}} \wedge \omega^{\hat{r}} \tag{6.23}$$

The remaining terms in (6.22) all vanish:

$$\Gamma^{\hat{r}}{}_{\hat{t}} \wedge \Gamma^{\hat{t}}{}_{\hat{t}} = 0 \quad (\text{since } \Gamma^{\hat{t}}{}_{\hat{t}} = 0)$$

$$\Gamma^{\hat{r}}{}_{\hat{r}} \wedge \Gamma^{\hat{r}}{}_{\hat{t}} = 0 \quad (\text{since } \Gamma^{\hat{r}}{}_{\hat{r}} = 0)$$

$$\Gamma^{\hat{r}}{}_{\hat{\phi}} \wedge \Gamma^{\hat{\phi}}{}_{\hat{t}} = -\frac{1}{R}\frac{\partial R}{\partial r}e^{-b(t,r)}\omega^{\hat{\phi}} \wedge \frac{1}{R}\frac{\partial R}{\partial t}\omega^{\hat{\phi}} = 0 \quad (\text{since } \omega^{\hat{\phi}} \wedge \omega^{\hat{\phi}} = 0)$$

Therefore (6.22) reduces to

$$\Omega^{\hat{r}}{}_{\hat{t}} = d\Gamma^{\hat{r}}{}_{\hat{t}}$$

$$\Rightarrow \Omega^{\hat{r}}{}_{\hat{t}} = \left[\frac{\partial^2 b}{\partial t^2} + \left(\frac{\partial b}{\partial t}\right)^2\right]\omega^{\hat{t}} \wedge \omega^{\hat{r}} \tag{6.24}$$

To find the components of the curvature tensor, we apply (6.21), which in this case gives

$$\Omega^{\hat{r}}_{\hat{t}} = \frac{1}{2}R^{\hat{r}}{}_{\hat{t}\hat{c}\hat{d}}\,\omega^{\hat{c}} \wedge \omega^{\hat{d}}$$

$$= \frac{1}{2}R^{\hat{r}}{}_{\hat{t}\hat{t}\hat{r}}\,\omega^{\hat{t}} \wedge \omega^{\hat{r}} + \frac{1}{2}R^{\hat{r}}{}_{\hat{t}\hat{r}\hat{t}}\,\omega^{\hat{r}} \wedge \omega^{\hat{t}}$$

$$= \frac{1}{2}R^{\hat{r}}{}_{\hat{t}\hat{t}\hat{r}}\,\omega^{\hat{t}} \wedge \omega^{\hat{r}} - \frac{1}{2}R^{\hat{r}}{}_{\hat{t}\hat{r}\hat{t}}\,\omega^{\hat{t}} \wedge \omega^{\hat{r}}$$

$$= \frac{1}{2}\left(R^{\hat{r}}{}_{\hat{t}\hat{t}\hat{r}} - R^{\hat{r}}{}_{\hat{t}\hat{r}\hat{t}}\right)\omega^{\hat{t}} \wedge \omega^{\hat{r}}$$

Now we use the flat space metric of the local frame $\eta_{\hat{a}\hat{b}} = \text{diag}(-1, \; 1, \; 1)$ to raise and lower indices, and then apply the symmetries of the curvature tensor to write

$$R^{\hat{r}}{}_{\hat{t}\hat{r}\hat{t}} = \eta^{\hat{r}\hat{r}}R_{\hat{r}\hat{t}\hat{r}\hat{t}} = R_{\hat{r}\hat{t}\hat{r}\hat{t}} = -R_{\hat{r}\hat{t}\hat{t}\hat{r}} = -\eta_{\hat{r}\hat{r}}R^{\hat{r}}{}_{\hat{t}\hat{t}\hat{r}} = -R^{\hat{r}}{}_{\hat{t}\hat{t}\hat{r}}$$

Therefore, we have

$$\Omega^{\hat{r}}{}_{\hat{t}} = \frac{1}{2}\left(R^{\hat{r}}{}_{\hat{t}\hat{t}\hat{r}} - R^{\hat{r}}{}_{\hat{t}\hat{r}\hat{t}}\right)\omega^{\hat{t}} \wedge \omega^{\hat{r}} = \frac{1}{2}\left(R^{\hat{r}}{}_{\hat{t}\hat{t}\hat{r}} + R^{\hat{r}}{}_{\hat{t}\hat{t}\hat{r}}\right)\omega^{\hat{t}} \wedge \omega^{\hat{r}}$$

$$= R^{\hat{r}}{}_{\hat{t}\hat{t}\hat{r}}\,\omega^{\hat{t}} \wedge \omega^{\hat{r}}$$

Comparison with (6.24) leads us to conclude that

$$R^{\hat{r}}_{\ \hat{t}\hat{t}\hat{r}} = \frac{\partial^2 b}{\partial t^2} + \left(\frac{\partial b}{\partial t}\right)^2$$

Let's calculate the curvature two form $\Omega^{\hat{t}}_{\ \hat{\phi}}$. Using (6.21), we have

$$\Omega^{\hat{t}}_{\ \hat{\phi}} = d\Gamma^{\hat{t}}_{\ \hat{\phi}} + \Gamma^{\hat{t}}_{\ \hat{c}} \wedge \Gamma^{\hat{c}}_{\ \hat{\phi}}$$

$$= d\left(\frac{1}{R}\frac{\partial R}{\partial t}\omega^{\hat{\phi}}\right) + \Gamma^{\hat{t}}_{\ \hat{t}} \wedge \Gamma^{\hat{t}}_{\ \hat{\phi}} + \Gamma^{\hat{t}}_{\ \hat{r}} \wedge \Gamma^{\hat{r}}_{\ \hat{\phi}} + \Gamma^{\hat{t}}_{\ \hat{\phi}} \wedge \Gamma^{\hat{\phi}}_{\ \hat{\phi}}$$

$$= d\left(\frac{1}{R}\frac{\partial R}{\partial t}\omega^{\hat{\phi}}\right) + \Gamma^{\hat{t}}_{\ \hat{r}} \wedge \Gamma^{\hat{r}}_{\ \hat{\phi}}$$

Now

$$d\left(\frac{1}{R}\frac{\partial R}{\partial t}\omega^{\hat{\phi}}\right) = d\left(\frac{1}{R}\frac{\partial R}{\partial t}R\,d\phi\right)$$

$$= d\left(\frac{\partial R}{\partial t}d\phi\right)$$

$$= \frac{\partial^2 R}{\partial t^2}dt \wedge d\phi + \frac{\partial^2 R}{\partial t\,\partial r}dr \wedge d\phi$$

$$= \frac{1}{R}\frac{\partial^2 R}{\partial t^2}\omega^{\hat{t}} \wedge \omega^{\hat{\phi}} + \frac{e^{-b(t,r)}}{R}\frac{\partial^2 R}{\partial t\,\partial r}\omega^{\hat{r}} \wedge \omega^{\hat{\phi}}$$

and

$$\Gamma^{\hat{t}}_{\ \hat{r}} \wedge \Gamma^{\hat{r}}_{\ \hat{\phi}} = \frac{\partial b}{\partial t}\omega^{\hat{r}} \wedge \left(-\frac{e^{-b(t,r)}}{R}\frac{\partial R}{\partial r}\omega^{\hat{\phi}}\right)$$

$$= -\frac{e^{-b(t,r)}}{R}\frac{\partial b}{\partial t}\frac{\partial R}{\partial r}\omega^{\hat{r}} \wedge \omega^{\hat{\phi}}$$

Putting these results together, we obtain

$$\Omega^{\hat{t}}{}_{\hat{\phi}} = \frac{1}{R}\frac{\partial^2 R}{\partial t^2}\omega^{\hat{t}} \wedge \omega^{\hat{\phi}} + \frac{e^{-b(t,r)}}{R}\frac{\partial^2 R}{\partial t\,\partial r}\omega^{\hat{r}} \wedge \omega^{\hat{\phi}} - \frac{e^{-b(t,r)}}{R}\frac{\partial b}{\partial t}\frac{\partial R}{\partial r}\omega^{\hat{r}} \wedge \omega^{\hat{\phi}}$$

$$= \frac{1}{R}\frac{\partial^2 R}{\partial t^2}\omega^{\hat{t}} \wedge \omega^{\hat{\phi}} + \frac{e^{-b(t,r)}}{R}\left(\frac{\partial^2 R}{\partial t\,\partial r} - \frac{\partial b}{\partial t}\frac{\partial R}{\partial r}\right)\omega^{\hat{r}} \wedge \omega^{\hat{\phi}}$$

Again, to find the components of the curvature tensor, we write out $\Omega^{\hat{t}}{}_{\hat{\phi}} = \frac{1}{2}R^{\hat{t}}{}_{\hat{\phi}\hat{c}\hat{d}}\omega^{\hat{c}} \wedge \omega^{\hat{d}}$ to obtain

$$\Omega^{\hat{t}}{}_{\hat{\phi}} = \frac{1}{2}R^{\hat{t}}{}_{\hat{\phi}\hat{r}\hat{\phi}}\,\omega^{\hat{r}} \wedge \omega^{\hat{\phi}} + \frac{1}{2}R^{\hat{t}}{}_{\hat{\phi}\hat{\phi}\hat{r}}\,\omega^{\hat{\phi}} \wedge \omega^{\hat{r}} + \frac{1}{2}R^{\hat{t}}{}_{\hat{\phi}\hat{t}\hat{\phi}}\,\omega^{\hat{t}} \wedge \omega^{\hat{\phi}}$$

$$+ \frac{1}{2}R^{\hat{t}}{}_{\hat{\phi}\hat{\phi}\hat{t}}\,\omega^{\hat{\phi}} \wedge \omega^{\hat{t}}$$

$$= \frac{1}{2}\left(R^{\hat{t}}{}_{\hat{\phi}\hat{r}\hat{\phi}} - R^{\hat{t}}{}_{\hat{\phi}\hat{\phi}\hat{r}}\right)\omega^{\hat{r}} \wedge \omega^{\hat{\phi}} + \frac{1}{2}\left(R^{\hat{t}}{}_{\hat{\phi}\hat{t}\hat{\phi}} - R^{\hat{t}}{}_{\hat{\phi}\hat{\phi}\hat{t}}\right)\omega^{\hat{t}} \wedge \omega^{\hat{\phi}}$$

$$= R^{\hat{t}}{}_{\hat{\phi}\hat{r}\hat{\phi}}\,\omega^{\hat{r}} \wedge \omega^{\hat{\phi}} + R^{\hat{t}}{}_{\hat{\phi}\hat{t}\hat{\phi}}\,\omega^{\hat{t}} \wedge \omega^{\hat{\phi}}$$

Therefore, we conclude that

$$R^{\hat{t}}{}_{\hat{\phi}\hat{t}\hat{\phi}} = \frac{1}{R}\frac{\partial^2 R}{\partial t^2} \quad \text{and} \quad R^{\hat{t}}{}_{\hat{\phi}\hat{r}\hat{\phi}} = \frac{e^{-b(t,r)}}{R}\left(\frac{\partial^2 R}{\partial t\,\partial r} - \frac{\partial b}{\partial t}\frac{\partial R}{\partial r}\right)$$

All together, the nonzero components of the curvature tensor are

$$R^{\hat{r}}{}_{\hat{t}\hat{t}\hat{r}} = \frac{\partial^2 b}{\partial t^2} + \left(\frac{\partial b}{\partial t}\right)^2 = -R^{\hat{r}}{}_{\hat{t}\hat{r}\hat{t}}$$

$$R^{\hat{t}}{}_{\hat{\phi}\hat{t}\hat{\phi}} = \frac{1}{R}\frac{\partial^2 R}{\partial t^2} = -R^{\hat{t}}{}_{\hat{\phi}\hat{\phi}\hat{t}}, \qquad R^{\hat{t}}{}_{\hat{\phi}\hat{r}\hat{\phi}} = \frac{e^{-b(t,r)}}{R}\left(\frac{\partial^2 R}{\partial t\,\partial r} - \frac{\partial b}{\partial t}\frac{\partial R}{\partial r}\right) = -R^{\hat{t}}{}_{\hat{\phi}\hat{\phi}\hat{r}}$$

$$R^{\hat{r}}{}_{\hat{\phi}\hat{r}\hat{\phi}} = \frac{e^{-2b(t,r)}}{R}\left(\frac{\partial b}{\partial t}\frac{\partial R}{\partial t}e^{2b(t,r)} - \frac{\partial^2 R}{\partial r^2} + \frac{\partial R}{\partial r}\frac{\partial b}{\partial r}\right) \tag{6.25}$$

**EXAMPLE 6-3**
Using the results of Example 6-1, find the components of the Einstein tensor in the coordinate basis.

**SOLUTION 6-3**
We begin by calculating the components of the Ricci tensor. For now, we continue working in the orthonormal basis. Therefore, we use

$$R_{\hat{a}\hat{b}} = R^{\hat{c}}{}_{\hat{a}\hat{c}\hat{b}} \tag{6.26}$$

The first nonzero component of the Ricci tensor is

$$R_{\hat{t}\hat{t}} = R^{\hat{c}}{}_{\hat{t}\hat{c}\hat{t}} = R^{\hat{t}}{}_{\hat{t}\hat{t}\hat{t}} + R^{\hat{r}}{}_{\hat{t}\hat{r}\hat{t}} + R^{\hat{\phi}}{}_{\hat{t}\hat{\phi}\hat{t}}$$

Noting that $R^{\hat{\phi}}{}_{\hat{t}\hat{\phi}\hat{t}} = \eta^{\hat{\phi}\hat{\phi}} R_{\hat{\phi}\hat{t}\hat{\phi}\hat{t}} = R_{\hat{\phi}\hat{t}\hat{\phi}\hat{t}} = -R_{\hat{t}\hat{\phi}\hat{\phi}\hat{t}} = -\eta_{\hat{t}\hat{t}} R^{\hat{t}}{}_{\hat{\phi}\hat{\phi}\hat{t}} = R^{\hat{t}}{}_{\hat{\phi}\hat{\phi}\hat{t}}$ and using the results of the previous example, we find that

$$R_{\hat{t}\hat{t}} = -\frac{\partial^2 b}{\partial t^2} - \left(\frac{\partial b}{\partial t}\right)^2 - \frac{1}{R}\frac{\partial^2 R}{\partial t^2} \tag{6.27}$$

Next we calculate

$$R_{\hat{t}\hat{r}} = R^{\hat{c}}{}_{\hat{t}\hat{c}\hat{r}} = R^{\hat{t}}{}_{\hat{t}\hat{t}\hat{r}} + R^{\hat{r}}{}_{\hat{t}\hat{r}\hat{r}} + R^{\hat{\phi}}{}_{\hat{t}\hat{\phi}\hat{r}}$$

The only nonzero term in this sum is $R^{\hat{\phi}}{}_{\hat{t}\hat{\phi}\hat{r}} = R^{\hat{t}}{}_{\hat{\phi}\hat{r}\hat{\phi}}$ and so

$$R_{\hat{t}\hat{r}} = \frac{e^{-b(t,r)}}{R}\left(-\frac{\partial^2 R}{\partial t\,\partial r} + \frac{\partial b}{\partial t}\frac{\partial R}{\partial r}\right) \tag{6.28}$$

Next, we find that

$$R_{\hat{r}\hat{r}} = R^{\hat{c}}{}_{\hat{r}\hat{c}\hat{r}} = R^{\hat{t}}{}_{\hat{r}\hat{t}\hat{r}} + R^{\hat{r}}{}_{\hat{r}\hat{r}\hat{r}} + R^{\hat{\phi}}{}_{\hat{r}\hat{\phi}\hat{r}}$$

$$= R^{\hat{t}}{}_{\hat{r}\hat{t}\hat{r}} + R^{\hat{r}}{}_{\hat{\phi}\hat{r}\hat{\phi}} \tag{6.29}$$

$$= \frac{\partial^2 b}{\partial t^2} + \left(\frac{\partial b}{\partial t}\right)^2 + \frac{e^{-2b(t,r)}}{R}\left(\frac{\partial b}{\partial t}\frac{\partial R}{\partial t}e^{2b(t,r)} - \frac{\partial^2 R}{\partial r^2} + \frac{\partial R}{\partial r}\frac{\partial b}{\partial r}\right)$$

Finally, using the same method it can be shown that

$$R_{\hat{\phi}\hat{\phi}} = \frac{1}{R}\frac{\partial^2 R}{\partial t^2} + \frac{e^{-2b(t,r)}}{R}\left(\frac{\partial b}{\partial t}\frac{\partial R}{\partial t}e^{2b(t,r)} - \frac{\partial^2 R}{\partial r^2} + \frac{\partial R}{\partial r}\frac{\partial b}{\partial r}\right) \qquad (6.30)$$

The next step is to find the Ricci scalar using $R = \eta^{\hat{a}\hat{b}}R_{\hat{a}\hat{b}}$ together with (6.27), (6.29), and (6.30):

$$R = -R_{\hat{t}\hat{t}} + R_{\hat{r}\hat{r}} + R_{\hat{\phi}\hat{\phi}}$$

$$= 2\frac{\partial^2 b}{\partial t^2} + 2\left(\frac{\partial b}{\partial t}\right)^2 + \frac{2}{R}\frac{\partial^2 R}{\partial t^2} + 2\frac{e^{-2b(t,r)}}{R}\left(\frac{\partial b}{\partial t}\frac{\partial R}{\partial t}e^{2b(t,r)} - \frac{\partial^2 R}{\partial r^2} + \frac{\partial R}{\partial r}\frac{\partial b}{\partial r}\right)$$

$$\qquad (6.31)$$

In the local frame, we can find the components of the Einstein tensor using

$$G_{\hat{a}\hat{b}} = R_{\hat{a}\hat{b}} - \frac{1}{2}\eta_{\hat{a}\hat{b}}R \qquad (6.32)$$

For example, using (6.29) together with (6.31), we find

$$G_{\hat{r}\hat{r}} = R_{\hat{r}\hat{r}} - \frac{1}{2}R$$

$$= \frac{\partial^2 b}{\partial t^2} + \left(\frac{\partial b}{\partial t}\right)^2 + \frac{e^{-2b(t,r)}}{R}\left(\frac{\partial b}{\partial t}\frac{\partial R}{\partial t}e^{2b(t,r)} - \frac{\partial^2 R}{\partial r^2} + \frac{\partial R}{\partial r}\frac{\partial b}{\partial r}\right)$$

$$\quad -\frac{1}{2}\left[2\frac{\partial^2 b}{\partial t^2} + 2\left(\frac{\partial b}{\partial t}\right)^2 + \frac{2}{R}\frac{\partial^2 R}{\partial t^2} + 2\frac{e^{-2b(t,r)}}{R}\left(\frac{\partial b}{\partial t}\frac{\partial R}{\partial t}e^{2b(t,r)} - \frac{\partial^2 R}{\partial r^2} + \frac{\partial R}{\partial r}\frac{\partial b}{\partial r}\right)\right]$$

$$= -\frac{1}{R}\frac{\partial^2 R}{\partial t^2}$$

Using (6.30) and (6.31), we obtain

$$G_{\hat{\phi}\hat{\phi}} = R_{\hat{\phi}\hat{\phi}} - \frac{1}{2}R$$

$$= \frac{1}{R}\frac{\partial^2 R}{\partial t^2} + \frac{e^{-2b(t,r)}}{R}\left(\frac{\partial b}{\partial t}\frac{\partial R}{\partial t}e^{2b(t,r)} - \frac{\partial^2 R}{\partial r^2} + \frac{\partial R}{\partial r}\frac{\partial b}{\partial r}\right)$$

$$-\frac{1}{2}\left[2\frac{\partial^2 b}{\partial t^2}+2\left(\frac{\partial b}{\partial t}\right)^2+\frac{2}{R}\frac{\partial^2 R}{\partial t^2}+2\frac{e^{-2b(t,r)}}{R}\left(\frac{\partial b}{\partial t}\frac{\partial R}{\partial t}e^{2b(t,r)}-\frac{\partial^2 R}{\partial r^2}+\frac{\partial R}{\partial r}\frac{\partial b}{\partial r}\right)\right]$$

$$=-\frac{\partial^2 b}{\partial t^2}-\left(\frac{\partial b}{\partial t}\right)^2$$

A similar exercise shows that the other nonzero components are

$$G_{\hat{t}\hat{t}}=\frac{1}{R}\frac{\partial b}{\partial t}\frac{\partial R}{\partial t}-\frac{e^{-2b(t,r)}}{R}\frac{\partial^2 R}{\partial r^2}+\frac{e^{-2b(t,r)}}{R}\frac{\partial R}{\partial r}\frac{\partial b}{\partial r}$$

$$G_{\hat{t}\hat{r}}=\frac{e^{-b(t,r)}}{R}\frac{\partial^2 R}{\partial t\,\partial r}-\frac{e^{-b(t,r)}}{R}\frac{\partial b}{\partial t}\frac{\partial R}{\partial r}$$

To write the components of the Einstein tensor in the coordinate basis, we need to write down the transformation matrix $\Lambda^{\hat{a}}{}_b$. Using the metric $ds^2=-dt^2+e^{2b(t,r)}\,dr^2+R(t,r)\,d\phi^2$, this is easy enough:

$$\Lambda^{\hat{a}}{}_b=\begin{pmatrix}-1 & 0 & 0\\ 0 & e^{b(t,r)} & 0\\ 0 & 0 & R(t,r)\end{pmatrix} \tag{6.33}$$

The transformation is given by

$$G_{ab}=\Lambda^{\hat{c}}{}_a\Lambda^{\hat{d}}{}_bG_{\hat{c}\hat{d}} \tag{6.34}$$

Note that the Einstein summation convention is being used on the right side of (6.34). However, since (6.33) is diagonal, each expression will use only one term from the sum. Considering each term in turn, using (6.33) and (6.34) we find

$$G_{tt}=\Lambda^{\hat{t}}{}_t\Lambda^{\hat{t}}{}_tG_{\hat{t}\hat{t}}=(-1)(-1)G_{\hat{t}\hat{t}}$$

$$=G_{\hat{t}\hat{t}}=\frac{e^{-2b(t,r)}}{R}\left(\frac{\partial b}{\partial t}\frac{\partial R}{\partial t}e^{2b(t,r)}-\frac{\partial^2 R}{\partial r^2}+\frac{\partial R}{\partial r}\frac{\partial b}{\partial r}\right) \tag{6.35}$$

$$G_{tr}=\Lambda^{\hat{t}}{}_t\Lambda^{\hat{r}}{}_rG_{\hat{t}\hat{r}}$$

$$=(-1)(e^{b(t,r)})\left(\frac{e^{-b(t,r)}}{R}\frac{\partial^2 R}{\partial t\,\partial r}-\frac{e^{-b(t,r)}}{R}\frac{\partial b}{\partial t}\frac{\partial R}{\partial r}\right) \tag{6.36}$$

$$=\frac{1}{R}\left(\frac{\partial b}{\partial t}\frac{\partial R}{\partial r}-\frac{\partial^2 R}{\partial t\,\partial r}\right)$$

$$G_{rr} = \Lambda^{\hat{r}}{}_r \Lambda^{\hat{r}}{}_r G_{\hat{r}\hat{r}}$$

$$= -\frac{e^{2b(t,r)}}{R}\frac{\partial^2 R}{\partial t^2} \tag{6.37}$$

and finally

$$G_{\phi\phi} = \Lambda^{\hat{\phi}}{}_\phi \Lambda^{\hat{\phi}}{}_\phi G_{\hat{\phi}\hat{\phi}}$$

$$= R^2 \left[ -\frac{\partial^2 b}{\partial t^2} - \left(\frac{\partial b}{\partial t}\right)^2 \right] \tag{6.38}$$

$$= -R^2 \left[ \frac{\partial^2 b}{\partial t^2} + \left(\frac{\partial b}{\partial t}\right)^2 \right]$$

### EXAMPLE 6-4

Using the results of Example 6-2, use the Einstein equations with nonzero cosmo-logical constant $\Lambda < 0$ to find the functional form of $e^{b(t,r)}$ and $R(t, r)$.

### SOLUTION 6-4

As stated earlier, the energy-momentum tensor for dust is given by $T_{ab} = \rho u_a u_b$. Since we are working in 2+1 dimensions, we will call $u^a$ the *three velocity*. It will be easiest to work in the co-moving frame. In this case, the three velocity takes on the simple form $u^a = \left(u^t,\ u^r,\ u^\phi\right) = (1,\ 0,\ 0)$. We can also use the local flat space metric

$$\eta_{\hat{a}\hat{b}} = \begin{pmatrix} -1 & 0 & 0 \\ 0 & 1 & 0 \\ 0 & 0 & 1 \end{pmatrix} \tag{6.39}$$

Einstein's equations can then be written as

$$G_{\hat{a}\hat{b}} + \Lambda \eta_{\hat{a}\hat{b}} = \kappa T_{\hat{a}\hat{b}} \tag{6.40}$$

where $\kappa$ is a constant. Since we are taking $\Lambda < 0$, we will make this more explicit by writing $\Lambda = -\lambda^2$ for some $\lambda^2 > 0$.

Earlier we found that $G_{\hat{r}\hat{r}} = -\frac{1}{R}\frac{\partial^2 R}{\partial t^2}$. Using $\eta_{\hat{r}\hat{r}} = 1$ and $T_{\hat{r}\hat{r}} = 0$ with (6.40), we find

$$\frac{\partial^2 R}{\partial t^2} + \lambda^2 R = 0 \tag{6.41}$$

This well-known equation has the solution

$$R = A \cos (\lambda t) + B \sin (\lambda t)$$

where $A = A(r)$ and $B = B(r)$. Turning to $G_{\hat\phi\hat\phi}$, using $\eta_{\hat\phi\hat\phi} = 1$ and $T_{\hat\phi\hat\phi} = 0$, we find

$$\frac{\partial^2 b}{\partial t^2} + \left( \frac{\partial b}{\partial t} \right)^2 + \lambda^2 = 0 \qquad (6.42)$$

To find a solution to this equation, we let $f = e^{b(t,r)}$. Therefore

$$\frac{\partial f}{\partial t} = \frac{\partial b}{\partial t} e^b$$

$$\frac{\partial^2 f}{\partial t^2} = \frac{\partial}{\partial t} \left( \frac{\partial b}{\partial t} e^b \right) = \frac{\partial^2 b}{\partial t^2} e^b + \left( \frac{\partial b}{\partial t} \right)^2 e^b = \left[ \frac{\partial^2 b}{\partial t^2} + \left( \frac{\partial b}{\partial t} \right)^2 \right] f$$

And so we can write

$$\frac{\partial^2 b}{\partial t^2} + \left( \frac{\partial b}{\partial t} \right)^2 = \frac{1}{f} \frac{\partial^2 f}{\partial t^2}$$

and (6.42) becomes

$$\frac{\partial^2 f}{\partial t^2} + \lambda^2 f = 0$$

Once again, we have a harmonic oscillator-type equation with solution

$$e^b = C \cos (\lambda t) + D \sin (\lambda t) \qquad (6.43)$$

where, as in the previous case, the "constants" of integration are functions of $r$.

There are two more Einstein's equations in this example that could be used for further analysis. We simply state them here:

$$\frac{\partial R}{\partial r}\frac{\partial b}{\partial t} - \frac{\partial^2 R}{\partial t \, \partial r} = 0$$

$$\frac{e^{-2b(t,r)}}{R}\left(\frac{\partial b}{\partial t}\frac{\partial R}{\partial t}e^{2b(t,r)} - \frac{\partial^2 R}{\partial r^2} + \frac{\partial R}{\partial r}\frac{\partial b}{\partial r}\right) + \lambda^2 = \kappa\rho$$

# Energy Conditions

Later, we will have use for the energy conditions. We state three of them here:

- The weak energy condition states that for any timelike vector $u^a$, $T_{ab}u^a u^b \geq 0$.
- The null energy condition states that for any null vector $l^a$, $T_{ab}l^a l^b \geq 0$.
- The strong energy condition states that for any timelike vector $u^a$, $T_{ab}u^a u^b \geq \frac{1}{2}T^c{}_c u^d u_d$.

# Quiz

1. Using the Bianchi identities, $\nabla_a R_{debc} + \nabla_c R_{deab} + \nabla_b R_{deca} = 0$, it can be shown that the contracted Bianchi identities for the Einstein tensor are
   (a) $\nabla_b R^{ab} = 0$
   (b) $\nabla_b G^{ab} = -T^{ac}$
   (c) $\nabla_b G^{ab} = 0$
   (d) $\nabla_b G^{ab} = \kappa\rho$

2. Consider Example 6-2. Using Cartan's equations
   (a) $R^{\hat{\phi}}{}_{\hat{t}\hat{t}\hat{\phi}} = \frac{1}{R^2}\frac{\partial^2 R}{\partial t^2}$
   (b) $R^{\hat{\phi}}{}_{\hat{t}\hat{t}\hat{\phi}} = \frac{1}{R}\frac{\partial^2 R}{\partial t^2}$
   (c) $R^{\hat{\phi}}{}_{\hat{t}\hat{t}\hat{\phi}} = -\frac{1}{R^2}\frac{\partial^2 R}{\partial t^2}$

3. The best statement of the strong equivalence principle is
   (a) the laws of physics are the same in an accelerated reference frame and in a uniform, static gravitational field

(b) tidal forces cannot be detected

(c) inertial and accelerated reference frames cannot be differentiated

4. The Einstein equations are

   (a) $\nabla^2 \phi = 4\pi G \rho$

   (b) $\frac{D^2 \eta^a}{D\tau^2} = R^a{}_{bcd} u^b u^c \eta^d$

   (c) $\nabla_b T^{ab} = 0$

   (d) $G_{ab} = \kappa T_{ab}$

   Consider the following metric:

   $$ds^2 = -dt^2 + L^2(t, r)\, dr^2 + B^2(t, r)\, d\phi^2 + M^2(t, r)\, dz^2$$

5. The Ricci rotation coefficient $\Gamma_{\hat{t}\hat{r}\hat{r}}$ is

   (a) $\frac{1}{L} \frac{\partial L}{\partial t}$

   (b) $-\frac{1}{L} \frac{\partial L}{\partial t}$

   (c) $\frac{1}{L^2} \frac{\partial L}{\partial t}$

   (d) $\frac{\partial B}{\partial t}$

6. $\Gamma_{\hat{r}\hat{\phi}\hat{\phi}}$ is given by

   (a) $\frac{1}{B} \frac{\partial B}{\partial t}$

   (b) $\frac{1}{M} \frac{\partial M}{\partial t}$

   (c) $-\frac{1}{LB} \frac{\partial B}{\partial r}$

   (d) $-\frac{1}{LB} \frac{\partial B}{\partial t}$

   Taking $T_{\hat{t}\hat{t}} = \rho$ and setting the cosmological constant equal to zero, show that the Einstein equation for $G_{\hat{t}\hat{t}}$ becomes

7. (a) $-\frac{B''L - B'L'}{BL^3} - \frac{M''L - M'L'}{ML^3} - \frac{B'M'}{BML^2} + \frac{\dot{B}\dot{L}}{BL} + \frac{\dot{M}\dot{L}}{ML} + \frac{\dot{B}\dot{M}}{BM} = \kappa\rho$

   (b) $\frac{B''L - B'L'}{BL^3} - \frac{M''L - M'L'}{ML^3} = \kappa\rho$

   (c) $-\frac{B''L - B'L'}{BL^3} - \frac{M''L - M'L'}{ML^3} - \frac{B'M'}{BML^2} + \frac{\dot{B}\dot{L}}{BLBL} + \frac{\dot{M}\dot{L}}{ML} + \frac{\dot{B}\dot{M}}{BM} = 0$

8.  The Ricci scalar is given by

(a)
$$R = \frac{2}{L}\frac{\partial^2 L}{\partial t^2} + \frac{2}{B}\frac{\partial^2 B}{\partial t^2} + \frac{2}{M}\frac{\partial^2 M}{\partial t^2} + \frac{2}{LB}\frac{\partial L}{\partial t}\frac{\partial B}{\partial t} + \frac{2}{LM}\frac{\partial L}{\partial t}\frac{\partial M}{\partial t}$$
$$+ \frac{2}{BM}\frac{\partial M}{\partial t}\frac{\partial B}{\partial t} - \frac{2}{L^2 B}\frac{\partial^2 B}{\partial r^2} - \frac{2}{L^2 M}\frac{\partial^2 M}{\partial r^2} + \frac{2}{L^3 B}\frac{\partial L}{\partial r}\frac{\partial B}{\partial r}$$
$$+ \frac{2}{L^3 M}\frac{\partial L}{\partial r}\frac{\partial M}{\partial r} - \frac{2}{L^2 BM}\frac{\partial B}{\partial r}\frac{\partial M}{\partial r}$$

(b)  $$R = \frac{2}{L}\frac{\partial^2 L}{\partial t^2} + \frac{2}{B}\frac{\partial^2 B}{\partial t^2} + \frac{2}{M}\frac{\partial^2 M}{\partial t^2} + \frac{2}{LB}\frac{\partial L}{\partial t}\frac{\partial B}{\partial t} - \frac{2}{LM}\frac{\partial L}{\partial t}\frac{\partial M}{\partial t}$$

(c)  $$R = \frac{2}{L^2}\frac{\partial^2 L}{\partial t^2} - \frac{2}{B}\frac{\partial^2 B}{\partial t^2} + \frac{2}{M}\frac{\partial^2 M}{\partial t^2} - \frac{2}{LB}\frac{\partial L}{\partial t}\frac{\partial B}{\partial t} - \frac{2}{LM}\frac{\partial L}{\partial t}\frac{\partial M}{\partial t}$$

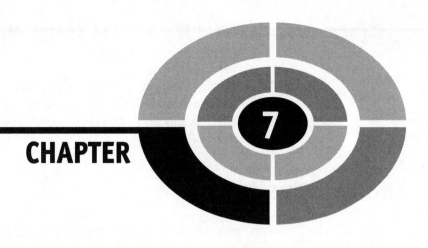

# The Energy-Momentum Tensor

In general relativity, the *stress-energy* or *energy-momentum* tensor $T^{ab}$ acts as the source of the gravitational field. It is related to the Einstein tensor and hence to the curvature of spacetime via the Einstein equation (with no cosmological constant)

$$G^{ab} = 8\pi\, G T^{ab}$$

The components of the stress-energy tensor can be arranged into a matrix with the property that $T^{ab} = T^{ba}$; i.e., the stress-energy tensor is symmetric. In the following and throughout the book, we will use the terms stress-energy tensor and energy-momentum interchangeably.

Let's see how to describe each component of the energy-momentum tensor. To understand the meaning of the component $T^{ab}$, consider the surface defined by constant $x^b$. Then $T^{ab}$ is the flux or flow of the $a$ component of momentum crossing the interface defined by constant $x^b$. In this case we're talking about the momentum four vector, so if $a = t$ then we are talking about the flow of *energy* across a surface. Let's describe each "type" of component we can have in turn. These are $T^{tt}$, $T^{it}$, $T^{ti}$, and $T^{ij}$.

# Energy Density

The $T^{tt}$ component represents energy density. To see why, consider the momentum four vector such that $\vec{p} = (E, \vec{p})$. Using the definition we gave above, we see that in this case $T^{tt}$ is the $p^0$ component of the momentum four vector, or simply the energy, crossing a surface of *constant time*. This is energy density.

In relativity, energy and mass are equivalent, and so we should really think of this as the mass-energy density.

In most applications energy density is denoted by $u$; however, we don't want to confuse that with the four velocity and so we denote the density of mass-energy by $\rho$. Therefore, for the stress-energy tensor, we can write

$$\rho = T^{tt}$$

# Momentum Density and Energy Flux

Momentum density is *momentum per unit volume*. If we call momentum density $\pi$, then the momentum density in the $i$ direction is

$$\pi^i = T^{it}$$

This is the flow of momentum crossing a surface of constant time.

Now consider $T^{ti}$. This term (which is actually equal to $T^{it}$ since the energy-momentum tensor is symmetric) represents the energy flow across the surface $x^i$.

# Stress

The final piece of the stress-energy tensor is given by the purely spatial components. These represent the flux of force per unit area—which is *stress*. We have

$$T^{ij}$$

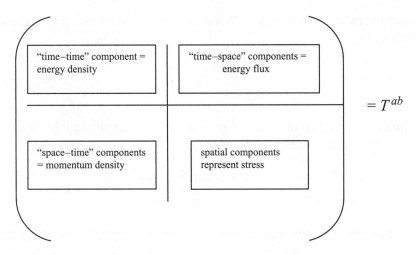

**Fig. 7-1.** A schematic representation of the stress-energy tensor. $T^{00}$ is the energy density. Terms of the form $T^{0j}$ (where $j$ is a spatial index) are energy flux. Terms with $T^{j0}$ are momentum density, while purely spatial components $T^{ij}$ are stress.

This term is the $i$th component of force per unit area (which is stress) across a surface with normal direction given by the basis vector $e_j$. Analogously, $T^{ji}$ is the $j$th component of force per unit area across a surface with normal given by the basis vector $e_i$. Returning to the view of a tensor that maps vectors and one forms to the real numbers, we obtain these components of the stress-energy tensor by passing as argument the basis vectors; i.e.,

$$T_{ij} = T\left(e_i, \ e_j\right)$$

The organization of the components of the stress-energy tensor into a matrix is shown schematically in Fig. 7-1. We will consider two types of stress-energy tensor seen frequently in relativity: perfect fluids and dust.

# Conservation Equations

Conservation equations can be derived from the stress-energy tensor using

$$\nabla_b T^{ab} = 0 \tag{7.1}$$

This equation means that energy and momentum are conserved. In a local frame, this reduces to

$$\frac{\partial T^{ab}}{\partial x^b} = 0 \tag{7.2}$$

In the local frame, when the conservation law (7.2) is applied to the time coordinate we obtain the familiar relation:

$$\frac{\partial T^{00}}{\partial t} + \frac{\partial T^{0i}}{\partial x^i} = \frac{\partial T^{00}}{\partial t} + \frac{\partial T^{0i}}{\partial x^i} = \frac{\partial \varepsilon}{\partial t} + \nabla \cdot \pi = 0$$

which is the conservation of energy.

# Dust

Later we will describe a perfect fluid which is characterized by pressure and density. If we start with a perfect fluid but let the pressure go to zero, we have *dust*. This is the simplest possible energy-momentum tensor that we can have.

It might seem that dust is too simple to be of interest. However, consider that the dust particles carry energy and momentum. The energy and momentum of the moving dust particles give rise to a gravitational field.

In this case, there are only two quantities that can be used to describe the matter field in the problem—the energy density—and how fast (and in what direction) the dust is moving. The simplest way to obtain the first quantity, the energy density, is to jump over to the co-moving frame. If you're in the co-moving frame, then you're moving along with the dust particles. In that case there is a number of dust particles per unit volume $n$, and each dust particle has energy $m$. So the energy density is given by $\rho = mn$.

The second item of interest is none other than the velocity four vector $\vec{u}$. This of course will give us the momentum carried by the dust. Generally speaking, to get the stress-energy tensor for dust, we put this together with the energy density. So for dust, the stress-energy tensor is given by

$$T^{ab} = \rho u^a u^b \tag{7.3}$$

For a co-moving observer, the four velocity reduces to $\vec{u} = (1, \ 0, \ 0, \ 0)$. In this case, the stress-energy tensor takes on the remarkably simple form

$$T^{ab} = \begin{pmatrix} \rho & 0 & 0 & 0 \\ 0 & 0 & 0 & 0 \\ 0 & 0 & 0 & 0 \\ 0 & 0 & 0 & 0 \end{pmatrix} \tag{7.4}$$

Now consider the case of a stationary observer seeing the dust particles go by with four velocity $\vec{u}$. In that case, we have $\vec{u} = (\gamma, \gamma u^x, \gamma u^y, \gamma u^z)$, where the $u^i$ are the ordinary components of three velocity and $\gamma = \frac{1}{\sqrt{1-v^2}}$. Looking at (7.3), we see that in this case the stress-energy tensor is

$$T^{ab} = \rho\gamma^2 \begin{pmatrix} 1 & u^x & u^y & u^z \\ u^x & (u^x)^2 & u^x u^y & u^x u^z \\ u^y & u^y u^x & (u^y)^2 & u^y u^z \\ u^z & u^z u^x & u^z u^y & (u^z)^2 \end{pmatrix} \tag{7.5}$$

**EXAMPLE 7-1**

Show that the conservation equations for the energy-momentum tensor in the case of dust lead to the equation of continuity of a fluid.

**SOLUTION 7-1**

The conservation equation is given by

$$\frac{\partial T^{ab}}{\partial x^b} = 0$$

Setting $a = t$, we obtain

$$\frac{\partial T^{tb}}{\partial x^b} = \frac{\partial T^{tt}}{\partial t} + \frac{\partial T^{tx}}{\partial x} + \frac{\partial T^{ty}}{\partial y} + \frac{\partial T^{tz}}{\partial z} = 0$$

Using (7.5), this becomes

$$\frac{\partial T^{tt}}{\partial t} + \frac{\partial T^{tx}}{\partial x} + \frac{\partial T^{ty}}{\partial y} + \frac{\partial T^{tz}}{\partial z} = \frac{\partial \rho}{\partial t} + \frac{\partial (\rho u^x)}{\partial x} + \frac{\partial (\rho u^y)}{\partial y} + \frac{\partial (\rho u^z)}{\partial z}$$

$$= \frac{\partial \rho}{\partial t} + \nabla \cdot (\rho \vec{u})$$

and so we have

$$\frac{\partial \rho}{\partial t} + \nabla \cdot (\rho \vec{u}) = 0$$

where $\vec{u}$ is the ordinary three-dimensional velocity. This is the equation of continuity.

# Perfect Fluids

A *perfect fluid* is a fluid that has no heat conduction or viscosity. As such the fluid is characterized by its mass density $\rho$ and the pressure $P$. The stress-energy tensor that describes a perfect fluid in the local frame is

$$T^{ab} = \begin{pmatrix} \rho & 0 & 0 & 0 \\ 0 & P & 0 & 0 \\ 0 & 0 & P & 0 \\ 0 & 0 & 0 & P \end{pmatrix} \tag{7.6}$$

To find the form of the stress-energy tensor in a general frame, we first consider the flat space of special relativity and boost to a frame of an observer with four velocity $\vec{u}$. The stress-energy transforms to a general frame by

$$T^{a'b'} = \Lambda^{a'}{}_c \Lambda^{b'}{}_d T^{cd} \tag{7.7}$$

However, we note that we can construct the most general form of the stress-energy tensor from the four velocity $\vec{u}$, the metric tensor $\eta_{ab}$ along with $\rho$ and $p$. Furthermore, the tensor is symmetric. This tells us that the general form of the stress energy tensor is

$$T^{ab} = A u^a u^b + B \eta_{ab} \tag{7.8}$$

where $A$ and $B$ are scalars. For this example, we assume that the metric is $\eta_{ab} = \text{diag}(1, -1, -1, -1)$. Looking at (7.6), we notice that the only spatial components are $T^{ii} = P$. Another way to write this is

$$T^{ij} = \delta^i_j P \tag{7.9}$$

In the rest frame, we have $u^0 = 1$ and all other components vanish. Therefore, (7.8) takes the form

$$T^{ij} = B\eta_{ij}$$

Comparison with (7.9) leads us to take $B = -P$. Now we consider the time component. In the local frame it is given by $T^{00} = \rho$ and so

$$T^{00} = \rho = Au^0u^0 + B\eta_{00} = Au^0u^0 - P = A - P$$

Therefore, we conclude that $A = P + \rho$ and write the general form of the stress-energy tensor for a perfect fluid in Minkowski spacetime as

$$T^{ab} = (\rho + P)u^au^b - P\eta_{ab} \qquad (7.10)$$

For any metric $g_{ab}$, this immediately generalizes to

$$T^{ab} = (\rho + P)u^au^b - Pg_{ab} \qquad (7.11)$$

Note that the form of the stress-energy tensor in general will change if we take $\eta_{ab} = \text{diag}(-1,\ 1,\ 1,\ 1)$. In that case the equations become

$$T^{ab} = (\rho + P)u^au^b + P\eta_{ab}$$
$$T^{ab} = (\rho + P)u^au^b + Pg_{ab} \qquad (7.12)$$

## EXAMPLE 7-2
Consider the Robertson-Walker metric used in Example 5-3:

$$ds^2 = -dt^2 + \frac{a^2(t)}{1 - kr^2}\,dr^2 + a^2(t)r^2\,d\theta^2 + a^2(t)r^2\sin^2\theta\,d\phi^2$$

Suppose we take the Einstein equation with nonzero cosmological constant. Find the Friedmann equations in this case.

## SOLUTION 7-2
In the last chapter we found that in the local frame, the components of the Einstein tensor for this metric were given by

$$G_{\hat{t}\hat{t}} = \frac{3}{a^2}\left(k + \dot{a}^2\right)$$
$$G_{\hat{r}\hat{r}} = G_{\hat{\theta}\hat{\theta}} = G_{\hat{\phi}\hat{\phi}} = -2\frac{\ddot{a}}{a} - \frac{1}{a^2}\left(k + \dot{a}^2\right) \qquad (7.13)$$

In the local frame, the components of the stress-energy tensor are simply given by (7.6), and so we have $T^{\hat{a}\hat{b}} = \text{diag}(\rho, p, p, p)$. Now with nonzero cosmological constant, the Einstein equation can be written as (using units with $c = G = 1$)

$$G_{ab} - \Lambda g_{ab} = 8\pi T_{ab} \qquad (7.14)$$

In the local frame, the way we have written the line element, we have $\eta_{\hat{a}\hat{b}} = \text{diag}(-1, 1, 1, 1)$. We use this to lower the indices of the stress-energy tensor

$$T_{\hat{a}\hat{b}} = \eta_{\hat{a}\hat{c}}\eta_{\hat{b}\hat{d}}T^{\hat{c}\hat{d}} \qquad (7.15)$$

In this case, this is easy since everything is diagonal, and it turns out any minus signs cancel. But you should be aware that in general you need to be careful about raising and lowering the indices. Anyway, we obtain

$$T_{\hat{a}\hat{b}} = \text{diag}(\rho, P, P, P) \qquad (7.16)$$

Putting this together with (7.14) and (7), we have

$$G_{\hat{t}\hat{t}} - \Lambda\eta_{\hat{t}\hat{t}} = 8\pi T_{\hat{t}\hat{t}}$$

$$\Rightarrow \frac{3}{a^2}\left(k + \dot{a}^2\right) + \Lambda = 8\pi\rho \qquad (7.17)$$

Since $G_{\hat{r}\hat{r}} = G_{\hat{\theta}\hat{\theta}} = G_{\hat{\phi}\hat{\phi}}$ and all of the spatial components of the stress-energy tensor are also identical, we need consider only one case. We find that

$$G_{\hat{r}\hat{r}} - \Lambda\eta_{\hat{r}\hat{r}} = 8\pi T_{\hat{r}\hat{r}} \qquad (7.18)$$

$$\Rightarrow 2\frac{\ddot{a}}{a} + \frac{1}{a^2}\left(k + \dot{a}^2\right) + \Lambda = -8\pi P$$

Equations (7.17) and (7.18) are *the Friedmann equations*.

# Relativistic Effects on Number Density

We now take a slight digression to investigate the effects of motion on the density of particles within the context of special relativity.

Consider a rectangular volume $V$ containing a set of particles. We can define a *number density* of particles which is simply the number of particles per unit volume. If we call the total number of particles in the volume $N$, then the number density is given by

$$n = \frac{N}{V}$$

In relativity, this is true only if we are in a frame that is at rest with respect to the volume. If we are not, then length contraction effects will change the number density that the observer sees. Suppose that we have two frames $F$ and $F'$ in the standard configuration, with $F'$ moving at velocity $v$ along the $x$-axis. The number of particles in the volume is a scalar, and so this does not change when viewed from a different frame. However, length contraction along the direction of motion means that the volume will change. In Fig. 7-2, we show motion along the $x$-axis.

Lengths along the $y$ and $z$ axes are unchanged under a Lorentz transformation under these conditions. If the volume of the box in a co-moving rest frame is $V$, then the volume of the box as seen by a stationary observer is

$$V' = \sqrt{1 - v^2}V = \frac{1}{\gamma}V$$

Therefore, the number density in a volume moving at speed $v$ as seen by a stationary observer is given by

$$n' = \frac{N}{V'} = \gamma n$$

**EXAMPLE 7-3**
Consider a box of particles. In the rest frame of the box, the volume $V = 1 \text{ m}^3$ and the total number of particles is $N = 2.5 \times 10^{25}$. Compare the number density of particles in the rest frame of the box and in a rest frame where the box has velocity $v = 0.9$. The box moves in the $x$-direction with respect to the stationary observer.

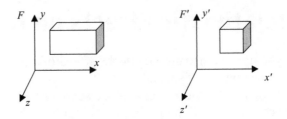

**Fig. 7-2.** A volume $V$, which we take in this example to be a rectangular box, is shortened along the direction of motion by the length contraction effect. This will change the number density of particles contained in $V$.

### SOLUTION 7-3
In the rest frame of the box, the number density $n = 2.5 \times 10^{25}$ particles per cubic meter. Now

$$\gamma = \frac{1}{\sqrt{1 - v^2}} = \frac{1}{\sqrt{1 - (0.9)^2}} \approx 2.29$$

A stationary observer who sees the box moving at velocity $v$ sees the number density of particles in the box as

$$n' = \gamma n = (2.3)\left(2.5 \times 10^{25}\right) = 5.75 \times 10^{25}$$

particles per cubic meter. Along the $x$-direction, the length of the box is

$$x = \frac{1}{\gamma} x' = \frac{1}{2.3} \text{ m} \approx 0.43 \text{ m}$$

The total number of particles is the same as viewed from both frames.

# More Complicated Fluids

The most general form that the stress-energy tensor can assume in the case of a fluid is that for a "nonperfect" fluid that can have viscosity and shear. This is beyond the scope of this book, but we will describe it here so that you will have seen it before and can see how viscosity is handled. The stress-energy tensor in this case is

$$T^{ab} = \rho\left(1 + \varepsilon\right)u^a u^b + (P - \zeta\theta)h^{ab} - 2\eta\sigma^{ab} + q^a u^b + q^b u^a$$

The quantities defined here are as follows:

$\varepsilon$ — specific energy density of the fluid in its rest frame

$P$ — pressure

$h^{ab}$ — the spatial projection tensor, $= u^a u^b + g^{ab}$

$\eta$ — shear viscosity

$\zeta$ — bulk viscosity

$\theta$ — expansion

$\sigma^{ab}$ — shear tensor

$q^a$ — energy flux vector

The expansion describes the divergence of the fluid worldliness. Therefore, it is given by

$$\theta = \nabla_a u^a$$

The shear tensor is

$$\sigma^{ab} = \frac{1}{2}\left(\nabla_c u^a h^{cb} + \nabla_c u^b h^{ca}\right) - \frac{1}{3}\theta h^{ab}$$

# Quiz

1. The $T^{tt}$ component of the stress-energy tensor
   (a) describes energy density
   (b) vanishes in most cases
   (c) represents conservation of momentum

2. The conservation equations are given by
   (a) $\nabla_b T^{ab} = -\rho$
   (b) $\nabla_b T^{ab} = 0$
   (c) $\nabla_b T^{ab} = \rho$

3. The Friedmann equations can be manipulated to obtain which of the following relationships?
   (a) $\frac{d}{dt}\left(\rho a^3\right) + P\frac{d}{dt}\left(a^3\right) = 0$
   (b) $\rho^2 \frac{d}{dt}\left(a^3\right) + P\frac{d}{dt}\left(a^3\right) = 0$
   (c) $\frac{d}{dt}\left(\rho a^3\right) = 0$

4. Using the correct result in Problem 3, if $a^3$ is taken to be volume $V$ and $E = \rho a^3$, which of the following is found to be correct?
   (a) $dE + P\,dV = 0$
   (b) $dE + T\,dS = 0$
   (c) $dE - P\,dV = 0$

5. Given a perfect fluid, we can write the spatial components of the stress-energy tensor as
   (a) $T^{ij} = \delta^i_j P$
   (b) $T^{ij} = \delta^i_j \rho$
   (c) $T^{ij} = \delta^i_j (P + \rho)$

# CHAPTER 8

# Killing Vectors

## Introduction

We've all heard the modern physics mantra and it's true: symmetries lead to conservation laws. So one thing you might be wondering is how can we find symmetries in relativity when the theory is so geometric? Geometrically speaking, a symmetry is hiding somewhere when we find that the metric is the same from point to point. Move from over here to over there, and the metric remains the same. That's a symmetry.

It turns out that there is a systematic way to tease out symmetries by finding a special type of vector called a *Killing vector*. A Killing vector $X$ satisfies *Killing's equation*, which is given in terms of covariant derivatives as

$$\nabla_b X_a + \nabla_a X_b = 0 \qquad (8.1)$$

Note that this equation also holds for contravariant components; i.e., $\nabla_b X^a + \nabla_a X^b = 0$. Killing vectors are related to symmetries in the following way: if $X$ is a vector field and a set of points is displaced by $X^a \, dx_a$ and all distance relationships remain the same, then $X$ is a Killing vector. This kind of distance preserving mapping is called an *isometry*. In a nutshell, if you move along the direction of a Killing vector, then the metric does not change. This is important because as we'll see in later chapters, this will lead us to conserved quantities. A free particle moving in a direction where the metric does not change will not feel any forces. This leads to momentum conservation. Specifically, if $X$ is a Killing vector, then

$$X \cdot u = \text{const}$$
$$X \cdot p = \text{const}$$

along a geodesic, where $u$ is the particle four velocity and $p$ is the particle four momentum.

Killing's equation can be expressed in terms of the Lie derivative of the metric tensor, as we show in this example.

**EXAMPLE 8-1**
Show that if the Lie derivative of the metric tensor vanishes, then

$$L_X g_{ab} = 0$$

This implies Killing's equation for $X$, given in (8.1).

**SOLUTION 8-1**
The Lie derivative of the metric is

$$L_X g_{ab} = X^c \partial_c g_{ab} + g_{cb} \partial_a X^c + g_{ac} \partial_b X^c$$

Let's recall the form of the covariant derivative. It's given by

$$\nabla_c X^a = \partial_c X^a + \Gamma^a{}_{bc} X^b \qquad (8.2)$$

Now, the covariant derivative of the metric tensor vanishes, $\nabla_c g_{ab} = 0$. The covariant derivative of the metric tensor is given by $\nabla_c g_{ab} = \partial_c g_{ab} - \Gamma^d{}_{ac} g_{db} - \Gamma^d{}_{bc} g_{ad}$. Since this vanishes, we can write

$$\partial_c g_{ab} = \Gamma^d{}_{ac} g_{db} + \Gamma^d{}_{bc} g_{ad}$$

Let's use this to rewrite the Lie derivative of the metric tensor. We have

$$L_X g_{ab} = X^c \partial_c g_{ab} + g_{cb} \partial_a X^c + g_{ac} \partial_b X^c$$
$$= X^c \left( \Gamma^d{}_{ac} g_{db} + \Gamma^d{}_{bc} g_{ad} \right) + g_{cb} \partial_a X^c + g_{ac} \partial_b X^c$$
$$= g_{db} \Gamma^d{}_{ac} X^c + g_{ad} \Gamma^d{}_{bc} X^c + g_{cb} \partial_a X^c + g_{ac} \partial_b X^c$$

Some manipulation can get this in the form we need to write down covariant derivatives of $X$. Note that

$$\nabla_b X^c = \partial_b X^c + \Gamma^c{}_{bd} X^d$$

In our expression we just derived for the Lie derivative the last term we have is $g_{ac} \partial_b X^c$. Can we find the other term needed to write down a covariant derivative? Yes, we can. Remember, a repeated index is a dummy index, and we are free to call it whatever we want. Looking at the last line we got for the Lie derivative, consider the second term

$$g_{ad} \Gamma^d{}_{bc} X^c$$

The indices $c$ and $d$ are repeated, so they are dummy indices. Let's switch them $c \leftrightarrow d$ and rewrite this term as

$$g_{ac} \Gamma^c{}_{bd} X^d$$

First, let's write down the result for the Lie derivative and rearrange the terms so that they are in the order we want for a covariant derivative.

$$L_X g_{ab} = g_{db} \Gamma^d{}_{ac} X^c + g_{ad} \Gamma^d{}_{bc} X^c + g_{cb} \partial_a X^c + g_{ac} \partial_b X^c$$
$$= g_{ac} \partial_b X^c + g_{db} \Gamma^d{}_{ac} X^c + g_{ad} \Gamma^d{}_{bc} X^c + g_{cb} \partial_a X^c$$
$$= g_{ac} \partial_b X^c + g_{ad} \Gamma^d{}_{bc} X^c + g_{cb} \partial_a X^c + g_{db} \Gamma^d{}_{ac} X^c$$

At this point, we put in the change of indices we used on what is now the second term in this expression.

$$L_X g_{ab} = g_{ac} \partial_b X^c + g_{ad} \Gamma^d{}_{bc} X^c + g_{cb} \partial_a X^c + g_{db} \Gamma^d{}_{ac} X^c$$
$$= g_{ac} \partial_b X^c + g_{ac} \Gamma^c{}_{bd} X^d + g_{cb} \partial_a X^c + g_{db} \Gamma^d{}_{ac} X^c$$
$$= g_{ac} \left( \partial_b X^c + \Gamma^c{}_{bd} X^d \right) + g_{cb} \partial_a X^c + g_{db} \Gamma^d{}_{ac} X^c$$
$$= g_{ac} \nabla_b X^c + g_{cb} \partial_a X^c + g_{db} \Gamma^d{}_{ac} X^c$$

Now we switch indices on the last term and get

$$
\begin{aligned}
L_X g_{ab} &= g_{ac}\nabla_b X^c + g_{cb}\,\partial_a X^c + g_{db}\Gamma^d{}_{ac}X^c \\
&= g_{ac}\nabla_b X^c + g_{cb}\,\partial_a X^c + g_{cb}\Gamma^c{}_{ad}X^d \\
&= g_{ac}\nabla_b X^c + g_{cb}\left(\partial_a X^c + \Gamma^c{}_{ad}X^d\right) \\
&= g_{ac}\nabla_b X^c + g_{cb}\nabla_a X^c
\end{aligned}
$$

Since the covariant derivative of the metric vanishes, we move the metric tensor inside the derivative and lower indices. For the first term we find

$$
g_{ac}\nabla_b X^c = \nabla_b\left(g_{ac}X^c\right) = \nabla_b X_a
$$

and for the second term we obtain

$$
g_{cb}\nabla_a X^c = \nabla_a\left(g_{cb}X^c\right) = \nabla_a X_b
$$

Therefore, we have

$$
L_X g_{ab} = g_{ac}\nabla_b X^c + g_{cb}\nabla_a X^c = \nabla_b X_a + \nabla_a X_b
$$

Since we are given that $L_X g_{ab} = 0$, this implies (8.1).

Often, we need to find the Killing vectors for a specific metric. We consider an explicit example by finding the Killing vectors for the 2-sphere.

**EXAMPLE 8-2**

Use Killing's equation to find the Killing vectors for the 2-sphere:

$$
ds^2 = a^2\,d\theta^2 + a^2\sin^2\theta\,d\phi^2
$$

**SOLUTION 8-2**

Killing's equation involves covariant derivatives. Therefore we need to recall the affine connection for this metric. In an earlier chapter we found

$$
\Gamma^\theta{}_{\theta\theta} = \Gamma^\phi{}_{\theta\theta} = \Gamma^\theta{}_{\phi\theta} = \Gamma^\phi{}_{\phi\phi} = 0
$$

$$
\Gamma^\phi{}_{\phi\theta} = \Gamma^\phi{}_{\theta\phi} = \cot\theta
$$

$$
\Gamma^\theta{}_{\phi\phi} = -\sin\theta\,\cos\theta
$$

Now, we recall that the covariant derivative is given by

$$
\nabla_b V_a = \partial_b V_a - \Gamma^c{}_{ab}V_c
$$

Starting with $a = b = \theta$ in Killing's equation, we find

$$\nabla_\theta X_\theta + \nabla_\theta X_\theta = 0$$
$$\Rightarrow \nabla_\theta X_\theta = 0$$

Using the equation for the covariant derivative, and recalling the Einstein summation convention, we obtain

$$\nabla_\theta X_\theta = \partial_\theta X_\theta - \Gamma^c{}_{\theta\theta} V_c = \partial_\theta X_\theta - \Gamma^\theta{}_{\theta\theta} V_\theta - \Gamma^\phi{}_{\theta\theta} V_\phi$$

Since $\Gamma^\theta{}_{\theta\theta} = \Gamma^\phi{}_{\theta\theta} = 0$, this reduces to the simple equation

$$\partial_\theta X_\theta = 0$$

More explicitly, we have

$$\frac{\partial X_\theta}{\partial \theta} = 0$$

Integrating, we find that the $X_\theta$ component of our Killing vector is some function of the $\phi$ variable:

$$X_\theta = f(\phi) \tag{8.3}$$

Next, we consider $a = b = \phi$. Using Killing's equation, we obtain

$$\nabla_\phi X_\phi = 0$$

Working out the left-hand side by writing out the covariant derivative, we find

$$\nabla_\phi X_\phi = \partial_\phi X_\phi - \Gamma^c{}_{\phi\phi} = \partial_\phi X_\phi - \Gamma^\theta{}_{\phi\phi} X_\theta - \Gamma^\phi{}_{\phi\phi} X_\phi$$

Now, $\Gamma^\phi{}_{\phi\phi} = 0$ and $\Gamma^\theta{}_{\phi\phi} = -\sin\theta\cos\theta$, and so using (8.3), this equation becomes

$$\frac{\partial X_\phi}{\partial \phi} = -\sin\theta\cos\theta\, X_\theta = -\sin\theta\cos\theta\, f(\phi)$$

Integrating, we obtain

$$X_\phi = -\sin\theta\cos\theta \int f(\phi')\,d\phi' + g(\theta) \tag{8.4}$$

Now, returning to Killing's equation, by setting $a = \theta$ and $b = \phi$, we have the last equation for this geometry, namely

$$\nabla_\theta X_\phi + \nabla_\phi X_\theta = 0$$

Let's write down each term separately. The first term is

$$\nabla_\theta X_\phi = \partial_\theta X_\phi - \Gamma^c{}_{\phi\theta} X_c = \partial_\theta X_\phi - \Gamma^\theta{}_{\phi\theta} X_\theta - \Gamma^\phi{}_{\phi\theta} X_\phi$$

Looking at the Christoffel symbols, we see that this leads to

$$\nabla_\theta X_\phi = \partial_\theta X_\phi - \cot\theta\, X_\phi$$

Now we consider the second term in $\nabla_\theta X_\phi + \nabla_\phi X_\theta = 0$. We obtain

$$\nabla_\phi X_\theta = \partial_\phi X_\theta - \Gamma^c{}_{\theta\phi} X_c = \partial_\phi X_\theta - \Gamma^\theta{}_{\theta\phi} X_\theta - \Gamma^\phi{}_{\theta\phi} X_\phi = \partial_\phi X_\theta - \cot\theta\, X_\phi$$

and so, the equation $\nabla_\theta X_\phi + \nabla_\phi X_\theta = 0$ becomes

$$\partial_\theta X_\phi + \partial_\phi X_\theta - 2\cot\theta\, X_\phi = 0$$
$$\Rightarrow \partial_\theta X_\phi + \partial_\phi X_\theta = 2\cot\theta\, X_\phi \tag{8.5}$$

We can refine this equation further using our previous results. Using (8.3) together with (8.4), we find

$$\partial_\theta X_\phi = \partial_\theta \left[ \sin\theta\cos\theta \int f(\phi')\,d\phi' + g(\theta) \right]$$

$$= (\sin^2\theta - \cos^2\theta) \int f(\phi')\,d\phi' + \partial_\theta g(\theta)$$

We also have

$$\partial_\phi X_\theta = \partial_\phi f(\phi)$$

Now, we can put all this together and obtain a solution. Adding these terms together, we get

$$\partial_\theta X_\phi + \partial_\phi X_\theta = (\sin^2\theta - \cos^2\theta)\int f(\phi')\,d\phi' + \partial_\theta g(\theta) + \partial_\phi f(\phi)$$

Next we want to set this equal to the right-hand side of (8.5). But let's work on that a little bit. We get

$$2\cot\theta\, X_\phi = 2\cot\theta\left[-\sin\theta\cos\theta\int f(\phi')\,d\phi' + g(\theta)\right]$$

Now we know that

$$\cot\theta\,(\sin\theta\cos\theta) = \frac{\cos\theta}{\sin\theta}(\sin\theta\cos\theta) = \cos^2\theta$$

Therefore, we can write

$$2\cot\theta\, X_\phi = -2\cos^2\theta\int f(\phi')\,d\phi' + 2\cot\theta g(\theta)$$

Finally, we equate both sides of (8.5) and we have

$$(\sin^2\theta - \cos^2\theta)\int f(\phi')\,d\phi' + \partial_\theta g(\theta) + \partial_\phi f(\phi)$$

$$= -2\cos^2\theta\int f(\phi')\,d\phi' + 2\cot\theta g(\theta)$$

Our goal is to get all $\theta$ terms on one side and all $\phi$ terms on the other. We can do this by adding $2\cos^2\theta\int f(\phi')\,d\phi'$ to both sides and then move the $\partial_\theta g(\theta)$ on the left-hand side over to the right. When we do this, we get this equation

$$\int f(\phi')\,d\phi' + \partial_\phi f(\phi) = 2\cot\theta g(\theta) - \partial_\theta g(\theta)$$

If you think back to your studies of partial differential equations, the kind where you used separation of variables, you will recall that when you have an equation in one variable equal to an equation in another variable, they must both be constant. So we will do that here. Let's call that constant $k$. Looking at the $\theta$

equation, we have

$$\partial_\theta g\,(\theta) - 2\cot\theta g\,(\theta) = -k$$

We multiplied through by $-1$ so that the derivative term would be positive. We can solve this kind of equation using the integrating factor method. Let's quickly review what that is. Consider your basic differential equation of the form

$$\frac{dy}{dt} + p\,(t)\,y = r\,(t)$$

First, we integrate the multiplying term $p(t)$:

$$p\,(t) = \int p(s)\,ds$$

Then we can solve the ordinary differential equation by writing

$$y\,(t) = e^{-p(t)} \int e^{p(s)} r(s)\,ds + Ce^{-p(t)}$$

Here $C$ is our constant of integration. Looking at our equation $\partial_\theta g\,(\theta) - 2\cot\theta g\,(\theta) = -k$, we make the following identifications. We set $p\,(\theta) = -2\cot\theta$ and $r\,(\theta) = -k$. First, we integrate $p$:

$$p\,(\theta) = \int -2\cot\theta\,d\theta = -2\int\frac{\cos\theta}{\sin\theta}\,d\theta = -2\ln\,(\sin\theta)$$

Now let's plug this into the exponential and we get

$$e^{-p(\theta)} = e^{2\ln(\sin\theta)} = e^{\ln\left(\sin^2\theta\right)} = \sin^2\theta$$

From this we deduce that $e^{p(\theta)} = \frac{1}{\sin^2\theta}$. Now we can use the integrating factor formula to write down a solution for the function $g$. The formula together with what we've just found gives us

$$g\,(\theta) = \sin^2\theta \int \frac{(-k)}{\sin^2 t}\,dt + C\sin^2\theta$$

In this case, $t$ is just a dummy variable of integration. When we integrate we will write the functions in terms of the $\theta$ variable. It's the easiest to look up the

integral in a table or use a program like Mathematica, and if you can't remember
you'll find out that

$$\int \frac{1}{\sin^2 t}\, dt = -\cot\theta$$

and so, we get the following:

$$g(\theta) = \sin^2\theta \int \frac{(-k)}{\sin^2 t}\, dt + C\sin^2\theta = \sin^2\theta\, k\cot\theta + C\sin^2\theta$$

$$= \sin^2\theta\,(k\cot\theta + C)$$

The final piece is to get a solution for the $\phi$ term. Earlier, we had
$\int f(\phi')\, d\phi' + \partial_\phi f(\phi) = 2\cot\theta g(\theta) - \partial_\theta g(\theta)$ and we decided to set this equal
to some constant that we called $k$. So looking at the $\phi$ piece, we obtain

$$\int f(\phi')d\phi' + \partial_\phi f(\phi) = k$$

Getting a solution to this one is easy. Let's differentiate it. That will get us rid
of the integral and turn the constant to zero, giving us the familiar equation

$$\frac{d^2 f}{d\phi^2} + f(\phi) = 0$$

This is a familiar equation to most of us, and we know that the solution is
given in terms of trignometric functions. More specifically, we have $f(\phi) = A\cos\phi + B\sin\phi$. We quickly see that $\partial_\phi f = -A\sin\phi + B\cos\phi$. Even better,
we can explicitly calculate the integral that has been hanging around since we
started this example. We get the following:

$$\int f(\phi')\, d\phi' = \int (A\cos\phi' + B\sin\phi')\, d\phi' = A\sin\phi - B\cos\phi$$

This is a really nice result. Remember, we had defined the constant $k$ in the
following way:

$$\int f(\phi')\, d\phi' + \partial_\phi f(\phi) = k$$

What this tells us is that the constant $k$ must be zero. Using what we have just found, we have

$$\int f(\phi')\, d\phi' + \partial_\phi f(\phi) = A \sin\phi - B \cos\phi - A \sin\phi + B \cos\phi = 0$$

This is a real headache, but we're almost done. Remember, for our $\theta$ function we found

$$g(\theta) = \sin^2\theta\,(k \cot\theta + C)$$

Since $k = 0$, we obtain the simple result

$$g(\theta) = C \sin^2\theta$$

Finally, at this point we have all the pieces we need to write down our Killing vector. Looking at (8.3), we recall that we had $X_\theta = f(\phi)$. With the results we've obtained, we have

$$X_\theta = A \cos\phi + B \sin\phi$$

In (8.4) we determined that

$$X_\phi = -\sin\theta \cos\theta \int f(\phi')\, d\phi' + g(\theta)$$

And using the results we've obtained, we get the final form for this component of the vector:

$$X_\phi = -\sin\theta \cos\theta\,(A \sin\phi - B \cos\phi) + C \sin^2\theta$$

The contravariant components of the Killing vector can be found by raising indices with the metric. It turns out that

$$X^\theta = X_\theta$$

and

$$\sin^2\theta\, X^\phi = X_\phi$$

It is going to turn out that for the 2-sphere, we are going to be able to write this Killing vector in terms of the angular momentum operators. Let's start by writing the entire vector out instead of individual components, and then do some algebraic manipulation:

$$
\begin{aligned}
X &= X^\theta \partial_\theta + X^\phi \partial_\phi \\
&= (A \cos\phi + B \sin\phi)\, \partial_\theta + [C - \cot\theta\, (A \sin\phi - B \cos\phi)]\, \partial_\phi \\
&= A \cos\phi\, \partial_\theta - A \cot\theta \sin\phi\, \partial_\phi + B \sin\phi\, \partial_\theta + A \cot\theta \cos\phi\, \partial_\phi + C \partial_\phi \\
&= -A L_x + B L_y + C L_z
\end{aligned}
$$

where the angular momentum operators are given by

$$
L_x = -\cos\phi\, \frac{\partial}{\partial\theta} + \cot\theta \sin\phi\, \frac{\partial}{\partial\phi}
$$

$$
L_y = \sin\phi\, \frac{\partial}{\partial\theta} + \cot\theta \cos\phi\, \frac{\partial}{\partial\phi}
$$

$$
L_z = \frac{\partial}{\partial\phi}
$$

# Derivatives of Killing Vectors

We can differentiate Killing vectors to obtain some useful relations between Killing vectors and the components of the Einstein equation. For the Riemann tensor, we have

$$
\nabla_c \nabla_b X^a = R^a{}_{bcd} X^d \tag{8.6}
$$

The Ricci tensor can be related to Killing vectors via

$$
\nabla_b \nabla_a X^b = R_{ac} X^c \tag{8.7}
$$

For the Ricci scalar, we have

$$
X^a \nabla_a R = 0 \tag{8.8}
$$

# Constructing a Conserved Current with Killing Vectors

Let $X$ be a Killing vector and $T$ be the stress-energy tensor. Let's define the following quantity as a current:

$$J^a = T^{ab} X_b$$

We can compute the covariant derivative of this quantity. We have

$$\nabla_a J^a = \nabla_a(T^{ab} X_b) = (\nabla_a T^{ab})X_b + T^{ab}(\nabla_a X_b)$$

The stress-energy tensor is conserved; therefore, $\nabla_c T^{ab} = 0$ and we are left with

$$\nabla_a J^a = T^{ab}(\nabla_a X_b)$$

Since the stress-energy tensor is symmetric, the symmetry of its indices allows us to write

$$\nabla_a J^a = T^{ab}(\nabla_a X_b) = \frac{1}{2}(T^{ab}\nabla_a X_b + T^{ba}\nabla_b X_a)$$

$$= \frac{1}{2}T^{ab}(\nabla_a X_b + \nabla_b X_a) = 0$$

Therefore, $J$ is a conserved current.

# Quiz

1. Killing's equation is given by
   (a) $\nabla_b X_a - \nabla_a X_b = 0$
   (b) $\nabla_b X_a = 0$
   (c) $\nabla_b X_a + \nabla_a X_b = G_{ab}$
   (d) $\nabla_b X_a + \nabla_a X_b = 0$

2. Given a Killing vector $X$, the Riemann tensor satisfies
   (a) $\nabla_c \nabla_b X^a = -R^a{}_{bcd} X^d$

(b) $\nabla_c \nabla_b X^a = R^a{}_{bcd} X^d$

(c) $\nabla_c \nabla_b X^a = R^a{}_{bcd} X^d - R^a{}_{bcd} X^b$

3. Given a Killing vector $X$, the Ricci scalar satisfies
   (a) $X^a \nabla_a R = 0$
   (b) $X^a \nabla_a R = -R$
   (c) $X^a \nabla_a R = R^a{}_a$

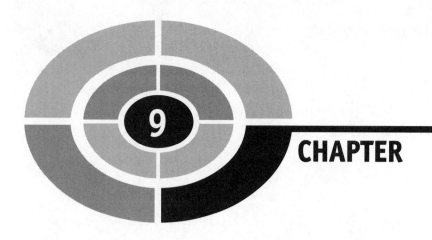

# CHAPTER 9

# Null Tetrads and the Petrov Classification

There is a viewpoint that the most fundamental entities that can be used to describe the structure of spacetime are light cones. After all, a light cone divides past and future, and in doing so defines which events are or can be causally related to one another. The light cone defines where in spacetime a particle with mass can move—nothing moves faster than the speed of light (see Fig. 9-1). We begin by reviewing a few concepts you've already seen. While they may already be familiar, they are important enough to be reviewed once again.

As described in Chapter 1, we can plot events in spacetime using a *spacetime diagram*. One or two spatial dimensions are suppressed, allowing us to represent space and time together graphically. By defining the speed of light $c = 1$, light rays move on 45° lines that define a cone. This light cone defines the structure of spacetime for some event $E$ that we have placed at the origin in the following way.

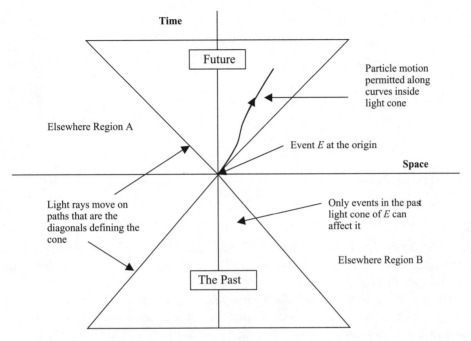

**Fig. 9-1.** The essence of spacetime structure is well described by the light cone, shown here with only one spatial dimension. Time is on the vertical axis. Particles with mass can move only on paths inside the light cone.

Events in the past that are causally related to $E$ are found in the lower light cone where $t < 0$ in the diagram. Events that can be affected by $E$ are inside the future light cone where $t > 0$. No object or particle with mass can move faster than the speed of light, so the motion of any massive particle is restricted to be inside the light cone.

In the diagram we have defined the two regions that are outside the light cone, regions A and B, as *elsewhere*. These regions are causally separated from each other. No event from region A can impact an event in region B because travel faster than the speed of light would be necessary for that to occur. This is all taking place in perfectly flat space, where light travels on straight lines.

Gravity manifests itself in the curvature of spacetime. As we will see when we examine the Schwarzschild solution in detail, gravity bends light rays. In a gravitational field, light no longer travels on perfectly straight lines. The more curvature there is, the more pronounced is the effect.

A remarkable consequence of this fact is that in strong gravitational fields, light cones begin to tip over. Inside a black hole we find the singularity, a point where the curvature of spacetime becomes infinite. As light cones get closer to

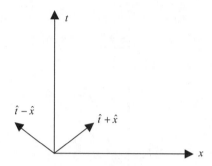

**Fig. 9-2.** Two null vectors pointing along the direction of light rays moving along $\hat{x}$ and $-\hat{x}$.

the singularity, the more they tip over. The future of any light cone is directed at the singularity. No matter what you do, if that's where you are, you are heading toward the singularity.

We will have more to say about this when we cover black holes in detail. The point of discussion right now is that the light cones themselves reveal the structure of spacetime. It was this notion that led Roger Penrose to consider a new way of doing relativity by introducing the *null tetrad*. This approach will be very useful in the study of black holes and also in the study of gravity waves, where gravitational disturbances propagate at the speed of light (and hence null vectors will be an appropriate tool).

The basic idea is the following: We would like to construct basis vectors that describe light rays moving in some direction. To simplify matters, for the moment consider the flat space of special relativity. Two vectors that describe light rays moving along $\hat{x}$ and $-\hat{x}$ are $\hat{t} + \hat{x}$ and $\hat{t} - \hat{x}$. These vectors are *null*, meaning that their lengths as defined by the dot product are zero. In order to have a basis for four-dimensional space, we need two more linearly independent vectors; this can be done by constructing them from the other spatial coordinates $\hat{y} \pm i\hat{z}$.

# Null Vectors

We have already seen how the introduction of an orthonormal basis can greatly simplify calculations and focus more on the geometry rather than being constrained to coordinates. In this chapter we will take that procedure a step further. As we mentioned in the last section, Penrose felt that the principal element of spacetime structure is the light cone. This consideration led him to introduce a set of *null* vectors to use as a tetrad basis (tetrad referring to the fact we have four basis vectors).

We recall from our studies in special relativity that an interval is *lightlike* if $(\Delta s)^2 = 0$, meaning that there is an equality of time and distance. We can carry this notion over to a vector by using its inner product to consider its length. A *null vector* $\vec{v}$ is one such that $\vec{v} \cdot \vec{v} = 0$.

### EXAMPLE 9-1

Classify the following vectors as timelike, spacelike, or null:

$$A^a = (-1,\ 4,\ 0,\ 1), \qquad B^a = (2, 0,\ -1,\ 1), \qquad C^a = (2,\ 0, -2,\ 0)$$

### SOLUTION 9-1

The Minkowski metric is

$$\eta_{ab} = \mathrm{diag}(1,\ -1,\ -1,\ -1)$$

Lowering indices for each vector, we have

$$A_a = \eta_{ab} A^b$$
$$\Rightarrow A_0 = \eta_{00} A^0 = (+1)(-1) = -1$$
$$A_1 = \eta_{11} A^1 = (-1)(4) = -4$$
$$A_2 = \eta_{22} A^2 = (-1)(0) = 0$$
$$A_3 = \eta_{33} A^3 = (-1)(1) = -1$$
$$\Rightarrow A_a = (-1,\ -4,\ 0,\ -1)$$

Applying a similar procedure to the other vectors gives

$$B_a = (2,\ 0,\ 1,\ -1) \quad \text{and} \quad C_a = (2,\ 0,\ 2,\ 0)$$

Computing the dot products $\vec{A} \cdot \vec{A}$, $\vec{B} \cdot \vec{B}$, and $\vec{C} \cdot \vec{C}$, we find

$$A_a A^a = (-1)(-1) + (-4)(4) + 0 + (-1)(1) = 1 - 16 - 1 = -16$$

Since $A_a A^a < 0$, with the convention we have used with the metric, $\vec{A}$ is spacelike. For the next vector, we find

$$B_a B^a = (2)(2) + 0 + (1)(-1) + (-1)(1) = 4 - 1 - 1 = 2$$
$$\Rightarrow B_a B^a > 0$$

Therefore $\vec{B}$ is timelike. For the last vector, we have

$$C_a C^a = (2)(2) + 0 + (2)(-2) + 0 = 4 - 4 = 0$$

Since $C_a C^a = 0$, this is a null vector.

# A Null Tetrad

Now that we've reminded ourselves about null vectors in Minkowski space, let's move on to the topic of this chapter. We begin by introducing a *tetrad* or set of four basis vectors that are labeled by $l$, $n$, $m$, and $\overline{m}$. In a departure from what you're used to in defining basis vectors, we allow two of these vectors to be *complex*. Specifically,

$$
\begin{array}{c}
l, n \text{ are real} \\
m, \overline{m} \text{ are complex}
\end{array}
\tag{9.1}
$$

As the notation makes obvious, $m$ and $\overline{m}$ are complex conjugates of one another. The second important observation is that these basis vectors are null vectors with respect to the Minkowski metric. Based on our quick review in the last section, this means that

$$l \cdot l = n \cdot n = m \cdot m = \overline{m} \cdot \overline{m} = 0 \tag{9.2}$$

or for example we can write $\eta_{ab} l^a l^b = 0$. In addition, these vectors satisfy a set of orthogonality relations that hold for inner products between a real and a complex vector:

$$l \cdot m = l \cdot \overline{m} = n \cdot m = n \cdot \overline{m} = 0 \tag{9.3}$$

We need two more pieces of information to characterize the tetrad. The first is we specify that the two real vectors satisfy

$$l \cdot n = 1 \tag{9.4}$$

and finally, the complex vectors satisfy

$$m \cdot \overline{m} = -1 \tag{9.5}$$

## BUILDING THE NULL TETRAD

It turns out that a null tetrad can be constructed rather easily with what we already know. We can do it using the basis of orthonormal one forms $\omega^{\hat{a}}$. First we state some relationships and then write down a matrix that will transform the $\omega^{\hat{a}}$ into a null tetrad.

Suppose that for a given spacetime we have defined an orthonormal tetrad as follows:

$$v^a \qquad \text{timelike vector}$$
$$i^a, \ j^a, \ k^a \quad \text{spacelike vectors}$$

We can then construct the null tetrad using a simple recipe. The two real vectors are given by

$$l^a = \frac{v^a + i^a}{\sqrt{2}} \quad \text{and} \quad n^a = \frac{v^a - i^a}{\sqrt{2}} \tag{9.6}$$

while the two complex vectors can be constructed using

$$m^a = \frac{j^a + ik^a}{\sqrt{2}} \quad \text{and} \quad \overline{m}^a = \frac{j^a - ik^a}{\sqrt{2}} \tag{9.7}$$

### EXAMPLE 9-2
Show that the definition given in (9.6) leads to $\eta_{ab}l^a l^b = 0$ and $\eta_{ab}l^a n^b = 1$.

### SOLUTION 9-2
The vectors $v^a$, $i^a$, $j^a$, and $k^a$ form an orthonormal tetrad. Since $v^a$ is time-like, we have $\eta_{ab}v^a v^b = +1$, while $i^a$ being spacelike satisfies $\eta_{ab}i^a i^b = -1$. Orthonormality also tells us that $\vec{v} \cdot \vec{i} = 0$. Proceeding with this knowledge, we have

$$\eta_{ab}l^a l^b = \eta_{ab}\left(\frac{v^a + i^a}{\sqrt{2}}\right)\left(\frac{v^b + i^b}{\sqrt{2}}\right) = \frac{1}{2}\eta_{ab}\left(v^a v^b + v^a i^b + i^a v^b + i^a i^b\right)$$

$$= \frac{1}{2}(1 + 0 + 0 - 1) = 0$$

In the second case, we have

$$\eta_{ab}l^a n^b = \eta_{ab}\left(\frac{v^a + i^a}{\sqrt{2}}\right)\left(\frac{v^b - i^b}{\sqrt{2}}\right) = \frac{1}{2}\eta_{ab}\left(v^a v^b - v^a i^b + i^a v^b - i^a i^b\right)$$

$$= \frac{1}{2}(1 + 0 + 0 + 1) = 1$$

Now let's see how to construct the null tetrad in a few specific cases. Two relations that are helpful if you are using the metric in a coordinate basis as components $g_{ab}$ relate the null tetrad to the components of the metric tensor as

$$g_{ab} = l_a n_b + l_b n_a - m_a \overline{m}_b - m_b \overline{m}_a \tag{9.8}$$

$$g^{ab} = l^a n^b + l^b n^a - m^a \overline{m}^b - m^b \overline{m}^a \tag{9.9}$$

If we are using an orthonormal tetrad, which is the preferred method of this book, then we can relate the two bases in the following way:

$$\begin{pmatrix} l \\ n \\ m \\ \overline{m} \end{pmatrix} = \frac{1}{\sqrt{2}} \begin{pmatrix} 1 & 1 & 0 & 0 \\ 1 & -1 & 0 & 0 \\ 0 & 0 & 1 & i \\ 0 & 0 & 1 & -i \end{pmatrix} \begin{pmatrix} \omega^{\hat{0}} \\ \omega^{\hat{1}} \\ \omega^{\hat{2}} \\ \omega^{\hat{3}} \end{pmatrix} \tag{9.10}$$

## EXAMPLE 9-3

Consider the flat space Minkowski metric, written in spherical polar coordinates:

$$ds^2 = dt^2 - dr^2 - r^2 \, d\theta^2 - r^2 \sin^2 \theta \, d\phi^2$$

Construct a null tetrad for this metric.

## SOLUTION 9-3

Writing down the components of the metric, we have

$$g_{ab} = \begin{pmatrix} 1 & 0 & 0 & 0 \\ 0 & -1 & 0 & 0 \\ 0 & 0 & -r^2 & 0 \\ 0 & 0 & 0 & -r^2 \sin^2 \theta \end{pmatrix} \tag{9.11}$$

Referring back to Chapter 5, we can write down the orthonormal basis one forms. These are

$$\omega^{\hat{t}} = dt, \qquad \omega^{\hat{r}} = dr, \qquad \omega^{\hat{\theta}} = r \, d\theta, \qquad \omega^{\hat{\phi}} = r \sin \theta \, d\phi$$

Applying (9.10), we have the following relations:

$$l = \frac{\omega^{\hat{t}} + \omega^{\hat{r}}}{\sqrt{2}} = \frac{dt + dr}{\sqrt{2}}$$

$$n = \frac{\omega^{\hat{t}} - \omega^{\hat{r}}}{\sqrt{2}} = \frac{dt - dr}{\sqrt{2}}$$

$$m = \frac{\omega^{\hat{\theta}} + i\omega^{\hat{\phi}}}{\sqrt{2}} = \frac{r\, d\theta + ir\sin\theta\, d\phi}{\sqrt{2}}$$

$$\overline{m} = \frac{\omega^{\hat{\theta}} - i\omega^{\hat{\phi}}}{\sqrt{2}} = \frac{r\, d\theta - ir\sin\theta\, d\phi}{\sqrt{2}}$$

In many cases, you will see the vectors written in terms of components with respect to the coordinate basis, i.e., $v^a = \left(v^0, \ v^1, \ v^2, \ v^3\right)$ or in this case $v^a = \left(v^t, \ v^r, \ v^\theta, \ v^\phi\right)$ or $v_a = \left(v_t, \ v_r, \ v_\theta, \ v_\phi\right)$. Using this notation, we have

$$l_a = \frac{1}{\sqrt{2}}\,(1, \ 1, \ 0, \ 0), \qquad n_a = \frac{1}{\sqrt{2}}\,(1, \ -1, \ 0, \ 0)$$

$$m_a = \frac{1}{\sqrt{2}}\,(0, \ 0, \ r, \ ir\sin\theta), \qquad \overline{m}_a = \frac{1}{\sqrt{2}}\,(0, \ 0, \ r, \ -ir\sin\theta)$$

With this in mind, let's verify (9.8). Starting with $g_{tt}$ in (9.11), we have

$$g_{tt} = l_t n_t + l_t n_t - m_t \overline{m}_t - m_t \overline{m}_t = \left(\frac{1}{\sqrt{2}}\right)\left(\frac{1}{\sqrt{2}}\right) + \left(\frac{1}{\sqrt{2}}\right)\left(\frac{1}{\sqrt{2}}\right)$$

$$= \frac{1}{2} + \frac{1}{2} = 1$$

Moving on, we find

$$g_{rr} = l_r n_r + l_r n_r - m_r \overline{m}_r - m_r \overline{m}_r = \left(\frac{1}{\sqrt{2}}\right)\left(-\frac{1}{\sqrt{2}}\right) + \left(\frac{1}{\sqrt{2}}\right)\left(-\frac{1}{\sqrt{2}}\right)$$

$$= -\frac{1}{2} - \frac{1}{2} = -1$$

$$g_{\theta\theta} = l_\theta n_\theta + l_\theta n_\theta - m_\theta \overline{m}_\theta - m_\theta \overline{m}_\theta = -\left(\frac{r}{\sqrt{2}}\right)\left(\frac{r}{\sqrt{2}}\right) - \left(\frac{r}{\sqrt{2}}\right)\left(\frac{r}{\sqrt{2}}\right)$$

$$= -\frac{r^2}{2} - \frac{r^2}{2} = -r^2$$

$$g_{\phi\phi} = l_\phi n_\phi + l_\phi n_\phi - m_\phi \overline{m}_\phi - m_\phi \overline{m}_\phi = -2\left(\frac{ir\sin\theta}{\sqrt{2}}\right)\left(-\frac{ir\sin\theta}{\sqrt{2}}\right)$$

$$= r^2\sin^2\theta$$

We can raise and lower the indices of the null tetrad using the metric $g_{ab}$. To see this, we use (9.9) together with the inner product relations defined in (9.2), (9.3), (9.4), and (9.5). Let's consider one case. We have

$$g^{ab} = l^a n^b + l^b n^a - m^a \overline{m}^b - m^b \overline{m}^a$$

What we would like is that $l^a = g^{ab}l_b$. So let's just put it there on both sides:

$$g^{ab}l_b = l^a n^b l_b + l^b n^a l_b - m^a \overline{m}^b l_b - m^b \overline{m}^a l_b$$

Now we rearrange terms a bit and use the fact that $A \cdot B = A^b B_b$ to write

$$g^{ab}l_b = l^a n^b l_b + n^a l^b l_b - m^a \overline{m}^b l_b - \overline{m}^a m^b l_b$$
$$= l^a\,(l \cdot n) + n^a\,(l \cdot l) - m^a\,(\overline{m} \cdot l) - \overline{m}^a\,(m \cdot l)$$

The inner product relations tell us the only term that survives is $l \cdot n = 1$. Therefore, we obtain the desired result $l^a = g^{ab}l_b$.

**EXAMPLE 9-4**
Using $g^{ab}$, find $l^a$, $n^a$, and $m^a$, and verify the inner product relations given in (9.2), (9.4), and (9.5).

**SOLUTION 9-4**
Now the metric with raised indices is given by

$$g^{ab} = \begin{pmatrix} 1 & 0 & 0 & 0 \\ 0 & -1 & 0 & 0 \\ 0 & 0 & -\frac{1}{r^2} & 0 \\ 0 & 0 & 0 & -\frac{1}{r^2\sin^2\theta} \end{pmatrix}$$

We can proceed in the usual way. For $l^a = g^{ab}l_b$ and $l_a = \frac{1}{\sqrt{2}}(1,\ 1,\ 0,\ 0)$, we have

$$l^t = g^{tt}\left(\frac{1}{\sqrt{2}}\right) = (+1)\left(\frac{1}{\sqrt{2}}\right) = \frac{1}{\sqrt{2}}$$

$$l^r = g^{rr}\left(\frac{1}{\sqrt{2}}\right) = (-1)\left(\frac{1}{\sqrt{2}}\right) = -\frac{1}{\sqrt{2}}$$

$$\Rightarrow l^a = \frac{1}{\sqrt{2}}(1, \ -1, \ 0, \ 0)$$

Therefore, we have

$$l \cdot l = l^a l_a = \left(\frac{1}{\sqrt{2}}\right)\left(\frac{1}{\sqrt{2}}\right) + \left(-\frac{1}{\sqrt{2}}\right)\left(\frac{1}{\sqrt{2}}\right) = \frac{1}{2} - \frac{1}{2} = 0$$

For $n_a = \frac{1}{\sqrt{2}}(1, -1, 0, 0)$, we obtain

$$n^t = g^{tt}\left(\frac{1}{\sqrt{2}}\right) = (+1)\left(\frac{1}{\sqrt{2}}\right) = \frac{1}{\sqrt{2}}$$

$$n^r = g^{rr}\left(-\frac{1}{\sqrt{2}}\right) = (-1)\left(-\frac{1}{\sqrt{2}}\right) = \frac{1}{\sqrt{2}}$$

$$\Rightarrow n^a = \frac{1}{\sqrt{2}}(1, \ 1, \ 0, \ 0)$$

And so we have $n \cdot n = n^a n_a = \left(\frac{1}{\sqrt{2}}\right)\left(\frac{1}{\sqrt{2}}\right) + \left(\frac{1}{\sqrt{2}}\right)\left(-\frac{1}{\sqrt{2}}\right) = \frac{1}{2} - \frac{1}{2} = 0$.
Now we compute

$$l \cdot n = l^a n_a = \left(\frac{1}{\sqrt{2}}\right)\left(\frac{1}{\sqrt{2}}\right) + \left(-\frac{1}{\sqrt{2}}\right)\left(-\frac{1}{\sqrt{2}}\right) = \frac{1}{2} + \frac{1}{2} = 1$$

So far, so good. Now let's check the complex basis vectors. For $m^a$ we find

$$m^\theta = g^{\theta\theta}m_\theta = \left(-\frac{1}{r^2}\right)\left(\frac{r}{\sqrt{2}}\right) = -\frac{1}{\sqrt{2}r}$$

$$m^\phi = g^{\phi\phi}m_\phi = \left(-\frac{1}{r^2\sin^2\theta}\right)\left(\frac{ir\sin\theta}{\sqrt{2}}\right) = -\frac{i}{\sqrt{2}r\sin\theta}$$

$$\Rightarrow m^a = \frac{1}{\sqrt{2}}(0, \ 0, \ -1/r, \ -i/r\sin\theta)$$

and so we have

$$m \cdot \overline{m} = m^a m_a = \frac{1}{2}\left[\left(-\frac{1}{r}\right)(r) + \left(-\frac{i}{r \sin \theta}\right)(-ir \sin \theta)\right] = \frac{1}{2}(-1 - 1)$$
$$= -1$$

And so we see, everything works out as expected.

## CONSTRUCTING A FRAME METRIC USING THE NULL TETRAD

We can construct a frame metric $v_{\hat{a}\hat{b}}$ using the null tetrad. The elements of $v_{\hat{a}\hat{b}} = e_{\hat{a}} \cdot e_{\hat{b}}$, and so we make the following designations:

$$e_{\hat{0}} = l, \qquad e_{\hat{1}} = n, \qquad e_{\hat{2}} = m, \qquad e_{\hat{3}} = \overline{m}$$

Using the orthogonality relations together with (9.4) and (9.5), it is easy to see that

$$v_{\hat{a}\hat{b}} = v^{\hat{a}\hat{b}} = \begin{pmatrix} 0 & 1 & 0 & 0 \\ 1 & 0 & 0 & 0 \\ 0 & 0 & 0 & -1 \\ 0 & 0 & -1 & 0 \end{pmatrix} \tag{9.12}$$

# Extending the Formalism

We are now going to introduce a new notation and method for calculation of the curvature tensor and related quantities, the *Newman-Penrose* formalism. This formalism takes advantage of null vectors and uses them to calculate components of the curvature tensor directly, allowing us to extract several useful pieces of information about a given spacetime. We will also be able to tie it together with a classification scheme called the *Petrov classification* that proves to be very useful in the study of black holes and gravitational radiation.

First let's introduce some notation. As a set of basis vectors, the null tetrad can be considered as directional derivatives. Therefore, we define the following:

$$D = l^a \nabla_a, \qquad \Delta = n^a \nabla_a, \qquad \delta = m^a \nabla_a, \qquad \delta^* = \overline{m}^a \nabla_a \tag{9.13}$$

Next we will define a set of symbols called the *spin coefficients*. These scalars (which can be complex) can be defined in two ways. The first is to write them down in terms of the Ricci rotation coefficients.

$$\pi = \Gamma_{130}, \qquad \nu = \Gamma_{131}, \qquad \lambda = \Gamma_{133}, \qquad \mu = \Gamma_{132}$$

$$\tau = \Gamma_{201}, \qquad \sigma = \Gamma_{202}, \qquad \kappa = \Gamma_{200}, \qquad \rho = \Gamma_{203}$$

$$\varepsilon = \frac{1}{2}\left(\Gamma_{100} + \Gamma_{230}\right), \qquad \gamma = \frac{1}{2}\left(\Gamma_{101} + \Gamma_{231}\right) \tag{9.14}$$

$$\alpha = \frac{1}{2}\left(\Gamma_{103} + \Gamma_{233}\right), \qquad \beta = \frac{1}{2}\left(\Gamma_{102} + \Gamma_{232}\right)$$

Better yet, these coefficients can also be defined directly in terms of the null tetrad:

$$\pi = -\nabla_b n_a \overline{m}^a l^b, \quad \nu = -\nabla_b n_a \overline{m}^a n^b, \quad \lambda = -\nabla_b n_a \overline{m}^a \overline{m}^b,$$

$$\mu = -\nabla_b n_a \overline{m}^a m^b$$

$$\tau = \nabla_b l_a m^a n^b, \quad \sigma = \nabla_b l_a m^a m^b, \quad \kappa = \nabla_b l_a m^a l^b, \quad \rho = \nabla_b l_a m^a \overline{m}^b$$

$$\varepsilon = \frac{1}{2}\left(\nabla_b l_a n^a l^b - \nabla_b m_a \overline{m}^a l^b\right), \quad \gamma = \frac{1}{2}\left(\nabla_b l_a n^a n^b - \nabla_b m_a \overline{m}^a n^b\right) \tag{9.15}$$

$$\alpha = \frac{1}{2}\left(\nabla_b l_a n^a \overline{m}^b - \nabla_b m_a \overline{m}^a \overline{m}^b\right), \quad \beta = \frac{1}{2}\left(\nabla_b l_a n^a m^b - \nabla_b m_a \overline{m}^a m^b\right)$$

In the usual course of business after finding the Ricci rotation coefficients (or Christoffel symbols in a coordinate basis) we tackle the problem of finding the curvature tensor. We will take the same approach here, so once we have the spin coefficients in hand we can obtain components of the Ricci and curvature tensors. However, in the Newman-Penrose formalism, we will be focusing on the *Weyl* tensor. We recall the definition from Chapter 4.

In a coordinate basis, we defined the Weyl tensor in terms of the curvature tensor as

$$C_{abcd} = R_{abcd} + \frac{1}{2}\left(g_{ad}R_{cb} + g_{bc}R_{da} - g_{ac}R_{db} - g_{bd}R_{ca}\right)$$

$$+ \frac{1}{6}\left(g_{ac}g_{db} - g_{ad}g_{cb}\right)R$$

The Weyl tensor has 10 independent components that we will represent with scalars. These can be referred to as *Weyl scalars* and are given by

$$\Psi_0 = -C_{abcd}l^a m^b l^c m^d \tag{9.16}$$

$$\Psi_1 = -C_{abcd}l^a n^b l^c m^d \tag{9.17}$$

$$\Psi_2 = -C_{abcd}l^a m^b \overline{m}^c n^d \tag{9.18}$$

$$\Psi_3 = -C_{abcd}l^a n^b \overline{m}^c n^d \tag{9.19}$$

$$\Psi_4 = -C_{abcd}n^a \overline{m}^b n^c \overline{m}^d \tag{9.20}$$

In addition, we have the following relations:

$$\begin{aligned} C_{0101} = C_{2323} = -\left(\Psi_1 + \Psi_2^*\right) \\ C_{0123} = \Psi_2 - \Psi_2^* \end{aligned} \tag{9.21}$$

Lastly, we develop some notation to represent the Ricci tensor by a set of scalars. This requires four real scalars and three complex ones. These are

$$\Phi_{00} = -\frac{1}{2}R_{ab}l^a l^b, \qquad \Phi_{01} = -\frac{1}{2}R_{ab}l^a m^b, \qquad \Phi_{02} = -\frac{1}{2}R_{ab}m^a m^b$$

$$\Phi_{11} = -\frac{1}{4}R_{ab}\left(l^a n^b + m^a \overline{m}^b\right), \qquad \Phi_{12} = -\frac{1}{2}R_{ab}n^a m^b, \tag{9.22}$$

$$\Phi_{22} = -\frac{1}{2}R_{ab}n^a n^b$$

$$\Lambda = \frac{1}{24}R$$

While we could go about finding $R_{ab}$ and the components of the Weyl tensor and then using them to calculate the scalars, that would miss the point. We want to avoid using those other quantities and calculate everything from the null vectors directly. It won't come as much of a surprise that there is a way to do it. First we calculate the spin coefficients directly from the null tetrad using (9.15). The next step relies on a monstrous looking set of equations called the *Newman-Penrose identities*. These threatening equations use the directional derivatives (9.13) applied to the spin coefficients to obtain the Weyl and Ricci scalars.

There are 18 identities in all, so we won't state them all here. Instead we will list 2 that will be useful for a calculation in the next example. This will be enough to communicate the flavor of the method. The reader who is interested in understanding the method in detail can consult Griffiths (1991) or Chandrasekhar (1992).

Two equations that will be of use to us in calculating the Weyl scalars and Ricci scalars are

$$\Delta\lambda - \bar{\delta}\nu = -\lambda\left(\mu + \bar{\mu}\right) - \lambda\left(3\gamma - \bar{\gamma}\right) + \nu\left(3\alpha + \bar{\beta} + \pi - \bar{\tau}\right) - \Psi_4 \quad (9.23)$$

$$\delta\nu - \Delta\mu = \mu^2 + \lambda\bar{\lambda} + \mu\left(\gamma + \bar{\gamma}\right) - \bar{\nu}\pi + \nu\left(\tau - 3\beta - \bar{\alpha}\right) + \Phi_{22} \quad (9.24)$$

Before we work an example, let's take a moment to discuss some physical interpretation of these definitions.

# Physical Interpretation and the Petrov Classification

In vacuum, the curvature tensor and the Weyl tensor coincide. Therefore, in many cases we need to study only the Weyl tensor. The *Petrov classification*, which describes the algebraic symmetries of the Weyl tensor, can be very useful in light of these considerations. To understand the meaning of the classifications, think in terms of matrices and eigenvectors. The eigenvectors of a matrix can be degenerate and occur in multiplicities. The same thing happens here and the Weyl tensor has a set of "eigenbivectors" that can occur in multiplicities. An eigenbivector satisfies

$$\frac{1}{2}C^{ab}_{\phantom{ab}cd}V^{cd} = \alpha V^{ab}$$

where $\alpha$ is a scalar. For mathematical reasons, which are beyond the scope of our present investigation, the Weyl tensor can have at most four distinct eigenbivectors.

Physically, the Petrov classification is a way to characterize a spacetime by the number of *principal null directions* it admits. The multiplicities of the eigenbivectors correspond to the number of principal null directions. If an eigenbivector is unique, we will call it *simple*. We will refer to the other eigenbivectors (and therefore the null directions) by the number of times they are repeated. If we say that there is a triple null direction, this means that three null directions coincide.

There are six basic types by which a spacetime can be classified in the Petrov scheme which we now summarize:

**Type I.** All four principal null directions are distinct (there are four simple principal null directions). This is also known as an *algebraically general spacetime*. The remaining types are known as *algebraically special*.

**Type II.** There are two simple null directions and one double null direction.

**Type III.** There is a single distinct null direction and a triple null direction. This type corresponds to longitudinal gravity waves with shear due to tidal effects.

**Type D.** There are two double principal null directions.

The Petrov type D is associated with the gravitational field of a star or black hole (Schwarzschild or Kerr vacuum). The two principal null directions correspond to ingoing and outgoing congruences of light rays.

**Type N.** There is a single principal null direction of multiplicity 4. This corresponds to transverse gravity waves.

**Type O.** The Weyl tensor vanishes and the spacetime is conformally flat.

We can learn about the principal null directions and the number of times they are repeated by examining $\Psi_A$. There are three situations of note:

1. *A principal null direction is repeated two times:* The nonzero components of the Weyl tensor are $\Psi_2$, $\Psi_3$, and $\Psi_4$.
2. *A principal null direction is repeated three times (Type III):* The nonzero components of the Weyl tensor are $\Psi_3$ and $\Psi_4$.
3. *A principal null direction is repeated four times (Type N):* The nonzero component of the Weyl tensor is $\Psi_4$.

Vanishing components of the Weyl tensor also tell us about the null vectors $l^a$ and $n^a$. For example, if $\Psi_0 = 0$ then $l^a$ is parallel with the principal null directions, while if $\Psi_4 = 0$ then $n^a$ is parallel with the principal null directions. If $\Psi_0 = \Psi_1 = 0$ then $l^a$ is aligned with repeated principal null directions.

In the context of gravitational radiation, we have the following interpretations:

$\Psi_0$   transverse wave component in the $n^a$ direction
$\Psi_1$   longitudinal wave component in the $n^a$ direction
$\Psi_3$   transverse wave component in the $l^a$ direction
$\Psi_4$   longitudinal wave component in the $l^a$ direction

Finally, we mention the physical meaning of some of the spin coefficients. First we consider a congruence of null rays defined by $l^a$:

$\kappa = 0$   $l^a$ is tangent to a null congruence
$-\mathrm{Re}\,(\rho)$   expansion of a null congruence (rays are diverging)
$\mathrm{Im}\,(\rho)$   twist of a null congruence
$|\sigma|$   shear of a null congruence

For a congruence defined by $n^a$, these definitions hold for $-\nu$, $-\mu$, and $-\lambda$.

**EXAMPLE 9-5**
Consider the Brinkmann metric that describes plane gravitational waves:

$$ds^2 = H(u,\ x,\ y)\,du^2 + 2\,du\,dv - dx^2 - dy^2$$

Find the nonzero Weyl scalars and components of the Ricci tensor, and interpret them.

**SOLUTION 9-5**
One can calculate these quantities by defining a basis of orthonormal one forms and proceeding with the usual methods, and then using (9.22) and related equations to calculate the desired quantities directly. However, we will take a new approach and use the Newman-Penrose identities to calculate the Ricci and Weyl scalars directly from the null tetrad.

The coordinates are $(u,\ v,\ x,\ y)$ where $u$ and $v$ are null coordinates. Specifically, $u = t - x$ and $v = t + x$. The vector $\partial_v$ defines the direction of propagation of the wave. Therefore, it is convenient to take

$$l^a = (0,\ 1,\ 0,\ 0) = \partial_v$$

We can raise and lower the indices using the components of the metric tensor, so let's begin by identifying them in the coordinate basis. We recall that we can write the metric as

$$ds^2 = g_{ab}\, dx^a\, dx^b$$

Therefore we can just read off the desired terms. Note that the cross term is given by $2\, du\, dv = g_{uv}\, du\, dv + g_{vu}\, dv\, du$ and so we take $g_{uv} = g_{vu} = 1$. The other terms can be read off immediately and so ordering the coordinates as $[u,\ v,\ x,\ y]$, we have

$$(g_{ab}) = \begin{pmatrix} H & 1 & 0 & 0 \\ 1 & 0 & 0 & 0 \\ 0 & 0 & -1 & 0 \\ 0 & 0 & 0 & -1 \end{pmatrix}$$

The inverse of this matrix gives the components of the metric tensor in the coordinate basis with raised indices:

$$(g^{ab}) = \begin{pmatrix} 0 & 1 & 0 & 0 \\ 1 & -H & 0 & 0 \\ 0 & 0 & -1 & 0 \\ 0 & 0 & 0 & -1 \end{pmatrix}$$

With this information in hand, we can calculate the various components of the null vectors. We start with $l^a = (0,\ 1,\ 0,\ 0)$. Lowering the index, we find

$$l_a = g_{ab} l^b = g_{av} l^v$$

The only nonzero term in the metric of this form is $g_{uv} = 1$, and so we conclude that

$$l_u = g_{uv} l^v = 1$$
$$\Rightarrow l_a = (1,\ 0,\ 0,\ 0) \tag{9.25}$$

To find the components of the rest of the tetrad, we can apply (9.8) and (9.9). To avoid flipping back through the pages, let's restate the identity for $g_{ab}$ here:

$$g_{ab} = l_a n_b + l_b n_a - m_a \overline{m}_b - m_b \overline{m}_a$$

Once we have all the components with lowered indices, we can raise indices with $g^{ab}$ to obtain all the other terms we need. Since there are not many terms in the metric, this procedure will be relatively painless. Since we also have $l_u = 1$ as the only nonzero component of the first null vector, we can actually guess as to what all the terms are without considering every single case. Starting from the top, we have

$$g_{uu} = H = 2l_u n_u - 2m_u \overline{m}_u$$

$$g_{uv} = 1 = l_u n_v + l_v n_u - m_u \overline{m}_v - \overline{m}_u m_v = l_u n_v - m_u \overline{m}_v - \overline{m}_u m_v$$

$$g_{vu} = 1 = l_v n_u + l_u n_v - m_v \overline{m}_u - \overline{m}_v m_u$$

$$g_{xx} = -1 = 2l_x n_x - 2m_x \overline{m}_x = -2m_x \overline{m}_x$$

$$g_{yy} = -1 = 2l_y n_y - 2m_y \overline{m}_y = -2m_y \overline{m}_y$$

We are free to make some assumptions, since all we have to do is come up with a null tetrad that gives us back the metric. We take $m_u = m_v = \overline{m}_u = \overline{m}_v = 0$ and assume that the $x$ component of $m$ is real. Then from the first and second equations, since $l_u = 1$, we obtain

$$n_u = \frac{1}{2} H \quad \text{and} \quad n_v = 1$$

Now $n$ is a null vector so we set $n_x = n_y = 0$. The fourth equation yields

$$m_x = \overline{m}_x = \frac{1}{\sqrt{2}}$$

These terms are equal to each other since they are complex conjugates and this is the real part. The final equation gives us the complex part of $m$. If we choose $m_y = -i \frac{1}{\sqrt{2}}$ then the fifth equation will be satisfied if $\overline{m}_y = i \frac{1}{\sqrt{2}}$, which is what we expect anyway since it's the complex conjugate. Anyway, all together we have

$$l_a = (1, 0, 0, 0), \qquad n_a = \left( \frac{1}{2} H, 1, 0, 0 \right)$$

$$m_a = \frac{1}{\sqrt{2}} (0, 0, 1, -i), \qquad \overline{m}_a = \frac{1}{\sqrt{2}} (0, 0, 1, i) \tag{9.26}$$

Next we use $g^{ab}$ to raise indices. We already know what $l^a$ is. The components of $n^a$ are given by

$$n^a = g^{ab}n_b$$

$$\Rightarrow n^u = g^{ub}n_b = g^{uu}n_u + g^{uv}n_v = n_v = 1$$

$$n^v = g^{vb}n_b = g^{vu}n_u + g^{vv}n_v = n_u - Hn_v = \frac{1}{2}H - H = -\frac{1}{2}H$$

It can be easily verified that the remaining components are zero. The final result is that

$$n^a = \left( 1, \ -\frac{1}{2}H, \ 0, \ 0 \right) \tag{9.27}$$

A similar exercise shows that

$$m^a = \frac{1}{\sqrt{2}}(0, 0, -1, i) \quad \text{and} \quad \overline{m}^a = \frac{1}{\sqrt{2}}(0, 0, -1, -i) \tag{9.28}$$

We are going to need the Christoffel symbols for this metric so that we can calculate covariant derivatives. These are fairly easy to calculate; we will compute two of them explicitly. Now

$$\Gamma^v{}_{xu} = \frac{1}{2}g^{vd}\left( \partial_x g_{du} + \partial_u g_{dx} - \partial_d g_{xu} \right)$$

There is only one nonconstant term in the metric with lowered indices, $g_{uu} = H$. Therefore, we can drop the last two terms and this reduces to

$$\Gamma^v{}_{xu} = \frac{1}{2}g^{vu}\left( \partial_x g_{uu} \right) = \frac{1}{2}\frac{\partial H}{\partial x}$$

Substitution of $y$ for $x$ yields a similar result

$$\Gamma^v{}_{yu} = \frac{1}{2}g^{vu}\left( \partial_y g_{uu} \right) = \frac{1}{2}\frac{\partial H}{\partial y}$$

All together, the nonzero Christoffel symbols for the metric are

$$\Gamma^v{}_{xu} = \frac{1}{2}\frac{\partial H}{\partial x}, \qquad \Gamma^v{}_{yu} = \frac{1}{2}\frac{\partial H}{\partial y} \tag{9.29}$$

$$\Gamma^v{}_{uu} = -\frac{1}{2}\frac{\partial H}{\partial u}, \qquad \Gamma^x{}_{uu} = \frac{1}{2}\frac{\partial H}{\partial x}, \qquad \Gamma^y{}_{uu} = \frac{1}{2}\frac{\partial H}{\partial y}$$

With this information in hand we can calculate the spin coefficients and the Weyl scalars. The spin coefficients can be calculated from

$$\pi = -\nabla_b n_a \overline{m}^a l^b, \quad \nu = -\nabla_b n_a \overline{m}^a n^b, \quad \lambda = -\nabla_b n_a \overline{m}^a \overline{m}^b,$$

$$\mu = -\nabla_b n_a \overline{m}^a m^b$$

$$\tau = \nabla_b l_a m^a n^b, \quad \sigma = \nabla_b l_a m^a m^b, \quad \kappa = \nabla_b l_a m^a l^b, \quad \rho = \nabla_b l_a m^a \overline{m}^b$$

$$\varepsilon = \frac{1}{2} \left( \nabla_b l_a n^a l^b - \nabla_b m_a \overline{m}^a l^b \right), \qquad \gamma = \frac{1}{2} \left( \nabla_b l_a n^a n^b - \nabla_b m_a \overline{m}^a n^b \right)$$

$$\alpha = \frac{1}{2} \left( \nabla_b l_a n^a \overline{m}^b - \nabla_b m_a \overline{m}^a \overline{m}^b \right), \qquad \beta = \frac{1}{2} \left( \nabla_b l_a n^a m^b - \nabla_b m_a \overline{m}^a m^b \right)$$

Now $\nabla_b l^a = 0$ and so we can immediately conclude that $\kappa = \rho = \sigma = \tau = 0$. As a result, we see there is no shear, twisting, or divergence for the null congruence in this case. In addition, one can see that $\alpha = \beta = \gamma = \varepsilon = 0$. The problem is starting to look tractable; we are going to have to worry about only very few terms. Let's consider $\nu$, which will turn out to be nonzero.

We will start by writing out the summation implied by the formula. However, note that we aren't going to have to write down every term. We are saved by the fact that $m$ has only $x$ and $y$ components while $n$ has only $u$ and $v$ components. Proceeding, we have

$$\nu = -\nabla_b n_a \overline{m}^a n^b$$
$$= -\nabla_u n_x \overline{m}^x n^u - \nabla_u n_y \overline{m}^y n^u - \nabla_v n_x \overline{m}^x n^v - \nabla_v n_y \overline{m}^y n^v$$

The last two terms drop out. Let's look at the very last term to see why. Remembering that $n_y = 0$, we have

$$\nabla_v n_y = \frac{\partial n_y}{\partial v} - \Gamma^d{}_{yv} n_d = -\Gamma^u{}_{yv} n_u - \Gamma^v{}_{yv} n_v$$

Having a look at the Christoffel symbols (0.29), we see that $\Gamma^u{}_{yv} = \Gamma^v{}_{yv} = 0$. So we can ignore these terms. Now the first two terms work out to be

$$\nabla_u n_x = \frac{\partial n_x}{\partial u} - \Gamma^d{}_{xu} n_d = -\Gamma^v{}_{xu} n_v = -\frac{1}{2} \frac{\partial H}{\partial x}$$

$$\nabla_u n_y = \frac{\partial n_y}{\partial u} - \Gamma^d{}_{yu} n_d = -\Gamma^v{}_{yu} n_v = -\frac{1}{2} \frac{\partial H}{\partial y}$$

Putting these results back into the expression for $\nu$, we find that

$$\nu = -\nabla_u n_x \overline{m}^x n^u - \nabla_u n_y \overline{m}^y n^u$$

$$= -\left(-\frac{1}{2}\frac{\partial H}{\partial x}\right)\left(-\frac{1}{\sqrt{2}}\right) - \left(-\frac{1}{2}\frac{\partial H}{\partial y}\right)\left(-i\frac{1}{\sqrt{2}}\right) \qquad (9.30)$$

$$= -\frac{1}{2\sqrt{2}}\left(\frac{\partial H}{\partial x} + i\frac{\partial H}{\partial y}\right)$$

(when going through this calculation, be careful with the overall minus sign). A similar exercise shows that the spin coefficients $\lambda = \pi = 0$. With all the vanishing spin coefficients, the Newman-Penrose identity we need to calculate, $\Psi_4$, takes on a very simple form. The expression we need is given by (9.23):

$$\Delta\lambda - \overline{\delta}\nu = -\lambda\left(\mu + \overline{\mu}\right) - \lambda\left(3\gamma - \overline{\gamma}\right) + \nu\left(3\alpha + \overline{\beta} + \pi - \overline{\tau}\right) - \Psi_4$$

In this case, we have

$$\Psi_4 = \overline{\delta}\nu$$

Recalling (9.13), this equation becomes $\Psi_4 = \overline{m}^a \nabla_a \nu$. Note, however, that $\nu$ is a scalar—we need to calculate only partial derivatives. The end result is that

$$\Psi_4 = \overline{m}^a \partial_a \nu = \overline{m}^x \partial_x \nu + \overline{m}^y \partial_y \nu$$

$$= \left(-\frac{1}{\sqrt{2}}\right)\frac{\partial}{\partial x}\left[-\frac{1}{2\sqrt{2}}\left(\frac{\partial H}{\partial x} + i\frac{\partial H}{\partial y}\right)\right]$$

$$+ \left(-\frac{i}{\sqrt{2}}\right)\frac{\partial}{\partial y}\left[-\frac{1}{2\sqrt{2}}\left(\frac{\partial H}{\partial x} + i\frac{\partial H}{\partial y}\right)\right]$$

$$= \frac{1}{4}\left(\frac{\partial^2 H}{\partial x^2} + i\frac{\partial^2 H}{\partial x\,\partial y}\right) + \frac{i}{4}\frac{\partial}{\partial y}\left(\frac{\partial^2 H}{\partial x\,\partial y} + i\frac{\partial^2 H}{\partial y^2}\right)$$

$$= \frac{1}{4}\frac{\partial^2 H}{\partial x^2} + \frac{i}{2}\frac{\partial^2 H}{\partial x\,\partial y} - \frac{1}{4}\frac{\partial^2 H}{\partial y^2}$$

and so we conclude that the Weyl scalar is given by

$$\Psi_4 = \frac{1}{4}\left(\frac{\partial^2 H}{\partial x^2} - \frac{\partial^2 H}{\partial y^2} + 2i\,\frac{\partial^2 H}{\partial x\,\partial y}\right) \tag{9.31}$$

In the exercises, if you are so inclined you can show that the only nonzero component of the Ricci tensor is given by

$$\Phi_{22} = \frac{1}{4}\left(\frac{\partial^2 H}{\partial x^2} + \frac{\partial^2 H}{\partial y^2}\right) \tag{9.32}$$

The fact that $\Psi_4$ is the only nonzero Weyl scalar tells us that the Petrov type is Type N, meaning that the principal null direction is repeated four times. Therefore, this metric describes transverse gravity waves.

# Quiz

1.  Using the Minkowski metric $\eta_{ab} = \text{diag}(1, -1, -1, -1)$, $\vec{A} = (3, 0, -3, 0)$ is
    (a) timelike
    (b) not enough information has been given
    (c) null
    (d) spacelike

2.  Using the definition of the complex vectors in the null tetrad, i.e.,

    $$m^a = \frac{j^a + ik^a}{\sqrt{2}}, \qquad \overline{m}^a = \frac{j^a - ik^a}{\sqrt{2}}$$

    which one of the following statements is true?
    (a) These are null vectors and that $m \cdot \overline{m} = -1$.
    (b) These are timelike vectors and that $m \cdot \overline{m} = -1$.
    (c) These are null vectors and that $m \cdot \overline{m} = 1$.
    (d) These are null vectors and that $m \cdot \overline{m} = 0$.

3.  Consider Example 9-5. Which of the following relations is true?
    (a) $\pi = -\nabla_b n_a \overline{m}^a l^b = 0$
    (b) $\pi = -\nabla_b n_a \overline{m}^a l^b = -1$
    (c) $\pi = \nabla_b n_a m^a \overline{l}^b = 0$
    (d) $\pi = \nabla_b n_a m^a \overline{l}^b = 1$

4. Again, consider Example 9-5. Which of the following relations is true?

(a) $\lambda = -\nabla_b l_a m^a m^b = 0$

(b) $\lambda = -\nabla_b n_a m^a m^b = 0$

(c) $\lambda = \nabla_b n_a \overline{m}^a \overline{m}^b = 0$

(d) $\lambda = -\nabla_b n_a \overline{m}^a \overline{m}^b = 0$

5. A spacetime described as Petrov type I

(a) has no null directions

(b) has four distinct principal null directions

(c) has a single null direction

(d) has a vanishing Weyl tensor, but has two null directions

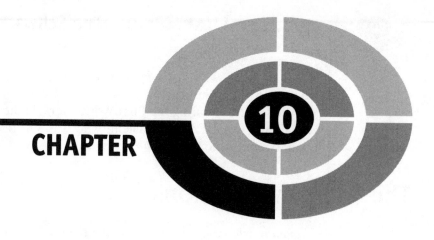

# The Schwarzschild Solution

When faced with a difficult set of mathematical equations, the first course of action one often takes is to look for special cases that are the easiest to solve. It turns out that such an approach often yields insights into the most interesting and physically relevant situations. This is as true for general relativity as it is for any other theory of mathematical physics.

Therefore for our first application of the theory, we consider a solution to the field equations that is time independent and spherically symmetric. Such a scenario can describe the gravitational field found outside of the Sun, for example. Since we might be interested only in the field *outside* of the matter distribution, we can simplify things even further by restricting our attention to the matter-free regions of space in the vicinity of some mass. Within the context of relativity, this means that one can find a solution to the problem using the *vacuum equations* and ignore the stress-energy tensor.

The solution we will obtain is known as the *Schwarzschild solution*. It was found in 1916 by the German physicist Karl Schwarzschild while he was serving on the Russian front during the First World War. He died from an illness soon after mailing his solution to Einstein, who was surprised that such a simple solution to his equations could be obtained.

# The Vacuum Equations

The *vacuum field equations* describe the metric structure of empty space surrounding a massive body. In the consideration of empty space where no matter or energy is present, we set $T_{ab} = 0$. In this case, the field equations become

$$R_{ab} = 0 \qquad (10.1)$$

# A Static, Spherically Symmetric Spacetime

To obtain the form of the metric that represents the field outside of a spherically symmetric body, we first consider the limiting form that it must take. Far away from the body (as $r$ becomes large), we expect that it will assume the form of the Minkowski metric. Since we are assuming spherical symmetry, we express the Minkowski metric in spherical coordinates

$$ds^2 = dt^2 - dr^2 - r^2 d\theta^2 - r^2 \sin^2 \theta \, d\phi^2 \qquad (10.2)$$

To obtain a general form of a time-independent, spherically symmetric metric that will reduce to (10.2) for large $r$, we first consider the requirement of time independence. If a metric is time independent, we should be able to change $dt \rightarrow -dt$ without affecting the metric. This tells us that the metric will not contain any mixed terms such as $dt\,dr$, $dt\,d\theta$, $dt\,d\phi$.

With no off-diagonal terms involving the time coordinate allowed in the metric, we can write the general form as

$$ds^2 = g_{tt} \, dt^2 + g_{ij} \, dx^i dx^j$$

We also require that the components of the metric are time independent; i.e.,

$$\frac{\partial g_{ab}}{\partial t} = 0$$

A metric that meets these conditions is called *static*.

Our next task is to think about how spherical symmetry affects the form of a metric. A spherically symmetric metric is one that has no preferred angular direction in space, meaning that we should be able to change $d\theta \rightarrow -d\theta$ and $d\phi \rightarrow -d\phi$ without changing the form of the metric. In the same way that time independence eliminated mixed terms involving $dt$ from consideration, we cannot have mixed terms such as $dr\,d\theta$, $dr\,d\phi$, $d\theta\,d\phi$ that would be affected by the changes $d\theta \rightarrow -d\theta$ and $d\phi \rightarrow -d\phi$.

So we have arrived at a metric that must have an entirely diagonal form. We have already eliminated any explicit time dependence from the terms of the metric. Since we are imposing radially symmetry, each term in the metric can be multiplied by a coefficient function that depends only on $r$. Using (10.2) as a guide, we can write this metric as

$$ds^2 = A(r)dt^2 - B(r)\,dr^2 - C(r)r^2 d\theta^2 - D(r)r^2 \sin^2 \theta\, d\phi^2 \qquad (10.3)$$

Spherical symmetry requires that the angular terms assume the normal form of $d\Omega^2$ that we are used to, and so we take $C = D$ and write this as

$$ds^2 = A(r)\,dt^2 - B(r)\,dr^2 - C(r)\left(r^2 d\theta^2 + r^2 \sin^2\theta\, d\phi^2\right) \qquad (10.4)$$

Now we can simplify matters even further by a change of radial coordinate to eliminate $C$. For the moment we will call the new radial coordinate $\rho$ and define it using $\rho = \sqrt{C(r)}r$. Then we have $\rho^2 = Cr^2$ and the angular part of the metric assumes the familiar form

$$C(r)\left(r^2\,d\theta^2 + r^2 \sin^2 \theta\, d\phi^2\right) = Cr^2\, d\theta^2 + Cr^2 \sin^2 \theta\, d\phi^2 \frac{1}{n}$$
$$= \rho^2\, d\theta^2 + \rho^2 \sin^2 \theta\, d\phi^2$$

From the definition $\rho = \sqrt{C(r)}r$, we see that we can write

$$d\rho = \frac{1}{2\sqrt{C}}\, dC\,r + \sqrt{C}\, dr$$

$$= \left(\frac{1}{2\sqrt{C}}\frac{dC}{dr}r + \sqrt{C}\right)dr$$

$$= \sqrt{C}\left(\frac{r}{2C}\frac{dC}{dr}r + 1\right)dr$$

Squaring both sides and solving for $dr^2$, we find that

$$dr^2 = \frac{1}{C}\left(1 + \frac{r}{2C}\frac{dC}{dr}\right)^{-2}d\rho^2$$

Now let's redefine the coefficient function $B' = \frac{1}{C}\left(1 + \frac{r}{2C}\frac{dC}{dr}\right)^{-2}B$ so that we have $B dr^2 = B'd\rho^2$. With these conditions in place, we can rewrite (10.4) as

$$ds^2 = A'dt^2 - B'd\rho^2 - \rho^2\left(d\theta^2 + \sin^2\theta\, d\phi^2\right)$$

where $A'$ is a function of $\rho$. Up to this point, other than the requirement that the coefficient functions go to unity as the radial coordinate gets large, we have not imposed any requirements on the form that they must take. Therefore we've basically been using nothing but labels. So we can relabel everything again, dropping primes on the coefficient functions and changing $\rho \to r$, allowing us to write the line element as

$$ds^2 = Adt^2 - Bdr^2 - r^2\left(d\theta^2 + \sin^2\theta\, d\phi^2\right)$$

Now we impose one last requirement. In order to correspond to the metric (10.2) at large $r$, we also need to preserve the signature. This can be done by writing the coefficient functions as exponentials, which are guaranteed to be positive functions. That is, we set $A = e^{2\nu(r)}$ and $B = e^{2\lambda(r)}$. This gives us the metric that is used to obtain the Schwarzschild solution:

$$ds^2 = e^{2\nu(r)}dt^2 - e^{2\lambda(r)}dr^2 - r^2\left(d\theta^2 + \sin^2\theta\, d\phi^2\right) \tag{10.5}$$

# The Curvature One Forms

We will proceed to find the solution, using the orthonormal tetrad method. By now the reader is aware that the first step along this path is to calculate the curvature one forms and the Ricci rotation coefficients. To do so, we define the following basis one forms:

$$\omega^{\hat{t}} = e^{\nu(r)}dt, \quad \omega^{\hat{r}} = e^{\lambda(r)}dr, \quad \omega^{\hat{\theta}} = r\,d\theta, \quad \omega^{\hat{\phi}} = r\sin\theta\, d\phi \tag{10.6}$$

Therefore, we have

$$dt = e^{-\nu(r)}\omega^{\hat{t}}, \quad dr = e^{-\lambda(r)}\omega^{\hat{r}}, \quad d\theta = \frac{1}{r}\omega^{\hat{\theta}}, \quad d\phi = \frac{1}{r\sin\theta}\omega^{\hat{\phi}} \tag{10.7}$$

The exterior derivative of each of the basis one forms is given by

$$d\omega^{\hat{t}} = d\left(e^{\nu(r)}dt\right) = \frac{d\nu}{dr}e^{\nu(r)}dr \wedge dt = \frac{d\nu}{dr}e^{-\lambda(r)}\omega^{\hat{r}} \wedge \omega^{\hat{t}} \tag{10.8}$$

$$d\omega^{\hat{r}} = d\left(e^{\lambda(r)}dr\right) = \frac{d\lambda}{dr}e^{\lambda(r)}dr \wedge dr = 0 \tag{10.9}$$

$$d\omega^{\hat{\theta}} = d\left(r\,d\theta\right) = dr \wedge d\theta = \frac{e^{-\lambda(r)}}{r}\omega^{\hat{r}} \wedge \omega^{\hat{\theta}} \tag{10.10}$$

$$d\omega^{\hat{\phi}} = d\left(r\sin\theta\,d\phi\right) = \sin\theta\,dr \wedge d\phi + r\cos\theta\,d\theta \wedge d\phi$$

$$= \frac{e^{-\lambda(r)}}{r}\omega^{\hat{r}} \wedge \omega^{\hat{\phi}} + \frac{\cot\theta}{r}\omega^{\hat{\theta}} \wedge \omega^{\hat{\phi}} \tag{10.11}$$

Once again recall Cartan's first structure equation, which for our coordinates will assume the form

$$d\omega^{\hat{a}} = -\Gamma^{\hat{a}}_{\ \hat{b}} \wedge \omega^{\hat{b}} = -\Gamma^{\hat{a}}_{\ \hat{t}} \wedge \omega^{\hat{t}} - \Gamma^{\hat{a}}_{\ \hat{r}} \wedge \omega^{\hat{r}} - \Gamma^{\hat{a}}_{\ \hat{\theta}} \wedge \omega^{\hat{\theta}} - \Gamma^{\hat{a}}_{\ \hat{\phi}} \wedge \omega^{\hat{\phi}}$$

Taking each basis one form in turn gives

$$d\omega^{\hat{t}} = -\Gamma^{\hat{t}}_{\ \hat{t}} \wedge \omega^{\hat{t}} - \Gamma^{\hat{t}}_{\ \hat{r}} \wedge \omega^{\hat{r}} - \Gamma^{\hat{t}}_{\ \hat{\theta}} \wedge \omega^{\hat{\theta}} - \Gamma^{\hat{t}}_{\ \hat{\phi}} \wedge \omega^{\hat{\phi}} \tag{10.12}$$

$$d\omega^{\hat{r}} = -\Gamma^{\hat{r}}_{\ \hat{t}} \wedge \omega^{\hat{t}} - \Gamma^{\hat{r}}_{\ \hat{r}} \wedge \omega^{\hat{r}} - \Gamma^{\hat{r}}_{\ \hat{\theta}} \wedge \omega^{\hat{\theta}} - \Gamma^{\hat{r}}_{\ \hat{\phi}} \wedge \omega^{\hat{\phi}} \tag{10.13}$$

$$d\omega^{\hat{\theta}} = -\Gamma^{\hat{\theta}}_{\ \hat{t}} \wedge \omega^{\hat{t}} - \Gamma^{\hat{\theta}}_{\ \hat{r}} \wedge \omega^{\hat{r}} - \Gamma^{\hat{\theta}}_{\ \hat{\theta}} \wedge \omega^{\hat{\theta}} - \Gamma^{\hat{\theta}}_{\ \hat{\phi}} \wedge \omega^{\hat{\phi}} \tag{10.14}$$

$$d\omega^{\hat{\phi}} = -\Gamma^{\hat{\phi}}_{\ \hat{t}} \wedge \omega^{\hat{t}} - \Gamma^{\hat{\phi}}_{\ \hat{r}} \wedge \omega^{\hat{r}} - \Gamma^{\hat{\phi}}_{\ \hat{\theta}} \wedge \omega^{\hat{\theta}} - \Gamma^{\hat{\phi}}_{\ \hat{\phi}} \wedge \omega^{\hat{\phi}} \tag{10.15}$$

Comparing (10.12) with (10.8), note that the only nonzero term in (10.8) of the form $\omega^{\hat{r}} \wedge \omega^{\hat{t}}$. Therefore, we conclude that

$$\Gamma^{\hat{t}}_{\ \hat{t}} = \Gamma^{\hat{t}}_{\ \hat{\theta}} = \Gamma^{\hat{t}}_{\ \hat{\phi}} = 0 \quad \text{and} \quad \Gamma^{\hat{t}}_{\ \hat{r}} = \frac{d\nu}{dr}e^{-\lambda(r)}\omega^{\hat{t}} \tag{10.16}$$

Using $\Gamma^{\hat{a}}_{\ \hat{b}} = \Gamma^{\hat{a}}_{\ \hat{b}\hat{c}}\omega^{\hat{c}}$, we set $\Gamma^{\hat{t}}_{\ \hat{r}\hat{t}} = \frac{d\nu}{dr}e^{-\lambda(r)}$. Since (10.9) vanishes and gives no direct information, we move on to compare (10.14) and (10.10). Again, with only a single term which this time involves $\omega^{\hat{r}} \wedge \omega^{\hat{\theta}}$, we take the nonzero term in (10.14) to be $\Gamma^{\hat{\theta}}_{\ \hat{r}} \wedge \omega^{\hat{r}}$ and conclude that

$$\Gamma^{\hat{\theta}}_{\ \hat{t}} = \Gamma^{\hat{\theta}}_{\ \hat{\theta}} = \Gamma^{\hat{\theta}}_{\ \hat{\phi}} = 0 \quad \text{and} \quad \Gamma^{\hat{\theta}}_{\ \hat{r}} = \frac{e^{-\lambda(r)}}{r}\omega^{\hat{\theta}} \tag{10.17}$$

Again using $\Gamma^{\hat{a}}_{\ \hat{b}} = \Gamma^{\hat{a}}_{\ \hat{b}\hat{c}}\omega^{\hat{c}}$, we conclude that $\Gamma^{\hat{\theta}}_{\ \hat{r}\hat{\theta}} = \frac{e^{-\lambda(r)}}{r}$.

Finally, We compare (10.15) with (10.11) using the same procedure, which gives

$$\Gamma^{\hat{\phi}}_{\ \hat{t}} = \Gamma^{\hat{\phi}}_{\ \hat{\phi}} = 0, \quad \Gamma^{\hat{\phi}}_{\ \hat{r}} = \frac{e^{-\lambda(r)}}{r}\omega^{\hat{\phi}}, \quad \Gamma^{\hat{\phi}}_{\ \hat{\theta}} = \frac{\cot\theta}{r}\omega^{\hat{\phi}} \tag{10.18}$$

where $\Gamma^{\hat{\phi}}_{\ \hat{r}\hat{\phi}} = \frac{e^{-\lambda(r)}}{r}$ and $\Gamma^{\hat{\phi}}_{\ \hat{\theta}\hat{\phi}} = \frac{\cot\theta}{r}$.

Now we return to the case of (10.9), which gave us no information since it vanished. To find the curvature one forms in this case, we will use symmetry considerations. Looking at the metric (10.5), we see that in this case we can define

$$\eta_{\hat{a}\hat{b}} = \begin{pmatrix} 1 & 0 & 0 & 0 \\ 0 & -1 & 0 & 0 \\ 0 & 0 & -1 & 0 \\ 0 & 0 & 0 & -1 \end{pmatrix}$$

to raise and lower indices. Therefore, we have the following relationships:

$$\Gamma^{\hat{t}}_{\ \hat{r}} = \eta^{\hat{t}\hat{t}}\Gamma_{\hat{t}\hat{r}} = \Gamma_{\hat{t}\hat{r}} = -\Gamma_{\hat{r}\hat{t}} = -\eta_{\hat{r}\hat{r}}\Gamma^{\hat{r}}_{\ \hat{t}} = \Gamma^{\hat{r}}_{\ \hat{t}}$$

$$\Gamma^{\hat{\theta}}_{\ \hat{r}} = \eta^{\hat{\theta}\hat{\theta}}\Gamma_{\hat{\theta}\hat{r}} = -\Gamma_{\hat{\theta}\hat{r}} = \Gamma_{\hat{r}\hat{\theta}} = \eta_{\hat{r}\hat{r}}\Gamma^{\hat{r}}_{\ \hat{\theta}} = -\Gamma^{\hat{r}}_{\ \hat{\theta}}$$

$$\Gamma^{\hat{\phi}}_{\ \hat{r}} = \eta^{\hat{\phi}\hat{\phi}}\Gamma_{\hat{\phi}\hat{r}} = -\Gamma_{\hat{\phi}\hat{r}} = \Gamma_{\hat{r}\hat{\phi}} = \eta_{\hat{r}\hat{r}}\Gamma^{\hat{r}}_{\ \hat{\phi}} = -\Gamma^{\hat{r}}_{\ \hat{\phi}}$$

And so, from (10.16) we conclude that

$$\Gamma^{\hat{r}}_{\ \hat{t}} = \Gamma^{\hat{t}}_{\ \hat{r}} = \frac{d\nu}{dr}e^{-\lambda(r)}\omega^{\hat{t}}$$

$$\Rightarrow \Gamma^{\hat{r}}_{\ \hat{t}\hat{t}} = \frac{d\nu}{dr}e^{-\lambda(r)} \tag{10.19}$$

Using (10.17), we find

$$\Gamma^{\hat{r}}_{\ \hat{\theta}} = -\Gamma^{\hat{\theta}}_{\ \hat{r}} = -\frac{e^{-\lambda(r)}}{r}\omega^{\hat{\theta}}$$

(10.20)

$$\Rightarrow \Gamma^{\hat{r}}_{\ \hat{\theta}\hat{\theta}} = -\frac{e^{-\lambda(r)}}{r}$$

and finally

$$\Gamma^{\hat{r}}_{\ \hat{\phi}} = -\Gamma^{\hat{\phi}}_{\ \hat{r}} = -\frac{e^{-\lambda(r)}}{r}\omega^{\hat{\phi}}$$

(10.21)

$$\Rightarrow \Gamma^{\hat{r}}_{\ \hat{\phi}\hat{\phi}} = -\frac{e^{-\lambda(r)}}{r}$$

# Solving for the Curvature Tensor

We now compute the components of the curvature tensor, using Cartan's second structure equation; i.e., $\Omega^{\hat{a}}_{\ \hat{b}} = d\Gamma^{\hat{a}}_{\ \hat{b}} + \Gamma^{\hat{a}}_{\ \hat{c}} \wedge \Gamma^{\hat{c}}_{\ \hat{b}} = \frac{1}{2}R^{\hat{a}}_{\ \hat{b}\hat{c}\hat{d}}\,\omega^{\hat{c}} \wedge \omega^{\hat{d}}$. We will explicitly compute one of the curvature two forms. Consider $\Omega^{\hat{r}}_{\ \hat{t}} = d\Gamma^{\hat{r}}_{\ \hat{t}} + \Gamma^{\hat{r}}_{\ \hat{c}} \wedge \Gamma^{\hat{c}}_{\ \hat{t}}$. For the summation, we have

$$\Gamma^{\hat{r}}_{\ \hat{c}} \wedge \Gamma^{\hat{c}}_{\ \hat{t}} = \Gamma^{\hat{r}}_{\ \hat{t}} \wedge \Gamma^{\hat{t}}_{\ \hat{t}} + \Gamma^{\hat{r}}_{\ \hat{r}} \wedge \Gamma^{\hat{r}}_{\ \hat{t}} + \Gamma^{\hat{r}}_{\ \hat{\theta}} \wedge \Gamma^{\hat{\theta}}_{\ \hat{t}} + \Gamma^{\hat{r}}_{\ \hat{\phi}} \wedge \Gamma^{\hat{\phi}}_{\ \hat{t}} = 0$$

since $\Gamma^{\hat{r}}_{\ \hat{t}} = \Gamma^{\hat{r}}_{\ \hat{r}} = \Gamma^{\hat{\theta}}_{\ \hat{t}} = \Gamma^{\hat{\phi}}_{\ \hat{t}} = 0$. This leaves

$$\Omega^{\hat{r}}_{\ \hat{t}} = d\Gamma^{\hat{r}}_{\ \hat{t}} = d\left(\frac{d\nu}{dr}e^{-\lambda(r)}\omega^{\hat{t}}\right)$$

$$= d\left(\frac{d\nu}{dr}e^{\nu(r)-\lambda(r)}dt\right)$$

$$= \frac{d^2\nu}{dr^2}e^{\nu(r)-\lambda(r)}dr \wedge dt + \left(\frac{d\nu}{dr}\right)^2 e^{\nu(r)-\lambda(r)}dr \wedge dt$$

$$- \left(\frac{d\nu}{dr}\right)\left(\frac{d\lambda}{dr}\right)e^{\nu(r)-\lambda(r)}dr \wedge dt$$

Using (10.7) to invert the differentials and write them in terms of the basis one forms, we arrive at the following expression:

$$\Omega^{\hat{r}}_{\ \hat{t}} = \left[\frac{d^2\nu}{dr^2} + \left(\frac{d\nu}{dr}\right)^2 - \left(\frac{d\nu}{dr}\right)\left(\frac{d\lambda}{dr}\right)\right]e^{-2\lambda(r)}\omega^{\hat{r}} \wedge \omega^{\hat{t}}$$

(10.22)

Now let's write out the curvature two form in terms of the components of the Riemann tensor, using the expression $\Omega^{\hat{a}}_{\ \hat{b}} = \frac{1}{2} R^{\hat{a}}_{\ \hat{b}\hat{c}\hat{d}}\, \omega^{\hat{c}} \wedge \omega^{\hat{d}}$. Remember that the Einstein summation convention is being used, so there is a summation over $\hat{c}$ and $\hat{d}$. However, noticing that in (10.22) there is only one term, $\omega^{\hat{r}} \wedge \omega^{\hat{t}}$, we have to consider only two terms in the sum, those involving $\omega^{\hat{r}} \wedge \omega^{\hat{t}}$ and $\omega^{\hat{t}} \wedge \omega^{\hat{r}}$. So we have

$$\Omega^{\hat{r}}_{\ \hat{t}} = \frac{1}{2} R^{\hat{r}}_{\ \hat{t}\hat{c}\hat{d}}\omega^{\hat{c}} \wedge \omega^{\hat{d}} = \frac{1}{2} R^{\hat{r}}_{\ \hat{t}\hat{r}\hat{t}}\omega^{\hat{r}} \wedge \omega^{\hat{t}} + \frac{1}{2} R^{\hat{r}}_{\ \hat{t}\hat{t}\hat{r}}\, \omega^{\hat{t}} \wedge \omega^{\hat{r}}$$

Now we use the fact that $\omega^{\hat{t}} \wedge \omega^{\hat{r}} = -\omega^{\hat{r}} \wedge \omega^{\hat{t}}$ to write this as

$$\Omega^{\hat{r}}_{\ \hat{t}} = \frac{1}{2} R^{\hat{r}}_{\ \hat{t}\hat{r}\hat{t}}\, \omega^{\hat{r}} \wedge \omega^{\hat{t}} - \frac{1}{2} R^{\hat{r}}_{\ \hat{t}\hat{t}\hat{r}}\, \omega^{\hat{r}} \wedge \omega^{\hat{t}} = \frac{1}{2} \left( R^{\hat{r}}_{\ \hat{t}\hat{r}\hat{t}} - R^{\hat{r}}_{\ \hat{t}\hat{t}\hat{r}} \right) \omega^{\hat{r}} \wedge \omega^{\hat{t}}$$

Using the same type of procedure we've seen in earlier chapters, we can simplify this even further using the symmetries of the Riemann tensor:

$$R^{\hat{r}}_{\ \hat{t}\hat{t}\hat{r}} = \eta^{\hat{r}\hat{r}} R_{\ \hat{r}\hat{t}\hat{t}\hat{r}} = -R_{\ \hat{r}\hat{t}\hat{r}\hat{t}} = R_{\ \hat{r}\hat{t}\hat{r}\hat{t}} = \eta_{\ \hat{r}\hat{r}} R^{\hat{r}}_{\ \hat{t}\hat{r}\hat{t}} = -R^{\hat{r}}_{\ \hat{t}\hat{r}\hat{t}}$$

and so we have

$$\Omega^{\hat{r}}_{\ \hat{t}} = \frac{1}{2} \left( R^{\hat{r}}_{\ \hat{t}\hat{r}\hat{t}} - R^{\hat{r}}_{\ \hat{t}\hat{t}\hat{r}} \right) \omega^{\hat{r}} \wedge \omega^{\hat{t}} = \frac{1}{2} \left( R^{\hat{r}}_{\ \hat{t}\hat{r}\hat{t}} + R^{\hat{r}}_{\ \hat{t}\hat{r}\hat{t}} \right) \omega^{\hat{r}} \wedge \omega^{\hat{t}} = R^{\hat{r}}_{\ \hat{t}\hat{r}\hat{t}}\omega^{\hat{r}} \wedge \omega^{\hat{t}}$$

Comparison with (10.22) shows that

$$R^{\hat{r}}_{\ \hat{t}\hat{r}\hat{t}} = \left[ \frac{d^2\nu}{dr^2} + \left( \frac{d\nu}{dr} \right)^2 - \left( \frac{d\nu}{dr} \right)\left( \frac{d\lambda}{dr} \right) \right] e^{-2\lambda(r)}$$

All together, the nonzero components of the Riemann tensor are given by

$$R^{\hat{r}}_{\ \hat{t}\hat{r}\hat{t}} = \left[ \frac{d^2\nu}{dr^2} + \left( \frac{d\nu}{dr} \right)^2 - \left( \frac{d\nu}{dr} \right)\left( \frac{d\lambda}{dr} \right) \right] e^{-2\lambda(r)}$$

$$R^{\hat{t}}_{\ \hat{\theta}\hat{t}\hat{\theta}} = R^{\hat{t}}_{\ \hat{\phi}\hat{t}\hat{\phi}} = -\frac{1}{r}\frac{d\nu}{dr}e^{-2\lambda}$$

$$R^{\hat{r}}_{\ \hat{\theta}\hat{r}\hat{\theta}} = R^{\hat{r}}_{\ \hat{\phi}\hat{r}\hat{\phi}} = \frac{1}{r}\frac{d\lambda}{dr}e^{-2\lambda} \qquad\qquad (10.23)$$

$$R^{\hat{\theta}}_{\ \hat{\phi}\hat{\theta}\hat{\phi}} = \frac{1 - e^{-2\lambda}}{r^2}$$

All other nonzero components can be found using the symmetries of the Riemann tensor.

# The Vacuum Equations

Now we can compute the components of the Ricci tensor to obtain the vacuum equations. This is done relatively easily by computing $R_{\hat{a}\hat{b}} = R^{\hat{c}}{}_{\hat{a}\hat{c}\hat{b}}$. The first term is easy to calculate

$$R_{\hat{t}\hat{t}} = R^{\hat{t}}{}_{\hat{t}\hat{t}\hat{t}} + R^{\hat{r}}{}_{\hat{t}\hat{r}\hat{t}} + R^{\hat{\theta}}{}_{\hat{t}\hat{\theta}\hat{t}} + R^{\hat{\phi}}{}_{\hat{t}\hat{\phi}\hat{t}}$$

$$= \left[ \frac{d^2 v}{dr^2} + \left( \frac{dv}{dr} \right)^2 - \left( \frac{dv}{dr} \right) \left( \frac{d\lambda}{dr} \right) + \frac{2}{r} \frac{dv}{dr} \right] e^{-2\lambda(r)} \qquad (10.24)$$

By showing that $R^{\hat{t}}{}_{\hat{r}\hat{t}\hat{r}} = -R^{\hat{r}}{}_{\hat{t}\hat{r}\hat{t}}$, you can see that

$$R_{\hat{r}\hat{r}} = -\left[ \frac{d^2 v}{dr^2} + \left( \frac{dv}{dr} \right)^2 - \left( \frac{dv}{dr} \right) \left( \frac{d\lambda}{dr} \right) - \frac{2}{r} \frac{d\lambda}{dr} \right] e^{-2\lambda(r)} \qquad (10.25)$$

Moving on we have

$$R_{\hat{\theta}\hat{\theta}} = R^{\hat{t}}{}_{\hat{\theta}\hat{t}\hat{\theta}} + R^{\hat{r}}{}_{\hat{\theta}\hat{r}\hat{\theta}} + R^{\hat{\theta}}{}_{\hat{\theta}\hat{\theta}\hat{\theta}} + R^{\hat{\phi}}{}_{\hat{\theta}\hat{\phi}\hat{\theta}}$$

$$= -\frac{1}{r} \frac{dv}{dr} e^{-2\lambda} + \frac{1}{r} \frac{d\lambda}{dr} e^{-2\lambda} + \frac{1 - e^{-2\lambda}}{r^2} \qquad (10.26)$$

$$R_{\hat{\phi}\hat{\phi}} = R^{\hat{t}}{}_{\hat{\phi}\hat{t}\hat{\phi}} + R^{\hat{r}}{}_{\hat{\phi}\hat{r}\hat{\phi}} + R^{\hat{\theta}}{}_{\hat{\phi}\hat{\theta}\hat{\phi}} + R^{\hat{\phi}}{}_{\hat{\phi}\hat{\phi}\hat{\phi}}$$

$$= -\frac{1}{r} \frac{dv}{dr} e^{-2\lambda} + \frac{1}{r} \frac{d\lambda}{dr} e^{-2\lambda} + \frac{1 - e^{-2\lambda}}{r^2} \qquad (10.27)$$

We obtain the vacuum equations by setting each component of the Ricci tensor equal to zero [recall (10.1)]. There are only two vacuum equations we need in order to find the functional form of $v(r)$ and $\lambda(r)$. We use (10.24) and (10.25) and set them equal to zero, which gives

$$\frac{d^2v}{dr^2} + \left(\frac{dv}{dr}\right)^2 - \left(\frac{dv}{dr}\right)\left(\frac{d\lambda}{dr}\right) + \frac{2}{r}\frac{dv}{dr} = 0 \qquad (10.28)$$

$$\frac{d^2v}{dr^2} + \left(\frac{dv}{dr}\right)^2 - \left(\frac{dv}{dr}\right)\left(\frac{d\lambda}{dr}\right) - \frac{2}{r}\frac{d\lambda}{dr} = 0 \qquad (10.29)$$

We subtract (10.29) from (10.28) and get

$$\frac{dv}{dr} + \frac{d\lambda}{dr} = 0$$

This implies that the sum of these functions is constant

$$v + \lambda = \text{ const} = k$$

We can use a trick so that the constant will vanish. We change the time coordinate to $t \to te^k$. This means that $dt \to dte^k$, $dt^2 \to dt^2\,e^{2k} \Rightarrow e^{2v}dt^2 \to e^{2(v+k)}dt^2$. In other words, we have transformed $v \to v + k$ and so

$$v + \lambda = 0$$
$$\Rightarrow \lambda = -v$$

Now we use this to replace $v$ with $-\lambda$ in (10.29) and get

$$\frac{d^2\lambda}{dr^2} - 2\left(\frac{d\lambda}{dr}\right)^2 + \frac{2}{r}\frac{d\lambda}{dr} = 0 \qquad (10.30)$$

To solve this equation, consider the second derivative of $re^{-2\lambda}$

$$(re^{-2\lambda})'' = \left(e^{-2\lambda} - 2r\frac{d\lambda}{dr}e^{-2\lambda}\right)'$$

$$= \left(-4\frac{d\lambda}{dr}e^{-2\lambda} - 2r\frac{d^2\lambda}{dr^2}e^{-2\lambda} + 4r\left(\frac{d\lambda}{dr}\right)^2 e^{-2\lambda}\right)$$

Let's quickly set this equal to zero to see that this is in fact another way of writing (10.30):

$$-4\frac{d\lambda}{dr}e^{-2\lambda} - 2r\frac{d^2\lambda}{dr^2}e^{-2\lambda} + 4r\left(\frac{d\lambda}{dr}\right)^2 e^{-2\lambda} = 0$$

Now divide through by $e^{-2\lambda}$ to get

$$-4\frac{d\lambda}{dr} - 2r\frac{d^2\lambda}{dr^2} + 4r\left(\frac{d\lambda}{dr}\right)^2 = 0$$

Next we divide through by $-2r$, which gives

$$\frac{d^2\lambda}{dr^2} - 2\left(\frac{d\lambda}{dr}\right)^2 + \frac{2}{r}\frac{d\lambda}{dr} = 0$$

Returning to $\left(re^{-2\lambda}\right)''$, since this vanishes because it is equivalent to (10.30), we can integrate once

$$\left(re^{-2\lambda}\right)' = \text{constant} \tag{10.31}$$

Now, recall that we had found that $\left(re^{-2\lambda}\right)'' = \left(e^{-2\lambda} - 2r\frac{d\lambda}{dr}e^{-2\lambda}\right)'$.
Let's turn to (10.26). We obtain another of the vacuum equations by setting this equal to zero:

$$R_{\hat{\theta}\hat{\theta}} = -\frac{1}{r}\frac{dv}{dr}e^{-2\lambda} + \frac{1}{r}\frac{d\lambda}{dr}e^{-2\lambda} + \frac{1 - e^{-2\lambda}}{r^2} = 0$$

We multiply through by $r^2$ to get

$$-r\frac{dv}{dr}e^{-2\lambda} + r\frac{d\lambda}{dr}e^{-2\lambda} + 1 - e^{-2\lambda} = 0$$

Now we use $v = -\lambda$ to obtain

$$2r\frac{d\lambda}{dr}e^{-2\lambda} + 1 - e^{-2\lambda} = 0$$

Moving 1 to the other side and then multiplying through by $-1$, we have

$$-2r\frac{d\lambda}{dr}e^{-2\lambda} + e^{-2\lambda} = 1$$

Now the left side is nothing other than $\left(re^{-2\lambda}\right)'$. And so we have found that $\left(re^{-2\lambda}\right)' = 1$. Integrating this equation, we find

$$re^{-2\lambda} = r - 2m$$

where $-2m$ is an unknown constant of integration. We choose this odd designation for the constant of integration because, as we will see, this term is related to the mass. Dividing through by $r$, we find that

$$e^{-2\lambda} = 1 - \frac{2m}{r}$$

Let's recall the original form of the metric. We had

$$ds^2 = e^{2\nu(r)}dt^2 - e^{2\lambda(r)}dr^2 - r^2\left(d\theta^2 + \sin^2\theta\, d\phi^2\right)$$

Using $\nu = -\lambda$, we obtain the coefficients in the metric we have been seeking

$$e^{2\nu} = 1 - \frac{2m}{r} \quad \text{and} \quad e^{2\lambda} = \left(1 - \frac{2m}{r}\right)^{-1} \tag{10.32}$$

# The Meaning of the Integration Constant

We find the constant appearing in the line element by seeking a correspondence between newtonian theory and general relativity. Later, we will discuss the weak field limit and find that relativity reduces to newtonian theory if we have (for the moment explicitly showing fundamental constants)

$$g_{tt} \approx 1 + \frac{2\Phi}{c^2}$$

where $\Phi$ is the gravitational potential in newtonian theory. For a point mass located at the origin,

$$\Phi = -G\frac{M}{r}$$

and so

$$g_{tt} \approx 1 - 2\frac{GM}{c^2 r}$$

Comparing this to (10.32), we set

$$m = \frac{GM}{c^2}$$

where $M$ is the mass of the body in kilograms (or whatever units are being used). Looking at this term you will see that the units of $m$ which appears in the metric has units of length. This constant $m$ is called the *geometric mass* of the body.

# The Schwarzschild Metric

With these results in hand, we can rewrite (10.5) in the familiar form of the Schwarzschild line element:

$$ds^2 = \left(1 - \frac{2m}{r}\right) dt^2 - \frac{dr^2}{\left(1 - \frac{2m}{r}\right)} - r^2 \left(d\theta^2 + \sin^2 \theta \, d\phi^2\right) \qquad (10.33)$$

Looking at the metric, we can see that as $r$ gets large, it approaches the form (10.2) of the flat space metric.

# The Time Coordinate

The correspondence between the Schwarzschild metric in (10.33) and the flat space metric tells us that the time coordinate $t$ used here is the time measured by a distant observer far from the origin.

# The Schwarzschild Radius

Notice that the metric defined in (10.33) becomes singular (i.e., blows sky high) when $r = 2m$. This value is known as the *Schwarzschild radius*. In terms of the mass of the object that is the source of the gravitational field, it is given by

$$r_s = \frac{2GM}{c^2} \tag{10.34}$$

For ordinary stars, the Schwarzschild radius lies deep in the stellar interior. Therefore we can use the metric given by (10.33) with confidence to describe the region found outside of your average star. As an example, the Schwarzschild radius for the Sun can be used by inserting the solar mass $1.989 \times 10^{30}$ kg into (10.34), and we find that $r_s \simeq 3$ km. Remember, we started out by solving the vacuum equations. Since this point lies inside the Sun where matter is present, the solutions we obtained in the previous section cannot apply in that region.

It should catch your attention that the metric blows up at a certain value of the radius. However, we need to investigate further to determine whether this is a real physical singularity, which means the curvature of spacetime becomes infinite, or whether this is just an artifact of the coordinate system we are using. While it seems like stating the obvious, it is important to note that the line element is written in terms of coordinates.

It may turn out that we can write the metric in a different way by a transformation to some other coordinates. To get a better idea of the behavior of the spacetime, the best course of action is to examine invariant quantities, i.e., scalars. If we can find a singularity that is present in a scalar quantity, then we know that this singularity is present in all coordinate systems and therefore represents something that is physically real. In the present case, the components of the Ricci tensor vanish, so the Ricci scalar vanishes as well. Instead we construct the following scalar using the Riemann tensor.

$$R_{abcd} R^{abcd} = \frac{48m^2}{r^6} \tag{10.35}$$

This quantity blows up at $r = 0$. Since this is a scalar, this is true in all coordinate systems, and therefore, we conclude that the point $r = 0$ is a genuine singularity.

# Geodesics in the Schwarzschild Spacetime

Now that we have the metric in hand, we can determine what the paths of particles and light rays are going to be in this spacetime. To find out we need to derive and solve the geodesic equations for each coordinate.

One way this can be done fairly easily is by deriving the Euler-Lagrange equations. We are not covering largrangian methods in this book, so we will simply demonstrate the method (interested readers, please see the references). To find the geodesic equations, we take the variation

$$\delta \int ds = \delta \int \left[ \left( 1 - \frac{2m}{r} \right) \dot{t}^2 - \frac{1}{\left( 1 - \frac{2m}{r} \right)} \dot{r}^2 - r^2 \dot{\theta}^2 - r^2 \sin^2 \theta \, \dot{\phi}^2 \right] ds = 0$$

Now we make the following definition:

$$F = \left( 1 - \frac{2m}{r} \right) \dot{t}^2 - \frac{1}{\left( 1 - \frac{2m}{r} \right)} \dot{r}^2 - r^2 \dot{\theta}^2 - r^2 \sin^2 \theta \, \dot{\phi}^2$$

The Euler-Lagrange equations are

$$\frac{d}{ds} \left( \frac{\partial F}{\partial \dot{x}^a} \right) - \frac{\partial F}{\partial x^a} = 0 \tag{10.36}$$

Starting with the time coordinate, we have

$$\frac{\partial F}{\partial \dot{t}} = 2 \left( 1 - \frac{2m}{r} \right) \dot{t}$$

$$\Rightarrow \frac{d}{ds} \left( \frac{\partial F}{\partial \dot{t}} \right) = \frac{d}{ds} \left[ 2 \left( 1 - \frac{2m}{r} \right) \dot{t} \right] = \frac{d}{ds} \left[ 2 \left( 1 - \frac{2m}{r} \right) \right] \dot{t} + 2 \left( 1 - \frac{2m}{r} \right) \ddot{t}$$

$$= \frac{4m}{r^2} \dot{r} \dot{t} + 2 \left( 1 - \frac{2m}{r} \right) \ddot{t}$$

Since there are no terms in $F$ that contain $t$, we set this to zero and obtain the first geodesic equation:

$$\ddot{t} + \frac{2m}{r \, (r - 2m)} \dot{r} \dot{t} = 0 \tag{10.37}$$

Now we consider the radial coordinate. We have

$$\frac{\partial F}{\partial \dot{r}} = -\frac{2}{\left( 1 - \frac{2m}{r} \right)} \dot{r}$$

$$\Rightarrow \frac{d}{ds} \left( \frac{\partial F}{\partial \dot{r}} \right) = -\frac{2}{\left( 1 - \frac{2m}{r} \right)} \ddot{r} - \frac{4m}{r^2} (\dot{r})^2$$

$F$ is $r$ dependent and so we calculate

$$\frac{\partial F}{\partial r} = \frac{2m}{r^2}\dot{t}^2 + \left(1 - \frac{2m}{r}\right)\frac{2m}{r}\dot{r}^2 - 2r\dot{\theta}^2 - 2r\sin^2\theta\,\dot{\phi}^2$$

Putting these results together using (10.36) gives the second geodesic equation

$$\ddot{r} + \frac{m}{r^3}(r - 2m)\dot{t}^2 - \frac{m}{r(r - 2m)}\dot{r}^2 - (r - 2m)\left(\dot{\theta}^2 + \sin^2\theta\,\dot{\phi}^2\right) = 0$$

$$(10.38)$$

Next we find

$$\frac{\partial F}{\partial\dot{\theta}} = -2r^2\dot{\theta} \Rightarrow \frac{d}{ds}\left(\frac{\partial F}{\partial\dot{\theta}}\right) = -4r\,\dot{r}\dot{\theta} - 2r^2\ddot{\theta}$$

$$\frac{\partial F}{\partial\theta} = -2r^2\sin\theta\cos\theta\,\dot{\phi}^2$$

and so the geodesic equation for $\theta$ is given by

$$\ddot{\theta} + \frac{2}{r}\dot{r}\dot{\theta} - \sin\theta\cos\theta\,\dot{\phi}^2 = 0 \qquad (10.39)$$

Finally, for the $\phi$ coordinate we find

$$\frac{\partial F}{\partial\dot{\phi}} = -2r^2\sin^2\theta\,\dot{\phi}$$

$$\Rightarrow \frac{d}{ds}\left(\frac{\partial F}{\partial\dot{\theta}}\right) = -4r\,\dot{r}\sin^2\theta\,\dot{\phi} - 4r^2\sin\theta\cos\theta\,\dot{\theta}\dot{\phi} - 2r^2\sin^2\theta\,\ddot{\phi}$$

$$\frac{\partial F}{\partial\phi} = 0$$

Therefore

$$\ddot{\phi} + 2\cot\theta\,\dot{\theta}\dot{\phi} + \frac{2}{r}\dot{r}\dot{\phi} = 0 \qquad (10.40)$$

# Particle Orbits in the Schwarzschild Spacetime

In the previous section we obtained a set of differential equations for the geodesics of the Schwarzschild spacetime. However, these equations are quite daunting and trying to solve them directly does not seem like a productive use of time. A different approach to learn about the motion of particles and light rays in the Schwarzschild spacetime is available, and is based on the use of Killing vectors. Remember, each symmetry in the metric corresponds to a Killing vector. Conserved quantities related to the motion of a massive particle can be found by forming the dot product of Killing vectors with the four velocity of the particle.

We will define vectors in terms of their components, using the coordinate ordering $(t, r, \theta, \phi)$. The four velocity of a particle in the Schwarzschild spacetime has components given by

$$\vec{u} = \left( \frac{dt}{d\tau}, \frac{dr}{d\tau}, \frac{d\theta}{d\tau}, \frac{d\phi}{d\tau} \right) \tag{10.41}$$

Here $\tau$ is the proper time. The four velocity satisfies

$$\vec{u} \cdot \vec{u} = g_{ab} u^a u^b = 1 \tag{10.42}$$

In a coordinate basis, the components of the metric tensor, which can be read off of (10.33), are

$$g_{tt} = \left( 1 - \frac{2m}{r} \right), \quad g_{rr} = -\frac{1}{\left( 1 - \frac{2m}{r} \right)}, \quad g_{\theta\theta} = -r^2, \quad g_{\phi\phi} = -r^2 \sin^2 \theta \tag{10.43}$$

Using (10.41), we find that (10.42) gives

$$\vec{u} \cdot \vec{u} = g_{ab} u^a u^b$$
$$= \left( 1 - \frac{2m}{r} \right) \left( \frac{dt}{d\tau} \right)^2 - \left( 1 - \frac{2m}{r} \right)^{-1} \left( \frac{dr}{d\tau} \right)^2 - r^2 \left( \frac{d\theta}{d\tau} \right)^2$$
$$- r^2 \sin^2 \theta \left( \frac{d\phi}{d\tau} \right)^2 = 1 \tag{10.44}$$

This equation forms the basis that we will use to find the orbits. It can be shown that in general relativity, as in classical mechanics, the orbit of a body in a central force field lies in a plane. Therefore we can choose our axes such that $\theta = \pi/2$ and set $\frac{d\theta}{d\tau} = 0$.

Before analyzing this equation further, we will define two Killing vectors for the Schwarzschild spacetime and use them to construct conserved quantities. This is actually quite easy. When the form of the metric was derived, we indicated two important criteria: the metric is time independent and spherically symmetric.

Looking at the metric

$$ds^2 = \left(1 - \frac{2m}{r}\right) dt^2 - \frac{dr^2}{\left(1 - \frac{2m}{r}\right)} - r^2 \left(d\theta^2 + \sin^2 \theta \, d\phi^2\right)$$

we see that there are no terms that explicitly depend on time $t$ or the angular variable $\phi$. Therefore we can define the following Killing vectors. The Killing vector that corresponds to the independence of the metric of $t$ is

$$\xi = (1, 0, 0, 0) \tag{10.45}$$

We associate the conservation of energy with the independence of the metric from time.

We can construct another Killing vector that corresponds to the independence of the metric from $\phi$. This is

$$\eta = (0, 0, 0, 1) \tag{10.46}$$

The independence of the metric with respect to $\phi$ is associated with conservation of angular momentum.

The conserved quantities related to these Killing vectors are found by taking the dot product of each vector with the four velocity. The *conserved energy per unit rest mass* is given by

$$e = -\vec{\xi} \cdot \vec{u} = -g_{ab}\xi^a u^b = -\left(1 - \frac{2m}{r}\right) \frac{dt}{d\tau} \tag{10.47}$$

The *conserved angular momentum per unit rest mass* is defined to be

$$l = \vec{\eta} \cdot \vec{u} = g_{ab}\eta^a u^b = r^2 \sin^2 \theta \, \frac{d\phi}{d\tau} = r^2 \frac{d\phi}{d\tau} \quad (\text{for } \theta = \frac{\pi}{2}) \tag{10.48}$$

With these definitions, and the choice of $\theta = \pi/2$, we can simplify (10.44) to read

$$\left(1 - \frac{2m}{r}\right)^{-1} e^2 - \left(1 - \frac{2m}{r}\right)^{-1} \left(\frac{dr}{d\tau}\right)^2 - \frac{l^2}{r^2} = 1 \tag{10.49}$$

Let's multiply through by $1 - \frac{2m}{r}$ and divide by 2. The result is

$$\frac{e^2}{2} - \frac{1}{2}\left(\frac{dr}{d\tau}\right)^2 - \frac{1}{2}\frac{l^2}{r^2}\left(1 - \frac{2m}{r}\right) = \frac{1}{2} - \frac{m}{r}$$

Now we isolate the energy per unit mass term, which gives

$$\frac{e^2 - 1}{2} = \frac{1}{2}\left(\frac{dr}{d\tau}\right)^2 + \frac{l^2}{2r^2}\left(1 - \frac{2m}{r}\right) - \frac{m}{r}$$

If we define the *effective potential* to be

$$V_{\text{eff}} = \frac{l^2}{2r^2}\left(1 - \frac{2m}{r}\right) - \frac{m}{r}$$

and set $E = \frac{e^2-1}{2}$, then we obtain an expression that corresponds to that used in classical mechanics

$$E = \frac{1}{2}\dot{r}^2 + V_{\text{eff}}$$

which describes a particle with energy $E$ and unit mass moving in the potential described by $V_{\text{eff}}$. However, a closer look at the effective potential is warranted. Multiplying through by the leading term, we have

$$V_{\text{eff}} = -\frac{m}{r} + \frac{l^2}{2r^2} - \frac{l^2 m}{r^3} \tag{10.50}$$

The first two terms are nothing more than what you would expect in the new-tonian case. In particular, the first term correlates to the gravitational potential while the second term is the angular momentum term we are familiar with from classical orbital mechanics. The final term in this expression is a modification of the potential that arises in general relativity.

We find the minimum and maximum values that $r$ can assume in the usual way. The first derivative is

$$\frac{dV_{\text{eff}}}{dr} = \frac{d}{dr}\left(-\frac{m}{r} + \frac{l^2}{2r^2} - \frac{l^2 m}{r^3}\right) = \frac{m}{r^2} - \frac{l^2}{r^3} + \frac{3l^2 m}{r^4}$$

Next we set this equal to zero:

$$\frac{m}{r^2} - \frac{l^2}{r^3} + \frac{3l^2m}{r^4} = 0$$

$$\Rightarrow mr^2 - l^2r + 3l^2m = 0$$

Applying the quadratic formula, we find the maximum and minimum values of $r$ to be given by

$$r_{1,2} = \frac{l^2 \pm \sqrt{l^4 - 12\,l^2m^2}}{2m} = \frac{l^2}{2m}\left(1 \pm \sqrt{1 - 12\frac{m^2}{l^2}}\right) \qquad (10.51)$$

These values correspond to circular orbits. We can use a binomial expansion of the term under the square root to rewrite this as

$$r_{1,2} = \frac{l^2}{2m}\left(1 \pm \sqrt{1 - 12\frac{m^2}{l^2}}\right) \cong \frac{l^2}{2m}\left[1 \pm \left(1 - 6\frac{m^2}{l^2}\right)\right]$$

The two values correspond to a stable and an unstable circular orbit, respectively. The stable circular orbit is one for which

$$r_1 \approx \frac{l^2}{m}$$

Meanwhile, for the unstable circular orbit, we take the minus sign and obtain

$$r_2 \cong \frac{l^2}{2m}\left[1 - \left(1 - 6\frac{m^2}{l^2}\right)\right] = \frac{l^2}{2m}\left(6\frac{m^2}{l^2}\right) = 3m$$

Using $m = \frac{GM}{c^2}$ the orbit is given by $r_2 = 3\frac{GM}{c^2}$. For the Sun, this value is

$$r_2^{\text{sun}} = 3\frac{(6.67 \times 10^{-11}\ \text{m}^3\text{s}^2/\text{kg})\,(1.989 \times 10^{30}\ \text{kg})}{(3 \times 10^8\ \text{m/s})^2} = 4422\ \text{km}$$

The equatorial radius of the Sun is 695,000 km, and so we see that the unstable circular orbit lies well inside the Sun.

Looking at (10.50), we can learn something about the behavior of the orbits for different values of $r$. First, consider the case where $r = 2m$, the Schwarzschild radius. In this case, we have

$$V_{\text{eff}}\,(r = 2m) = -\frac{m}{2m} + \frac{l^2}{2\,(2m)^2} - \frac{l^2 m}{(2m)^3} = -\frac{1}{2} + \frac{l^2}{8m^2} - \frac{l^2 m}{8m^3} = -\frac{1}{2}$$

For large $r$, the potential reduces to the newtonian potential

$$V_{\text{eff}}\,(r) \approx -\frac{m}{r}$$

Looking once again at (10.51), notice that the term under the square root becomes negative when the angular momentum $l^2$ is less than $12m^2$. We then get the unphysical result where the radius is a complex number. This tells us that in this case no stable circular orbit can exist. Physically speaking, if $l^2$ is less than $12m^2$, this indicates that the orbiting body will crash into the surface of the star. If the body happens to be approaching a black hole under these conditions, it will simply be swallowed up by the black hole.

These results are illustrated in Figs. 10-1 and 10-2. By comparing the curves in the two figures, you can see that as $r$ gets large the newtonian and relativistic cases converge. At small $r$, the differences are quite dramatic. This tells us that the best place to look for relativistic effects in the solar system is near the Sun.

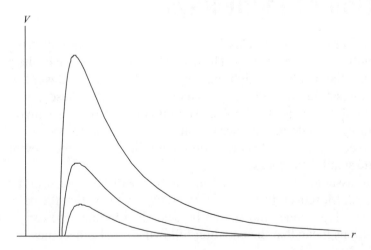

**Fig. 10-1.** Plots of the effective potential for general relativity. For simplicity, we used a unit mass to generate the plots that shows three different values of $l$. The behavior at small $r$ is significant; a comparison shows that the orbits differ markedly from the newtonian case in this region. This is a result of the $1/r^3$ term.

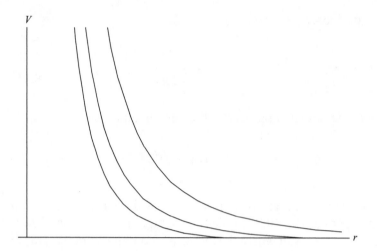

**Fig. 10-2.** The newtonian case for the same values of $l$. At large $r$, the orbits correspond to those in the previous plot. At small $r$, the behavior is quite different.

This is what Einstein did when he considered the precession in the orbit of Mercury.

# The Deflection of Light Rays

There are four standard or "classical tests" of general relativity that have been applied within the solar system. These include the precession of the perihelion of Mercury, the bending of light passing near the Sun, the travel time of light in a Schwarzschild field, and the gravitational red shift. These phenomena are covered in all the major textbooks. We will consider two tests involving light and begin by considering the derivation of an equation for the trajectory of a light ray (see Fig. 10-3). This derivation will follow that of the previous section with some small differences.

We can assume once again that the motion will take place in a plane and set $\theta = \pi/2$. Moreover, from special relativity we know that the path of a light ray lies on a light cone, and so can be described by the case $ds^2 = 0$. These considerations mean that the equation following (10.44) becomes

$$\left(1 - \frac{2m}{r}\right)\dot{t}^2 - \left(1 - \frac{2m}{r}\right)^{-1}\dot{r}^2 - r^2\dot{\phi}^2 = 0 \tag{10.52}$$

Proceeding as we did for massive particles, we use definitions (10.47) and (10.48) to write (with the difference that differentiation is not with respect to the proper time $\tau$, instead we take the derivative with respect to some parameter we denote by $\lambda$)

$$e^2 = \left(1 - \frac{2m}{r}\right)^2 \dot{t}^2 \quad \text{and} \quad l^2 = r^4 \dot{\phi}^2$$

where we have used dots to denote differentiation with respect to $\lambda$. Then (10.52) becomes

$$\left(1 - \frac{2m}{r}\right)^{-1} e^2 - \left(1 - \frac{2m}{r}\right)^{-1} \dot{r}^2 - \frac{l^2}{r^2} = 0 \qquad (10.53)$$

To find the trajectory, we are interested in obtaining an expression for $r = r(\phi)$. We will now use primes to denote differentiation with respect to $\phi$. With this in mind, let's rewrite $\dot{r}$ as

$$\dot{r} = \frac{dr}{d\lambda} = \frac{dr}{d\lambda}\left(\frac{d\phi}{d\phi}\right) = \frac{dr}{d\phi}\left(\frac{d\phi}{d\lambda}\right) = r'\dot{\phi}$$

Now we can use $r^2\dot{\phi} = l$ to write

$$\dot{r} = r'\dot{\phi} = \frac{r'l}{r^2}$$

To write the equation in a more convenient form, we introduce a new variable $u = 1/r$. Note that

$$u' = \left(\frac{1}{r}\right)' = -\frac{1}{r^2}r'$$

and so, we can write

$$\dot{r} = \frac{r'l}{r^2} = \left(-r^2 u'\right)\frac{l}{r^2} = -lu'$$

Returning to (10.53), we multiply through by $(1 - 2m/r)$ and set $u = 1/r$ to obtain

$$e^2 - l^2 u'^2 - l^2 u^2 (1 - 2mu) = 0$$

To obtain an equation for the trajectory of a light ray, we differentiate with respect to $\phi$ a second time to get rid of the constant $e^2$. This gives

$$u' \left( u'' + u - 3mu^2 \right) = 0$$

Dividing through by $u'$, we obtain the final result

$$u'' + u = 3mu^2 \tag{10.54}$$

The standard procedure is to solve this equation using perturbation methods. First we set $\varepsilon = 3m$ and then try a solution of the form

$$u = u_0 + \varepsilon u_1 + O\left(\varepsilon^2\right) \tag{10.55}$$

Ignoring the higher order terms, we have

$$u' = u'_0 + \varepsilon u'_1$$
$$u'' = u''_0 + \varepsilon u''_1$$

Now, ignoring terms that are second order and higher, we have $3mu^2 = \varepsilon u^2 \approx \varepsilon u_0^2$. Inserting these results into (10.54), we obtain

$$u''_0 + \varepsilon u''_1 + u_0 + \varepsilon u_1 = \varepsilon u_0^2$$

We now equate terms by their order in $\varepsilon$. We start with

$$u''_0 + u_0 = 0$$

The solution of this equation is given by $u_0 = A \sin \phi + B \cos \phi$. Without loss of generality, we can choose our initial conditions so that $B = 0$ and $u_0 = A \sin \phi$. This equation represents straight-line motion since $r = 1/u$, and using $y = r \sin \phi$ in polar coordinates, we can write this as $1/A = r \sin \phi = y$. Therefore, the constant $A$ represents the distance of closest approach to the origin. Next, we equate terms that are first order in $\varepsilon$.

$$u''_1 + u_1 = u_0^2 = A^2 \sin^2 \phi$$

The homogeneous equation is the same as we had before

$$u''_1 + u_1 = 0$$

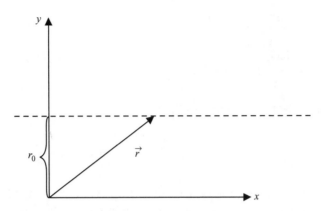

**Fig. 10-3.** Lowest order representation of the trajectory of a light ray. The light ray follows the straight path represented by the dashed line. The distance $r_0$ is the distance of closest approach to the origin.

with solution $u_1^H = B \sin \phi + C \cos \phi$. Without loss of generality, we take $B = 0$. For the particular solution, we guess that $u_p = D \sin^2 \phi + E \cos^2 \phi$; therefore, differentiating we have

$$u_p' = 2D \sin \phi \cos \phi - 2E \cos \phi \sin \phi$$
$$u_p'' = 2D \cos^2 \phi - 2D \sin^2 \phi - 2E \cos^2 \phi + 2E \sin^2 \phi$$

and so, we have

$$u_p'' + u_p = -D \sin^2 \phi + 4D \sin^2 \phi = 3D \sin^2 \phi$$
$$u_p'' + u_p = 2D \cos^2 \phi - 2D \sin^2 \phi - 2E \cos^2 \phi + 2E \sin^2 \phi$$
$$+ D \sin^2 \phi + E \cos^2 \phi$$
$$= 2D \cos^2 \phi - D \sin^2 \phi - E \cos^2 \phi + 2E \sin^2 \phi$$

Now the particular solution must satisfy $u_p'' + u_p = A^2 \sin^2 \phi$ and this only be true can if $2D - E = 0$, leaving us with

$$-D \sin^2 \phi + 4D \sin^2 \phi = 3D \sin^2 \phi$$

Using $u_p'' + u_p = A^2 \sin^2 \phi$, we conclude that $D = A^2/3$. And so the particular solution becomes

$$u_1^p = D \sin^2 \phi + E \cos^2 \phi = \frac{A^2}{3} \sin^2 \phi + 2\frac{A^2}{3} \cos^2 \phi$$

$$= \frac{A^2}{3}\left(1 - \cos^2 \phi\right) + 2\frac{A^2}{3}\cos^2 \phi$$

$$= \frac{A^2}{3} + \frac{A^2}{3}\cos^2 \phi$$

The complete first-order solution is then

$$u_1 = \frac{A^2}{3}\left(1 + K\cos\phi + \cos^2\phi\right) \tag{10.56}$$

where $K$ is another integration constant. Putting everything together and using $\varepsilon = 3m$, the complete solution (which is an approximation) for the trajectory of a light ray is

$$u = u_0 + \varepsilon u_1 = A\sin\phi + mA^2\left(1 + K\cos\phi + \cos^2\phi\right) \tag{10.57}$$

The $mA^2$ will cause a deflection of the trajectory from a straight-line path.

In an astrophysics situation, a light ray originating from a distant star approaches the Sun along an asymptotic straight line, is deflected a small amount by the Sun's gravitational field, and then heads off into the distance along another asymptotic straight line. The asymptotes correspond to $u = 0$ and are parallel to the $x$-axis. We can take $\phi = \delta$ (where $\delta$ is some small angle) and use the small angle approximation $\sin\delta \approx \delta$, $\cos\delta \approx 1$ to write (10.57) as

$$u \approx A\delta + mA^2\left(2 + K\right)$$

We can define a new constant $\kappa = 2 + K$, and setting this equal to zero, we find (see Fig. 10-4)

$$\delta = -m\kappa A$$

The minus sign indicates that the deflection is towards the Sun. Recalling that the constant $A$ is inversely related to the straight-line distance we found above, we can write

$$\delta = -\frac{m\kappa}{r_0}$$

The total deflection is given by

$$\Delta = \frac{4m}{r_0}$$

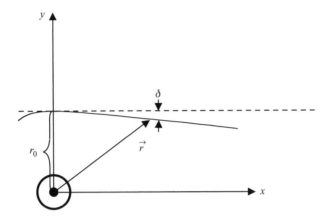

**Fig. 10-4.** A light ray deflected by the Sun.

In the case of the Sun, a deflection of 1.75 in. of arc is predicted. The interested reader can learn about the observational challenges and results in trying to measure this phenomenon, which has been done successfully.

# Time Delay

The final phenomenon that manifests itself in the Schwarzschild geometry is the time travel that is required for light to go between two points. The curvature induced in the spacetime surrounding a massive body like the Sun increases the travel time of light rays relative to what would be the case in flat space.

Again setting $\theta = \pi/2$ and using $ds^2 = 0$, we have

$$0 = \left(1 - \frac{2m}{r}\right) dt^2 - \left(1 - \frac{2m}{r}\right)^{-1} dr^2 - r^2 d\phi^2$$

Using the previous results for light rays, we write the last piece in terms of $dr$ and obtain the following result:

$$dt^2 = \frac{dr^2 \left(1 - 2mr_0^2/r^3\right)}{\left(1 - r_0^2/r^2\right)\left(1 - 2m/r\right)^2}$$

Taking the square root, expanding to first order, and using conventional units (so we put $ct$ in place of $t$), we obtain

$$c\,dt = \frac{dr}{\sqrt{1 - r_0^2/r^2}} \left(1 + \frac{2m}{r} - \frac{mr_0^2}{r^3}\right)$$

This result can be integrated. To consider the travel time of light between Earth and another planet in the solar system, we integrate between $r_0$ to $r_p$ and $r_0$ to $r_e$, where $r_p$ is the planet radius and $r_e$ is the Earth radius. The result is

$$ct = \sqrt{r_p^2 - r_0^2} + \sqrt{r_e^2 - r_0^2} + 2m \ln \frac{\left(\sqrt{r_p^2 - r_0^2} + r_p\right)\left(\sqrt{r_e^2 - r_0^2} + r_e\right)}{r_0^2}$$
$$- m\left(\frac{\sqrt{r_p^2 - r_0^2}}{r_p} + \frac{\sqrt{r_e^2 - r_0^2}}{r_e}\right)$$

The ordinary flat space distance between Earth and the planet is given by the first term, $\sqrt{r_p^2 - r_0^2} + \sqrt{r_e^2 - r_0^2}$. The remaining terms indicate the increased distance caused by the curvature of spacetime (i.e., by the gravitational field of the Sun). These terms cause a time delay that is measurable in the solar system. For example, radar reflections to Venus are delayed by about 200 $\mu$s.

Because of limited space our coverage of the Schwarzschild solution is incomplete. The reader is encouraged to consult the references listed at the end of the book for more extensive treatment.

# Quiz

1.  Using the variational method described in Example 4-10, the nonzero Christoffel symbols for the Schwarzschild metric are
    (a) $\Gamma^t_{rt} = d\nu/dr$
       $\Gamma^r_{tt} = e^{2(\nu-\lambda)}\frac{d\nu}{dr}$, $\Gamma^r_{rr} = \frac{d\lambda}{dr}$, $\Gamma^r_{\theta\theta} = -r\,e^{-2\lambda}$, $\Gamma^r_{\phi\phi} = -r\,e^{-2\lambda}\sin^2\theta$
       $\Gamma^\theta_{r\theta} = \frac{1}{r}$, $\Gamma^\theta_{\phi\phi} = -\sin\theta\,\cos\theta$
       $\Gamma^\phi_{r\phi} = \frac{1}{r}$, $\Gamma^\phi_{\theta\phi} = \cot\theta$

(b) $\Gamma^t{}_{rt} = d\nu/dr$

$\Gamma^r{}_{tt} = e^{2(\nu-\lambda)} \frac{d\nu}{dr}, \Gamma^r{}_{rr} = \frac{d\lambda}{dr}, \Gamma^r{}_{\theta\theta} = r\,e^{-2\lambda}, \Gamma^r{}_{\phi\phi} = -r\,e^{-2\lambda}\,\sin^2\theta$

$\Gamma^\theta{}_{r\theta} = \frac{1}{r}, \Gamma^\theta{}_{\phi\phi} = \sin\theta\,\cos\theta$

$\Gamma^\phi{}_{r\phi} = -\frac{1}{r}, \Gamma^\phi{}_{\theta\phi} = -\cot\theta$

(c) $\Gamma^r{}_{tt} = \frac{d\nu}{dr}, \Gamma^r{}_{rr} = -\frac{d\lambda}{dr}, \Gamma^r{}_{\theta\theta} = -r\,e^{-2\lambda}, \Gamma^r{}_{\phi\phi} = r\,e^{-2\lambda}\,\cos^2\theta$

$\Gamma^\theta{}_{r\theta} = \frac{1}{r}, \Gamma^\theta{}_{\phi\phi} = -\sin\theta\,\cos\theta$

$\Gamma^\phi{}_{r\phi} = \frac{1}{r}, \Gamma^\phi{}_{\theta\phi} = \cot\theta$

2. Suppose that we were to drop the requirement of time independence and wrote the line element as

$$ds^2 = e^{2\nu(r,t)}\,dt^2 - e^{2\lambda(r,t)}\,dr^2 - r^2\left(d\theta^2 + \sin^2\theta\,d\phi^2\right)$$

The $R_{rt}$ component of the Ricci tensor is given by

(a) $R_{rt} = \frac{1}{r}\frac{d\nu}{dt}$

(b) $R_{rt} = \frac{1}{r}\frac{d\lambda}{dt}$

(c) $R_{rt} = -\frac{1}{r^2}\frac{d\lambda}{dt}$

For the following set of problems, we consider a Schwarzschild metric with a nonzero cosmological constant. We make the following definition:

$$f(r) = 1 - \frac{2m}{r} - \frac{1}{3}\Lambda r^2$$

We write the line element as

$$ds^2 = -f(r)\,dt^2 + \frac{1}{f(r)}\,dr^2 + r^2\,d\theta^2 + r^2\,\sin^2\theta\,d\phi^2$$

3. When you calculate the Ricci rotation coefficients, you will find

(a) $\Gamma^{\hat{r}}{}_{\hat{t}\hat{t}} = \frac{-\Lambda r^3}{\sqrt{9r-18m-3\,\Lambda r^3}}$

(b) $\Gamma^{\hat{r}}{}_{\hat{t}\hat{t}} = \frac{3m-\Lambda r^3}{\sqrt{9r-18m-3\,\Lambda r^3}}$

(c) $\Gamma^{\hat{r}}{}_{\hat{t}\hat{t}} = \frac{3m-\Lambda r^3}{\sqrt{9r-18m}}$

4.  When you calculate the components of the Ricci tensor, you will find
    (a) $-R_{\hat{t}\hat{t}} = R_{\hat{r}\hat{r}} = R_{\hat{\theta}\hat{\theta}} = R_{\hat{\phi}\hat{\phi}} = \Lambda$
    (b) $-R_{\hat{t}\hat{t}} = R_{\hat{r}\hat{r}} = R_{\hat{\theta}\hat{\theta}} = R_{\hat{\phi}\hat{\phi}} = \Lambda r^3$
    (c) $-R_{\hat{t}\hat{t}} = R_{\hat{r}\hat{r}} = R_{\hat{\theta}\hat{\theta}} = R_{\hat{\phi}\hat{\phi}} = 0$

5.  The Petrov type of the Schwarzschild spacetime is best described as
    (a) type O
    (b) type I
    (c) type III
    (d) type D

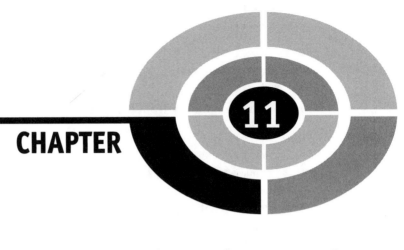

# CHAPTER 11

# Black Holes

A *black hole* is a region of spacetime where gravity is so strong that not even light can escape. In nature, it is believed that black holes form at the end of a stars lifetime, when a massive star runs out of fuel and ends its life in collapse. We shall begin our study of black holes by taking a closer look at the Schwarzschild solution. As we will see, according to classical general relativity, a black hole is completely characterized by just three parameters. These are

- Mass
- Charge
- Angular momentum

This characterization results in three general classes of black holes that are studied:

- Static black holes with no charge, described by the Schwarzschild solution
- Black holes with electrical charge, described by the Reissner-Nordstrøm solution
- Rotating black holes, described by the Kerr solution

In this chapter we will consider two cases: the Schwarzschild and Kerr black holes. To begin, let's review the problem of coordinate singularities and see how to remove the singularity from the Schwarzschild metric.

# Redshift in a Gravitational Field

When studying black holes you will often see an *infinite redshift* being discussed. Let's take a moment to see what happens to light as it is emitted upward in a gravitational field, that is from an observer located at some inner radius $r_i$ to an observer positioned at some outer radius $r_o$.

The key to seeing what happens to the light is to see how time passes for each observer. In other words we are interested in the period of the light wave as seen by each observer. Recall that the proper time $\tau$ is the time a given observer measures on his or her own clock. For the Schwarzschild metric, for a stationary observer the proper time relates to the time as measured by a distant observer via the relationship

$$d\tau = \sqrt{1 - \frac{2m}{r}}\, dt$$

It is a simple matter to calculate a redshift factor by comparing the proper time for observers located at two different values of $r$. This is best illustrated by an example.

**EXAMPLE 11-1**
Consider two fixed observers located nearby a Schwarzschild black hole. One observer, located at $r_1 = 3m$, emits a pulse of ultraviolet light to a second observer located at $r_2 = 8m$. Show that the second observer finds that the light has been redshifted to orange.

**SOLUTION 11-1**
To find the redshift factor, we simply calculate the ratio $d\tau_2/d\tau_1$ where

$$d\tau_i = \sqrt{1 - \frac{2m}{r_i}}\, dt$$

Let's denote the redshift factor by $\alpha$. Then we have

$$\alpha = \frac{d\tau_2}{d\tau_1} = \frac{\sqrt{1 - \frac{2m}{r_2}} \, dt}{\sqrt{1 - \frac{2m}{r_1}} \, dt} = \frac{\sqrt{1 - \frac{2m}{r_2}}}{\sqrt{1 - \frac{2m}{r_1}}}$$

$$= \frac{\sqrt{1 - \frac{2m}{8m}}}{\sqrt{1 - \frac{2m}{3m}}} = \frac{\sqrt{1 - \frac{1}{4}}}{\sqrt{1 - \frac{2}{3}}} = \frac{\sqrt{\frac{3}{4}}}{\sqrt{\frac{1}{3}}} = \sqrt{\frac{9}{4}} = \frac{3}{2}$$

We can take the wavelength of ultraviolet light to be $\lambda_1 \sim 400$ nm. The wavelength that the second observer sees is then

$$\lambda_2 = \alpha \lambda_1 = 1.5 \times 400 \text{ nm} = 600 \text{ nm}$$

This is in the orange region of the spectrum, which roughly runs from 542 nm to 620 nm.

# Coordinate Singularities

Let's step back for a moment and review the distinction between coordinate and curvature singularities. First recall the Schwarzschild metric given in (10.33):

$$ds^2 = \left(1 - \frac{2m}{r}\right) dt^2 - \frac{dr^2}{(1 - 2m/r)} - r^2(d\theta^2 + \sin^2\theta d\phi^2)$$

It's pretty clear that the Schwarzschild metric exhibits unusual behavior at $r = 2m$. For $r > 2m$, $g_{tt} > 0$ and $g_{rr} < 0$. However, notice that for $r < 2m$, the signs of these components of the metric reverse. This means that a world line along the $t$-axis has $ds^2 < 0$ and so describes a spacelike curve. Meanwhile, a world line along the $r$-axis has $ds^2 > 0$ and so describes a timelike curve. The time and space character of the coordinates has reversed. This indicates that a massive particle inside the Schwarzschild radius could not remain stationary at a constant value of $r$.

Now let's take a direct look at $r = 2m$. Considering $g_{tt}$ first, we see that at $r = 2m$

$$g_{tt} = 1 - \frac{2m}{2m} = 1 - 1 = 0$$

While this is well behaved mathematically, the fact that $g_{tt}$ vanishes means that the surface $r = 2m$ is a surface of infinite redshift. Something unusual is obviously going on, and we will examine this behavior again later. But for now let's consider the other components of the metric. While nothing unusual happens to $g_{\theta\theta}$ and $g_{\phi\phi}$, we see that $g_{rr}$ behaves very badly. In fact, this term blows up:

$$g_{rr} = -\frac{1}{(1 - 2m/r)} \to \infty \quad \text{as } r \to 2m$$

When a mathematical expression goes to infinity at some point, we call that point a *singularity*. However, in geometry and in physics and hence in general relativity, the presence of a singularity must be explored carefully. The first question to ask is whether or not the singularity is physically real or whether it is due to the choice of coordinates we have made.

In this case, we will find that while the surface $r = 2m$ has some unusual properties, the singularity is due to the choice of coordinates, and so is a *coordinate singularity*. Simply put, by using a different set of coordinates we can write the metric in such a way that the singularity at $r = 2m$ is removed. However, we will see that the point $r = 0$ is due to infinite curvature and cannot be removed by a change in coordinates.

We have already seen a way to investigate this question: construct invariant quantities—invariant quantities will not depend on our particular choice of coordinates. In Chapter 10, we found that

$$R_{abcd} R^{abcd} = \frac{48m^2}{r^6}$$

This invariant (it's a scalar) tells us that the curvature tensor *does* blow up at $r = 0$, but that at $r = 2m$, nothing unusual happens. This tells us that we can remove the singularity at $r = 2m$ by changing to an appropriate coordinate system.

# Eddington-Finkelstein Coordinates

We can study these problems further by examining the behavior of light cones near $r = 2m$. Consider paths along radial lines, which means we can set

$d\theta = d\phi = 0$. In this case, the Schwarzschild metric simplifies to

$$ds^2 = \left(1 - \frac{2m}{r}\right) dt^2 - \frac{dr^2}{\left(1 - 2m/r\right)}$$

To study the paths of light rays, we set $ds^2 = 0$. This leads to the following relationship, which gives the slope of a light cone:

$$\frac{dt}{dr} = \pm \left(1 - \frac{2m}{r}\right)^{-1} \tag{11.1}$$

The first thing to notice that far from $r = 2m$, that is as $r \to \infty$, this equation becomes

$$\frac{dt}{dr} = \pm 1$$

Therefore in this limit we recover the motion of light rays in flat space (integration gives $t = \pm r$ modulo a constant, just what we would expect for light cones in Minkowski space).

Now let's examine the behavior as we approach smaller $r$, specifically approaching $r = 2m$. It will be helpful to examine the positive sign, which corresponds to outgoing radial null curves. Then we can write (11.1) as

$$\frac{dt}{dr} = \frac{r}{r - 2m}$$

Notice that as $r \to 2m$, $dt/dr \to \infty$. This tells us that the light cones are becoming more narrow as we approach $r = 2m$ (at $r = 2m$, the lines become vertical). This effect is shown in Fig. 11-1.

We can find the key to getting rid of the singularity by integrating (11.1) to get time as a function of $r$. Once again, if we take the positive sign, which applies for outgoing radial null curves, then integration gives

$$t = r + 2m \ln |r - 2m|$$

(we are ignoring the integration constant). The form of $t(r)$ suggests a coordinate transformation that we can use. We now consider the *tortoise coordinate*, which will allow us to write down the metric in a new way that shows only the curvature singularity at the origin.

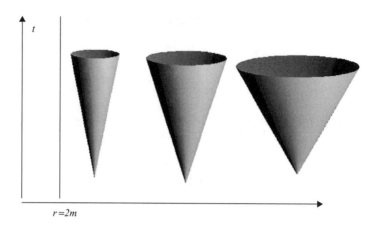

**Fig. 11-1.** Using Schwarzschild coordinates, light cones close up as they approach $r = 2m$.

# The Path of a Radially Infalling Particle

In the Schwarzschild geometry a radially infalling particle falling from infinity with vanishing initial velocity moves on a path described by

$$\left(1 - \frac{2m}{r}\right) \frac{dt}{d\tau} = 1 \quad \text{and} \quad \left(\frac{dr}{d\tau}\right)^2 = \frac{2m}{r}$$

From these equations, we find that

$$\frac{dr/d\tau}{dt/d\tau} = \frac{dr}{dt} = -\sqrt{\frac{2m}{r}} \left(1 - \frac{2m}{r}\right)$$

Integration yields

$$t - t_0 = \frac{2}{3\sqrt{2m}} \left(r^{3/2} - r_0^{3/2} + 6m\sqrt{r} - 6m\sqrt{r_0}\right)$$

$$+ 2m \ln \frac{\left(\sqrt{r} + \sqrt{2m}\right)\left(\sqrt{r_0} - \sqrt{2m}\right)}{\left(\sqrt{r_0} + \sqrt{2m}\right)\left(\sqrt{r} - \sqrt{2m}\right)}$$

In the limit that $r$ becomes close to 2m, this becomes

$$r - 2m = 8m\, e^{-(t-t_0)/2m}$$

This indicates that if we choose to use $t$ as the time coordinate, the surface $r = 2m$ is approached but never passed. We recall that $t$ corresponds to the proper time of a distant observer. Therefore for an outside observer far from the black hole, a falling test body will never reach $r = 2m$.

Revisiting the path of a radially infalling particle using the particle's proper time, we have

$$\frac{dr}{d\tau} = -\sqrt{\frac{2m}{r}}$$

We assume the particle starts from $r = r_0$ at proper time $\tau = \tau_0$. Cross multiplying terms and integrating, we have

$$\frac{1}{\sqrt{2m}} \int_r^{r_0} \sqrt{r'} dr' = \int_{\tau_0}^{\tau} d\tau'$$

where the primes denote dummy integration variables. Integrating both sides gives

$$\frac{2}{3\sqrt{2m}} \left( r_0^{3/2} - r^{3/2} \right) = \tau - \tau_0$$

Looking at this equation, the mysterious surface $r = 2m$ makes no appearance. The body falls continuously to $r = 0$ in finite proper time, in contrast to the result seen by an outside observer. In fact, one can say that the entire evolution of the physical universe has occurred by the time the body passes the surface $r = 2m$.

To study the spacetime inside $r = 2m$, we need to remove the coordinate singularity. We take this up in the next section.

# Eddington-Finkelstein Coordinates

The first attempt to get around the problem of the coordinate singularity was made with *Eddington-Finkelstein* coordinates. First we introduce a new coordinate $r^*$ called the tortoise coordinate

$$r^* = r + 2m \ln \left( \frac{r}{2m} - 1 \right) \tag{11.2}$$

along with two null coordinates

$$u = t - r^* \quad \text{and} \quad v = t + r^* \tag{11.3}$$

From (11.2) we find

$$dr^* = dr + \frac{2m}{(r/2m - 1)} \left( \frac{1}{2m} \right) dr = dr + \frac{dr}{(r/2m - 1)}$$

$$= \frac{(r/2m - 1)}{(r/2m - 1)} dr + \frac{dr}{(r/2m - 1)} = \left( \frac{r}{2m} \right) \frac{dr}{(r/2m - 1)}$$

$$= \frac{dr}{1 - 2m/r}$$

Now we use (11.3) to write

$$dt = dv - dr^*$$

$$= dv - \frac{dr}{\left( 1 - 2m/r \right)}$$

$$\Rightarrow dt^2 = dv^2 - 2 \frac{dv \, dr}{\left( 1 - 2m/r \right)} + \frac{dr^2}{\left( 1 - 2m/r \right)^2}$$

Substitution of this result into the Schwarzschild metric gives the Eddington-Finkelstein form of the metric

$$ds^2 = \left( 1 - \frac{2m}{r} \right) dv^2 - 2 \, dv \, dr - r^2 \left( d\theta^2 + \sin^2 \theta \, d\phi^2 \right) \tag{11.4}$$

While the curvature singularity at $r = 0$ is clearly evident, in these new coordinates the metric is no longer singular at $r = 2m$. Once again, let's consider the radial paths of light rays by setting $d\theta = d\phi = 0$ and $ds^2 = 0$. This time we find

$$\left( 1 - \frac{2m}{r} \right) dv^2 - 2 \, dv \, dr = 0$$

Dividing both sides by $dv^2$, we obtain

$$\left(1 - \frac{2m}{r}\right) - 2\frac{dr}{dv} = 0$$

If we set $r = 2m$, then we have $dr/dv = 0$; that is, the radial coordinate velocity of light has vanished. We can integrate to find that $r(v) = $ const, which describes light rays that stay right where they are, neither outgoing nor ingoing.

Rearranging terms gives

$$\frac{dv}{dr} = \frac{2}{\left(1 - 2m/r\right)}$$

Integrating, we find that

$$v(r) = 2\left(r + 2m\ \ln|r - 2m|\right) + \text{const}$$

This equation gives us the paths that radial light rays will follow using $(v, r)$ coordinates. If $r > 2m$, then as $r$ increases, $v$ increases. This describes the behavior we would expect for radial light rays that are outgoing. On the other hand, if $r < 2m$, as $r$ decreases, $v$ increases. So the light rays are ingoing.

In these coordinates, light cones no longer become increasingly narrow and they make it past the line $r = 2m$; however, the fact that the time and radial coordinates reverse their character inside $r = 2m$ means that light cones tilt over in this region (see Fig. 11-2).

In summary, we have found the following:

- The surface defined by $r = 2m$ is a coordinate singularity. We can find a suitable change of coordinates to remove the singularity.
- However, the surface $r = 2m$ defines a one-way membrane called the *event horizon*. It is possible for future-directed lightlike and timelike curves to cross from $r > 2m$ to $r < 2m$, but the reverse is not possible. Events inside the event horizon are hidden from external observers.
- Moving in the direction of smaller $r$, light cones begin to tip over. At $r = 2m$, outward traveling photons remain stationary.
- For $r < 2m$, future-directed lightlike and timelike curves are directed toward $r = 0$.
- The Schwarzschild coordinates are well suited for describing the geometry over the region $2m < r < \infty$ and $-\infty < t < \infty$; however, another coordinate system must be used to describe the point $r = 2m$ and the interior region.

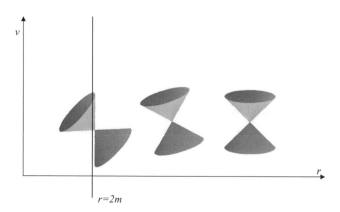

**Fig. 11-2.** Using Eddington-Finkelstein coordinates $(v, r)$ removes the coordinate singularity at $r = 2m$. As $r$ gets smaller, light cones tip over. For $r < 2m$, all geodesics directed toward the future head toward $r = 0$.

# Kruskal Coordinates

Kruskal-Szekeres coordinates allow us to extend the Schwarzschild geometry into the region $r < 2m$. Two new coordinates which we label $u$ and $v$ are introduced. They are related to the Schwarzschild coordinates $t, r$ in the following ways, depending on whether $r < 2m$ or $r > 2m$:

$r > 2m$:

$$u = e^{r/4m} \sqrt{\frac{r}{2m} - 1} \cosh \frac{t}{4m} \tag{11.5}$$

$$v = e^{r/4m} \sqrt{\frac{r}{2m} - 1} \sinh \frac{t}{4m}$$

$r < 2m$:

$$u = e^{r/4m} \sqrt{1 - \frac{r}{2m}} \sinh \frac{t}{4m} \tag{11.6}$$

$$v = e^{r/4m} \sqrt{1 - \frac{r}{2m}} \cosh \frac{t}{4m}$$

The Kruskal form of the metric is given by

$$ds^2 = \frac{32m^3}{r}e^{-r/2m}\left(du^2 - dv^2\right) + r^2\left(d\theta^2 + \sin^2\theta d\phi^2\right) \qquad (11.7)$$

These coordinates are illustrated in Fig. 11-3. These coordinates exhibit the following features:

- The "outside world" is labeled by O is the region $r \geq 2m$, which corresponds to $u \geq |v|$.
- The line $u = v$ corresponds to the Schwarzschild coordinate $t \to \infty$ while $u = -v$ corresponds to $t \to -\infty$.
- The region inside the event horizon, which is $r < 2m$, corresponds to $v > |u|$.
- In Kruskal coordinates, light cones are 45°.

We also have the following relationships:

$$u^2 - v^2 = \left(\frac{r}{2m} - 1\right)e^{r/2m} \quad \text{and} \quad \frac{v}{u} = \tanh\frac{t}{4m} \qquad (11.8)$$

The coordinate singularity at $r = 2m$ corresponds to $u^2 - v^2 = 0$. The real, curvature singularity $r = 0$ is a hyperbola that maps to

$$v^2 - u^2 = 1$$

Once again we can examine the paths of light rays by studying $ds^2 = 0$. For the Kruskal metric, we have

$$ds^2 = 0 = \frac{32m^3}{r}e^{-r/2m}\left(du^2 - dv^2\right)$$

This immediately leads to

$$\left(\frac{du}{dv}\right)^2 = 1$$

In Kruskal coordinates massive bodies move inside light cones and have slope

$$\left(\frac{dv}{du}\right)^2 > 1$$

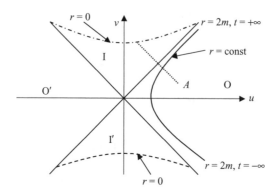

**Fig. 11-3.** An illustration of Kruskal coordinates. Regions O and O′ are outside the event horizon and so correspond to $r > 2m$. Regions I and I′ correspond to regions $r < 2m$. The hyperbola $r$ = const is some constant radius outside of $r = 2m$; it could, for example, represent the surface of a star.

which tells us that the velocity of light is 1 everywhere. Therefore there is no boundary to light propagation in these coordinates. Furthermore,

- $u$ serves as a global radial marker.
- $v$ serves as a global time marker.
- The metric is equivalent to the Schwarzschild solution but does not correspond to flat spherical coordinates at large distances.
- There is no coordinate singularity at $r = 2m$.
- It still has the real singularity at $r = 0$.

In Fig. 11-3, the dashed line indicated by $A$ represents a light ray traveling radially inward. The slope is $-1$ and in Kruskal coordinates it must hit the singularity at $r = 0$.

# The Kerr Black Hole

Observation of astronomical objects like the Earth, Sun, or a neutron star reveals one fact: most (if not all) of them rotate. While the Schwarzschild solution still works as a description of the spacetime around a slowly rotating object, to accurately describe a spinning black hole we need a new solution. Such a solution is given by the *Kerr metric*.

The Kerr metric reveals some interesting new phenomena that are wholly unexpected. For example, we will find that an object that is placed near a spinning black hole cannot avoid rotating along with the black hole—no matter what state of motion we give to the object. Put a rocket ship there. Fire the most powerful engines that can be constructed so that the rocket ship will move in a direction opposite to that in which the black hole is rotating. But the engines cannot help—no matter what we do—the rocket ship will be carried along in the direction of rotation. In fact, we will see below the rotation even effects light!

We're also going to see that a rotating black hole has two event horizons. In between these event horizons is a region called the *ergosphere*, and we will see that it is there where the effects of rotation are felt.

It is also possible to extract energy from a Kerr black hole using a method known as the *Penrose process*. Let's get started examining the Kerr black hole by making some definitions.

As you know a spinning object is characterized by its angular momentum. When describing a black hole, physicists and astronomers give the angular momentum the label $J$ and are usually concerned with angular momentum per unit mass. This is given by $a = J/M$, and if we are using the "gravitational mass" for $M$ then the units of $a$ are given in meters. With the Kerr metric, the effects of spinning on the spacetime around the black hole will be seen by the presence of angular momentum in the metric along with mixed or *cross terms*. These are terms of the form $dt\, d\phi$ that indicate a change in angle with time—a rotation.

The Kerr metric is a bit complicated, and so we are simply going to state what it is. To simplify notation, we make the following definitions:

$$\Delta = r^2 - 2mr + a^2$$
$$\Sigma = r^2 + a^2 \cos^2 \theta$$

where, as defined above, $a$ is the angular momentum per unit mass. With these definitions the metric describing the spacetime around a rotating black hole is

$$ds^2 = \left(1 - \frac{2mr}{\Sigma}\right) dt^2 \frac{4amr \sin^2 \theta}{\Sigma} dt\, d\phi - \frac{\Sigma}{\Delta} dr^2 - \Sigma d\theta^2$$
$$- \left(r^2 + a^2 + \frac{2a^2 mr \sin^2 \theta}{\Sigma}\right) \sin^2 \theta d\phi^2 \qquad (11.9)$$

We have written the metric in Boyer-Lindquist coordinates. The components of the metric tensor are given by

$$g_{tt} = \left(1 - \frac{2mr}{\Sigma}\right), \qquad g_{t\phi} = g_{\phi t} = \frac{2mar \sin^2 \theta}{\Sigma}$$

$$g_{rr} = -\frac{\Sigma}{\Delta}, \qquad g_{\theta\theta} = -\Sigma, \qquad g_{\phi\phi} = -\left(r^2 + a^2 + \frac{2ma^2 r \sin^2\theta}{\Sigma}\right)\sin^2\theta$$

(11.10)

Note that the metric components are independent of $t$ and $\phi$. This implies that two suitable killing vectors for the spacetime are $\partial_t$ and $\partial_\phi$. To invert this complicated metric, we first make the observation that the off-diagonal terms involve only $g_{t\phi}$. Therefore, we can invert the terms $g_{rr}$ and $g_{\theta\theta}$ using

$$g^{rr} g_{rr} = 1 \Rightarrow g^{rr} = -\frac{\Delta}{\Sigma}, \qquad g^{\theta\theta} g_{\theta\theta} = -1 \Rightarrow g^{\theta\theta} = -\frac{1}{\Sigma} \qquad (11.11)$$

To find the other components, we form the matrix

$$\begin{pmatrix} g_{tt} & g_{t\phi} \\ g_{\phi t} & g_{\phi\phi} \end{pmatrix} = \begin{pmatrix} \left(1 - \frac{2mr}{\Sigma}\right) & \frac{2mar\sin^2\theta}{\Sigma} \\ \frac{2mar\sin^2\theta}{\Sigma} & -\left(r^2 + a^2 + \frac{2ma^2 r \sin^2\theta}{\Sigma}\right)\sin^2\theta \end{pmatrix}$$

(11.12)

This matrix can be inverted with a great deal of tedious algebra, or using computer (the path we choose), which gives

$$g^{tt} = \frac{\left(r^2 + a^2\right)^2 - \Delta a^2 \sin^2\theta}{\Sigma\Delta}, \qquad g^{t\phi} = \frac{2mar}{\Sigma\Delta}$$

$$g^{\phi\phi} = -\frac{\Delta - a^2 \sin^2\theta}{\Sigma\Delta} \qquad (11.13)$$

The fact that there are mixed components of the metric tensor leads to some interesting results. For example, we can consider the four momentum of a particle with components $\left(p_t, \ p_r, \ p_\theta, \ p_\phi\right)$. Notice that

$$p^t = g^{ta} p_a = g^{tt} p_t + g^{t\phi} p_\phi$$
$$p^\phi = g^{\phi a} p_a = g^{\phi t} p_t + g^{\phi\phi} p_\phi$$

and so a particle will have a nonzero $p^\phi = g^{\phi t} p_t$ even when $p_\phi = 0$.

It is possible to simplify matters a bit and still get to the essential features of the Kerr metric. Let's consider the *equatorial plane*, which is a plane that cuts right through the equator of a sphere. If we imagine the sphere being the earth or some other rotating body, the plane is perpendicular to spin axis.

In the case of a black hole, we can also imagine a plane through the center of the black hole and perpendicular to the spin axis. In this case, $\theta = 0$, which means that $\cos\theta = 1$ and $d\theta = 0$. Looking at the metric in (11.9) together with the definitions of $\Delta$ and $\Sigma$, we see that the metric is greatly simplified. We can write

$$ds^2 = \left(1 - \frac{2m}{r}\right)dt^2 + \frac{4ma}{r}dt\,d\phi - \frac{1}{\left(1 - \frac{2m}{r} + \frac{a^2}{r^2}\right)}dr^2$$

$$- \left(1 + \frac{a^2}{r^2} + \frac{2ma^2}{r^3}\right)r^2 d\phi^2 \qquad (11.14)$$

With this simplified metric, some features of the spacetime about a spinning black hole jump out at you rather quickly. First let's note that the time coordinate used in the metric, $t$, is the time recorded by a distant observer as it was for the Schwarzschild metric. With this in mind, let's follow the same procedure used with the Schwarzschild metric and note where terms go to zero or blow up.

The first thing to note about this metric is that the coefficient $g_{tt}$ is the same as that we saw in the Schwarzschild metric (10.33). Let's set it to zero to get

$$1 - \frac{2m}{r} = 0$$

Solving for $r$, this term goes to zero at $r_s = 2m$. This is the same value we found in the Schwarzschild case, but since the present metric is more complicated we are going to find other values of $r$ where interesting things are happening. So we call this the *static limit*. More on this in a moment. For now, let's turn to the $g_{rr}$ term. It is here that we see the first impact of rotation. In the Schwarzschild case we were interested in seeing where this term blew up. We are in this case as well, and notice that now $g_{rr}$ depends on the angular momentum per unit mass $a$. We have

$$g_{rr} = -\frac{1}{\left(1 - \frac{2m}{r} + \frac{a^2}{r^2}\right)}$$

which means that this term will blow up when

$$1 - \frac{2m}{r} + \frac{a^2}{r^2} = 0 \tag{11.15}$$

Multiplying through by $r^2$ we obtain the following quadratic equation:

$$r^2 - 2mr + a^2 = 0$$

with solutions given by (using the quadratic formula)

$$r_\pm = \frac{-b \pm \sqrt{b^2 - 4ac}}{2a} = \frac{2m \pm \sqrt{4m^2 - 4a^2}}{2} \tag{11.16}$$

Looking at this result, we see that in the case of a rotating black hole, there are *two* horizons! First, it is reassuring to note that if we consider a nonrotating black hole by setting $a = 0$, we get back the Schwarzschild result; that is,

$$r_\pm = \frac{2m \pm \sqrt{4m^2 - 0}}{2} = m \pm m$$
$$\Rightarrow r = 2m \text{ or } r = 0$$

Now let's consider the case of interest here, where $a \neq 0$. If we take the positive sign we obtain an outer horizon while the minus sign gives us an inner horizon. The inner horizon goes by the name the *Cauchy horizon*.

Let's consider the maximum angular momentum that can be found. We can find this rather easily. Notice the term under the square root sign in (11.16). We have $\sqrt{4m^2 - 4a^2} = 2\sqrt{m^2 - a^2}$. This term is real only when $m^2 - a^2 \geq 0$. The inner radius is going to attain a maximum when $a = m$. In fact, looking at (11.16), we see that in this case we obtain $r_\pm = m$. The inner and outer horizons coincide.

Returning to the two horizons associated with the Kerr metric, let's label the inner and outer horizons by $r_\pm$. These are genuine horizons; that is, they are one-way membranes where you can cross going in, but can't come out. As we mentioned above, the two one-way membranes or horizons are given by

$$r_\pm = m \pm \sqrt{m^2 - a^2}$$

We take the outer horizon to be

$$r_+ = m + \sqrt{m^2 - a^2}$$

This horizon represents a boundary between the black hole and the outside world. It is analogous to the Schwarzschild horizon that we found in the case of a nonrotating black hole, and as we noted above, if you set $a = 0$ then you will get the familiar result $r = 2m$. Now, turning to the inner horizon represented by $r_- = m - \sqrt{m^2 - a^2}$, note that since it resides inside the outer horizon, it is inaccessible to an outside observer.

Earlier we noted that in the Kerr geometry at $r_s = 2m$, the time component of the metric vanishes, i.e., $g_{tt} = 0$. The solution $r_s = 2m$ can be described as an outer *infinite redshift* surface that lies outside of the outer horizon $r_+ = m + \sqrt{m^2 - a^2}$. Particles and light can cross the infinite redshift surface in either direction. But think of the surface represented by the horizon $r_+ = m + \sqrt{m^2 - a^2}$ as the "actual" black hole. It is the one-way membrane that represents the point of no return. If a particle or light beam passes it, escape is not possible. Interestingly, however, at $\theta = 0$, $\pi$ the horizon and the surface of infinite redshift coincide, and so at these points if light or massive bodies cross, they cannot escape.

The volume between these surfaces defined by the static limit and the horizon, that is, the region where $r_+ < r < r_s$, is called the ergosphere. Inside the ergosphere one finds the frame-dragging effect: an object inside this region is dragged along regardless of its energy or state of motion. More formally we can say that inside the ergosphere, all timelike geodesics rotate with the mass that is the source of the gravitational field.

In between the two horizons where $r_- < r < r_+$, $r$ becomes a timelike coordinate. This is just like the Schwarzschild case. This means that if we were to find ourselves in this region, no matter what we do, we would be pulled with inevitability to the Cauchy horizon $r = r_-$ in the same way that we all march through life to the future. It is believed that the Kerr solution describes the geometry accurately up to the Cauchy horizon.

# Frame Dragging

The rotational nature of the Kerr solution leads to an interesting effect known as *frame dragging*. We imagine dropping a particle in from infinity with zero angular momentum. This particle will acquire an angular velocity in the direction in which the source is rotating. An easy way to describe this phenomenon is to consider the momentum four vector and the components of the metric tensor. Looking at the inverse components of the metric (11.13) consider the ratio

$$\frac{g^{t\phi}}{g^{tt}} = \frac{2mar}{\left(r^2 + a^2\right)^2 - \Delta a^2 \sin^2 \theta} = \omega \qquad (11.17)$$

Now imagine a massive particle dropped in with zero angular momentum. The angular velocity is given by

$$\frac{d\phi}{dt} = \frac{d\phi/d\tau}{dt/d\tau} = \frac{p^\phi}{p^t}$$

With $p_\phi = 0$, we have $p^t = g^{tt} p_t$ and $p^\phi = g^{\phi t} p_t$ and so this expression becomes, using (11.17),

$$\frac{p^\phi}{p^t} = \frac{g^{\phi t}}{g^{tt}} = \omega = \frac{2mar}{\left(r^2 + a^2\right)^2 - \Delta a^2 \sin^2 \theta}$$

Note that the angular velocity is proportional to terms that make up the metric, and so think of it as being due to the gravitational field. Therefore, we see that if we drop a particle in from infinity with zero angular momentum, it will pick up an angular velocity from the gravitational field. This effect is called *frame dragging* and it causes a gyroscopic precession effect known as the *Lense-Thirring* effect. In the equatorial plane we have $\theta = 0$ and so $\sin^2 \theta = (1/2)(1 - \cos 2\theta) = 0$, giving a simplified expression for the angular velocity given by

$$\omega = \frac{2mar}{\left(r^2 + a^2\right)^2}$$

Let's consider what happens to light near a Kerr black hole. More specifically, we consider light that initially moves on a tangential path (so we set $dr = 0$). Recall that for a null ray $ds^2 = 0$, and so confining ourselves to the equatorial plane using (11.14) we find the following relation for light:

$$0 = \left(1 - \frac{2m}{r}\right) dt^2 + \frac{4ma}{r} dt\, d\phi - \left(1 + \frac{a^2}{r^2} + \frac{2ma^2}{r^3}\right) r^2 d\phi^2 \qquad (11.18)$$

To simplify notation, we follow Taylor and Wheeler (2000) and introduce the *reduced circumference*

$$R^2 = r^2 + m^2 + \frac{2m^3}{r} \tag{11.19}$$

Then (11.18) can be written in the more compact form

$$0 = \left(1 - \frac{2m}{r}\right) dt^2 + \frac{4ma}{r} dt\, d\phi - R^2\, d\phi^2 \tag{11.20}$$

Dividing through (11.20) by $dt^2$ and then by $-R^2$ gives us the following quadratic equation for $d\phi/dt$:

$$\left(\frac{d\phi}{dt}\right)^2 - \frac{4ma}{rR^2}\frac{d\phi}{dt} - \frac{1}{R^2}\left(1 - \frac{2m}{r}\right) = 0 \tag{11.21}$$

We wish to consider an important special case, the static limit where $r = r_s = 2m$. In this case notice that the last term in (11.21) vanishes:

$$1 - \frac{2m}{r} = 1 - \frac{2m}{2m} = 1 - 1 = 0$$

Meanwhile, at the static limit $R^2 = 6m^2$ and the middle coefficient in (11.21) becomes

$$\frac{4ma}{rR^2} = \frac{4ma}{(2m)6m^2} = \frac{4ma}{12m^3} = \frac{a}{3m^2}$$

Putting these results together, at the static limit (11.21) can be written as

$$\left(\frac{d\phi}{dt}\right)^2 - \frac{a}{3m^2}\frac{d\phi}{dt} = 0 \tag{11.22}$$

There are two solutions to this equation, given by

$$\frac{d\phi}{dt} = \frac{a}{3m^2} \quad \text{and} \quad \frac{d\phi}{dt} = 0 \tag{11.23}$$

The first solution, $\frac{d\phi}{dt} = \frac{a}{3m^2}$, represents light that is emitted in the same direction in which the black hole is spinning. This is a very interesting result indeed; note that the motion of the light is constrained by the angular momentum $a$ of the black hole! The second solution, however, represents an even more astonishing result. If the light is emitted in the direction *opposite* to that of the black hole's rotation, then $\frac{d\phi}{dt} = 0$; that is, the light is completely stationary! Since no material particle can attain a velocity that is faster than that of light, it is absolutely impossible to move in a direction opposite to that of the black hole's rotation. No rocket ship, probe, or elementary particle can do it.

# The Singularity

Following the process outlined in the Schwarzschild case, we wish to move beyond the coordinate singularity and consider any singularity we can find from invariant quantities. In this case we again consider the invariant quantity formed from the Riemann tensor $R^{abcd} R_{abcd}$, which leads to a genuine singularity described by

$$r^2 + a^2 \cos^2 \theta = 0$$

In the equatorial plane, again we have $\theta = 0$ and the singularity is described by the equation $r^2 + a^2 = 0$. This rather innocuous equation actually describes a *ring* of radius $a$ that lies in the $x-y$ plane. So—for a rotating black hole—the intrinsic singularity is not given by $r = 0$—but is instead a ring of radius $a$ that lies in the equatorial plane with $z = 0$.

# A Summary of the Orbital Equations for the Kerr Metric

The equations that govern the orbital motion of particles in the Kerr metric are given by

$$\sum \dot{r} = \pm\sqrt{V_r}$$
$$\sum \dot{\theta} = \pm\sqrt{V_\theta}$$
$$\sum \dot{\phi} = -\left(aE - L_z/\sin^2\theta\right) + \frac{a}{\Delta}P \qquad (11.24)$$

$$\sum i = -a\left(aE\,\sin^2\theta - L_z\right) + \frac{r^2 + a^2}{\Delta}P$$

where the derivative is in respect to the proper time or an affine parameter. The extra quantities defined in these equations are

$$P = E\left(r^2 + a^2\right) - L_z a$$

$$V_r = P^2 - \Delta\left[\mu^2 r^2 + (L_z - aE)^2 + A\right]$$

$$V_\theta = A - \cos^2\theta\left[a^2\left(\mu^2 - E^2\right) + L_z^2/\sin^2\theta\right]$$

where    $E$ = conserved energy

$L_z$ = conserved $z$ component of angular momentum

$A$ = conserved quantity associated with total angular momentum

$\mu$ = particles rest mass

# Further Reading

The study of black holes is an interesting, but serious and complicated topic. A great deal of this chapter was based on a very nice introduction to the subject, *Exploring Black Holes: An Introduction to General Relativity* by Edwin F. Taylor and John Archibald Wheeler, Addison-Wesley, 2000.

For a more technical and detailed introductory exposition on black holes consult D'Inverno (1992). There one can find a good description of black holes, charged black holes, and Kerr black holes.

One interesting phenomenon associated with rotating black holes we are not able to cover owing to space limitations is the *Penrose process*. This is a method that could be used to extract energy from the black hole. See Chapter 15 of Hartle (2002) or pages F21–F30 of Taylor and Wheeler (2000) for more information.

Our definition of the orbital equations was taken from Lightman, Press, Price, and Teukolsky (1975), which contains several solved problems related to black holes.

To see how to choose an orthonormal tetrad to use with this metric and the results of calculations of the curvature tensor, consult http://panda.unm.edu/courses/finley/p570/handouts/kerr.pdf.

# Quiz

1. Which of the following could not be used to characterize a black hole?
   (a) Mass
   (b) Electron density
   (c) Electric charge
   (d) Angular momentum

2. Using Eddington-Finkelstein coordinates, one finds that
   (a) the surface defined by $r = 2m$ is a genuine singularity
   (b) moving along the radial coordinate, in the direction of smaller $r$, light cones begin to tip over. At $r = 2m$, outward traveling photons remain stationary.
   (c) moving along the radial coordinate, in the direction of smaller $r$, light cones begin to become narrow. At $r = 2m$, outward traveling photons remain stationary.
   (d) moving along the radial coordinate, in the direction of smaller $r$, light cones remain stationary. At $r = 2m$, outward traveling photons remain stationary.

3. In Kruskal coordinates, there is a genuine singularity at
   (a) $r = 0$
   (b) $r = 2m$
   (c) $r = m$

4. Frame dragging can be best described as
   (a) an intertial effect
   (b) a particle giving up angular momentum
   (c) a particle, initially having zero angular momentum, will acquire an angular velocity in the direction in which the source is rotating
   (d) a particle, initially having zero angular momentum, will acquire an angular velocity in the direction opposite to that in which the source is rotating

5. The ergosphere can be described by saying
   (a) inside the ergosphere, all timelike geodesics rotate with the mass that is the source of the gravitational field

(b) the ergosphere is a region of zero gravitational field
 (c) inside the ergosphere, spacelike geodesics rotate with the mass that
   is the source of the gravitational field
 (d) no information can be known about the ergosphere

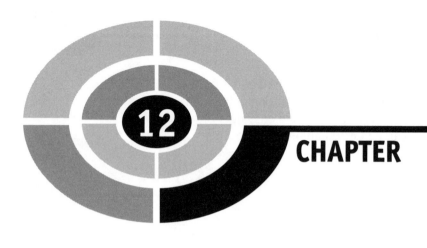

# 12

**CHAPTER**

# Cosmology

We now turn to the study of the dynamics of the entire universe, the science known as *cosmology*. The mathematical study of cosmology turns out to be relatively simple for two reasons. The first is that gravity dominates on large scales, so we don't need to consider the local complexity that arises from the nuclear and electromagnetic forces. The second reason is that on large enough scales, the universe is to good approximation *homogeneous* and *isotropic*. By large enough scales, we are talking about the level of clusters of galaxies. We apply the terms homogeneous and isotropic to the spatial part of the metric only.

By homogeneous, we mean that the geometry (i.e., the metric) is the same at any one point of the universe as it is at any other. Remember, we are talking about the universe on a large scale, so we are not considering local variations such as those in the vicinity of a black hole.

An isotropic space is one without any preferred directions. If you do a rotation, the space looks the same. Therefore, we can say an isotropic space is one for which the geometry is spherically symmetric about any point.

Incorporating these two characteristics into the spatial part of the metric allows us to consider spaces of constant curvature. The curvature in a space is denoted by $K$. Consider an $n$-dimensional space $R^n$. A result from differential geometry known as Schur's theorem tells us that if all points in some neighborhood $N$ about a point are isotropic, and the dimension of the space is $n \geq 3$, then the curvature $K$ is constant throughout $N$.

In our case we are considering a globally isotropic space, and therefore $K$ is constant everywhere. At an isotropic point in $R^n$, we can define the Riemann tensor in terms of the curvature and the metric using

$$R_{abcd} = K \left( g_{ac}g_{bd} - g_{ad}g_{bc} \right) \tag{12.1}$$

In our case, keep in mind that we will be able to apply this result only to the spatial part of the metric.

The observation that on large scales the universe is homogeneous and isotropic is embodied in a philosophical statement known as the *cosmological principle.*

# The Cosmological Principle

Copernicus told us that the Earth is not the center of the solar system. This idea can be generalized to basically say that the Earth is not the center of the universe. We call this statement the *cosmological principle*. Basically, we are saying that the universe is the same from point to point.

# A Metric Incorporating Spatial Homogeneity and Isotropy

As we mentioned earlier, the properties of homogeneity and isotropy apply only to the spatial part of the metric. Observation indicates that the universe is evolving in time and therefore we cannot extend these properties to include all of *spacetime*. This type of situation is described by using *gaussian normal coordinates*.

A detailed study of gaussian normal coordinates is beyond the scope of this book, but we will take a quick look to understand why the metric can be written in the general form:

$$ds^2 = dt^2 - a^2(t) \, d\sigma^2$$

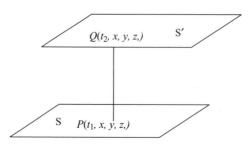

**Fig. 12-1.** A geodesic at the same spatial point, moving through time.

where $d\sigma^2$ is the spatial part of the metric and $a(t)$ the *scale factor*, a function that implements the evolution in time of the spatial part of the metric. Note that if a(t) > 0, we are describing an expanding universe.

We model the universe in the following way. At a given time, it is spatially isotropic and homogeneous, but it evolves in time. Mathematically we represent this by a set of three-dimensional spacelike hypersurfaces or slices S stacked one on top of the other. An observer who sits at a fixed point in space remains at that point but moves forward in time. This means that the observer moves along a geodesic that is parallel with the time coordinate.

Suppose that S' represents the spacelike hypersurface at some time $t_1$ and that S' is a spacelike hypersurface at a later time $t_2$. Let us denote two points on these slices as $P$ and $Q$, and consider a geodesic that moves between the two points (see Fig. 12-1).

Since the observer is sitting at the same point in space, the spatial coordinates of the points $P$ and $Q$ are unchanged as we move from S to S'. Therefore, the arc length of the geodesic is given by the time coordinate; i.e., $t_2 - t_1 =$ arc length of the geodesic. More precisely, we can write

$$ds^2 = dt^2$$

Therefore, the component of the metric must be $g_{tt} = 1$. To derive the form of the spatial component of the metric, we rely on our previous studies. The Schwarzschild metric had the property of spherical symmetry—which is exactly what we are looking for. Let's recall the general form of the Schwarzschild metric:

$$ds^2 = e^{2\nu(r)} dt^2 - e^{2\lambda(r)} dr^2 - r^2 \left(d\theta^2 + \sin^2\theta\, d\phi^2\right)$$

Using this as a guide, we write the spatial component of the metric in the present case in the following way:

$$d\sigma^2 = e^{2f(r)} dr^2 + r^2 d\theta^2 + r^2 \sin^2 \theta \, d\phi^2$$

We could go through the long and tedious process of deriving the Christoffel symbols and the components of the Ricci tensor to find $e^{2f(r)}$. However, there is an easier method available. The similarity in form to the Schwarzschild metric means that we can quickly derive the function $e^{2f(r)}$ by using our previous results. Recalling our results for the Schwarzschild metric

$$R_{\hat{t}\hat{t}} = \left[\frac{d^2\nu}{dr^2} + \left(\frac{d\nu}{dr}\right)^2 - \left(\frac{d\nu}{dr}\right)\left(\frac{d\lambda}{dr}\right) + \frac{2}{r}\frac{d\nu}{dr}\right] e^{-2\lambda(r)}$$

$$R_{\hat{r}\hat{r}} = -\left[\frac{d^2\nu}{dr^2} + \left(\frac{d\nu}{dr}\right)^2 - \left(\frac{d\nu}{dr}\right)\left(\frac{d\lambda}{dr}\right) - \frac{2}{r}\frac{d\lambda}{dr}\right] e^{-2\lambda(r)}$$

$$R_{\hat{\theta}\hat{\theta}} = -\frac{1}{r}\frac{d\nu}{dr}e^{-2\lambda} + \frac{1}{r}\frac{d\lambda}{dr}e^{-2\lambda} + \frac{1-e^{-2\lambda}}{r^2}$$

$$R_{\hat{\phi}\hat{\phi}} = -\frac{1}{r}\frac{d\nu}{dr}e^{-2\lambda} + \frac{1}{r}\frac{d\lambda}{dr}e^{-2\lambda} + \frac{1-e^{-2\lambda}}{r^2}$$

we can solve the problem by focusing on one term. It is the easiest to choose $R_{\hat{r}\hat{r}}$ and work in the coordinate basis. We can write $R_{\hat{r}\hat{r}}$ in the coordinate basis as

$$R_{rr} = -\left[\frac{d^2\nu}{dr^2} + \left(\frac{d\nu}{dr}\right)^2 - \left(\frac{d\nu}{dr}\right)\left(\frac{d\lambda}{dr}\right) - \frac{2}{r}\frac{d\lambda}{dr}\right] \qquad (12.2)$$

We can obtain the equations in the present case by setting $\nu \to 0 \, and \, \lambda \to f$. So for $R_{rr}$, we have

$$R_{rr} = \frac{2}{r}\frac{df}{dr} \qquad (12.3)$$

Using (12.1), $R_{ijkl} = K\left(g_{ik}g_{jl} - g_{il}g_{jk}\right)$, we can obtain the Ricci tensor in terms of $K$ and the metric. We have labeled the indices using $(i, j, k, l)$ because we are considering only the spatial part of the metric. The Ricci tensor is defined as

$$R_{jl} = R^k{}_{jkl}$$

We obtain the contraction via

$$R^k{}_{jkl} = g^{ki} R_{ijkl}$$
$$= g^{ki} K \left( g_{ik}g_{jl} - g_{il}g_{jk} \right)$$
$$= K \left( g^{ki} g_{ik}g_{jl} - g^{ki} g_{il}g_{jk} \right)$$
$$= K \left( \delta^k_k g_{jl} - \delta^k_l g_{jk} \right)$$
$$= K \left( \delta^k_k g_{jl} - g_{jl} \right)$$

We are working in three dimensions, and the Einstein summation convention is being used, and so $\delta^k_k = 1 + 1 + 1 = 3$. This means that the Ricci tensor for three dimensions in the constant curvature case is given by $R_{jl} = K \left( 3g_{jl} - g_{jl} \right) = 2Kg_{jl}$. We can use this with our results we obtained by comparison to the Schwarzschild solution to quickly find the function $e^{2f}$.

Now, looking at $d\sigma^2 = e^{2f(r)} dr^2 + r^2 d\theta^2 + r^2 \sin^2 \theta \, d\phi^2$, we can write the metric as

$$g_{ij} = \begin{pmatrix} e^{2f} & 0 & 0 \\ 0 & r^2 & 0 \\ 0 & 0 & r^2 \sin^2 \theta \end{pmatrix}$$

Using $R_{jl} = 2Kg_{jl}$ together with (12.3) and $g_{rr} = e^{2f}$, we obtain the following differential equation:

$$\frac{2}{r} \frac{df}{dr} = 2K \, e^{2f}$$

Cross multiplying we obtain

$$e^{-2f} \, df = Kr \, dr$$

Integration yields

$$-\frac{1}{2} e^{-2f} = K \frac{r^2}{2} + C$$

where $C$ is a constant of integration. Solving we find

$$e^{2f} = \frac{1}{C - Kr^2} \tag{12.4}$$

To find the constant $C$, we can use the other components of the Ricci tensor. In the coordinate basis, $R_{\theta\theta}$ for the Schwarzschild metric is $R_{\theta\theta} = 1 - e^{-2\lambda} + r\,e^{-2\lambda}\,(\lambda' - \nu')$. Therefore in the present case we have, using $\nu \to 0$ *and* $\lambda \to f$,

$$R_{\theta\theta} = 1 - e^{-2f} + r\,e^{-2f}\,\frac{df}{dr}$$

Let's rewrite this, using our previous results. We found that $\frac{2}{r}\frac{df}{dr} = 2K\,e^{2f}$; therefore,

$$r\,e^{-2f}\,\frac{df}{dr} = r\,e^{-2f}\left(Kr\,e^{2f}\right) = Kr^2$$

Using this along with (12.4), we can rewrite $R_{\theta\theta}$ as

$$R_{\theta\theta} = 1 - e^{-2f} + r\,e^{-2f}\,\frac{df}{dr} = 1 - C + Kr^2 + Kr^2 = 1 - C + 2Kr^2$$

Now we use $R_{jl} = 2Kg_{jl}$ together with $g_{\theta\theta} = r^2$. We have $R_{\theta\theta} = 2Kg_{\theta\theta} = 2Kr^2$. Therefore, we have $1 - C + 2Kr^2 = 2Kr^2$ or $1 - C = 0 \Rightarrow C = 1$. The final result is then

$$e^{2f} = \frac{1}{1 - Kr^2}$$

and we can write the spatial part of the metric as

$$d\sigma^2 = \frac{dr^2}{1 - kr^2} + r^2\,d\theta^2 + r^2 \sin^2\theta\,d\phi^2 \tag{12.5}$$

The curvature constant $K$ is normalized and denoted by $k$ (we can absorb any constants into the scale factor). There are three cases to consider. If $k = +1$, this corresponds to positive curvature, while $k = -1$ corresponds to negative curvature, and $k = 0$ is flat. We consider each of these cases in turn.

# Spaces of Positive, Negative, and Zero Curvature

With the normalized curvature $k$ there are three possibilities to consider: positive, negative, and zero curvature. To describe these three surfaces, we write (12.5) in the more general form

$$d\sigma^2 = d\chi^2 + r^2(\chi)\, d\theta^2 + r^2(\chi) \sin^2\theta\, d\phi^2 \qquad (12.6)$$

A space with *positive* curvature is specified by setting $k = 1$ in (12.5) or by setting $r(\chi) = \sin\chi$ in (12.6). Doing so we obtain the metric

$$d\sigma^2 = d\chi^2 + \sin^2(\chi)\, d\theta^2 + \sin^2(\chi) \sin^2\theta\, d\phi^2$$

In order to understand this space, we examine the surface we obtain if we set $\theta$ to some constant value: we take $\theta = \pi/2$. The surface then turns out to be a two sphere (see Fig. 12-2). The line element with $\theta = \pi/2$ is

$$d\sigma^2 = d\chi^2 + \sin^2(\chi)\, d\phi^2$$

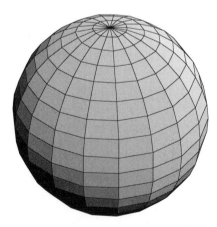

**Fig. 12-2.** An embedding diagram for a space of positive curvature. We take $\theta = \pi/2$, for which the two surface is a sphere.

A three-dimensional space that has constant curvature has two analogies with the surface of a sphere. If we start at some point and travel in a *straight* line on the sphere, we end up at the same point eventually. This would also be true moving in the three-dimensional space of a universe with positive curvature. Second, if we add up the angles of a triangle drawn on the surface, we would find that the sum was greater than $180°$.

Next we consider a space of negative curvature, which means that we take $k = -1$. In this case, we set $r = \sinh \chi$ and the line element becomes

$$d\sigma^2 = d\chi^2 + \sinh^2(\chi)\,d\theta^2 + \sinh^2(\chi)\sin^2\theta\,d\phi^2$$

When we use this as the spatial line element for the universe $dt^2 - a^2(t)\,d\sigma^2$, spatial slices have the remarkable property that they have infinite volume. In this case, the sum of the angles of a triangle add up to *less* than $180°$.

Once again considering $\theta = \pi/2$ in order to obtain an embedding diagram, we obtain

$$d\sigma^2 = d\chi^2 + \sinh^2(\chi)\,d\phi^2$$

The embedding diagram for a surface with negative curvature is a saddle (see Fig. 12-3).

Finally, we consider the case of zero curvature for which $k = 0$. It turns out that current observations indicate that this is the closest approximation to the real universe we live in. In this final case, we set $r = \chi$ and we can write the line element as

$$d\sigma^2 = d\chi^2 + \chi^2\,d\theta^2 + \chi^2\sin^2\theta\,d\phi^2$$

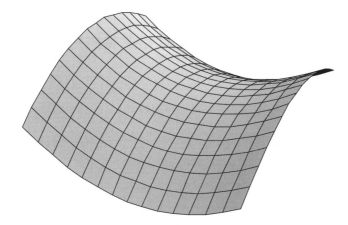

**Fig. 12-3.** A surface of negative curvature is a saddle.

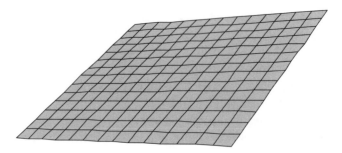

**Fig. 12-4.** When $k = 0$, space is perfectly flat.

which is nothing but good old flat Euclidean space. Setting $\theta = \pi/2$, we obtain a flat plane for the embedding diagram (see Fig. 12-4).

# Useful Definitions

We now list several definitions that you will come across when reading about cosmology.

## THE SCALE FACTOR

The universe is expanding and therefore its size changes with time. The spatial size of the universe at a given time $t$ is called the *scale factor*. This quantity is variously labeled $R(t)$ and $a(t)$ by different authors. In this chapter we will denote the scale factor by $a(t)$. Observation indicates that the universe is expanding as time moves forward and therefore $a(t) > 0$.

## THE GENERAL ROBERTSON-WALKER METRIC

The Robertson-Walker metric is defined by

$$\mathrm{d}s^2 = \mathrm{d}t^2 - a^2(t)\,\mathrm{d}\sigma^2$$

where $\mathrm{d}\sigma^2$ is given by one of the constant curvature metrics described by (12.6) in *spaces of positive, negative, and zero curvature.*

## MATTER DENSITY

Matter is all the stuff in the universe—stars, planets, comets, asteroids, and galaxies (and whatever else there may be). When constructing the stress-energy tensor, we model matter as *dust*. We will indicate the energy density of matter as $\rho_m$. The pressure $P_m = 0$. Just as a model to keep in your head for the moment, think of the universe as a box filled with a diffuse gas, with the particles that make up the gas representing galaxies. The expansion of the universe is represented by the volume of the box expanding. But the number of particles in the box remains the same—therefore, the number density decreases with the expansion. In the real universe, we model this behavior by saying as the universe expands, the energy density of matter $\rho_m$ decreases.

## RADIATION DENSITY

We indicate the energy density of radiation by $\rho_R$ and the pressure by $P_R$ and treat radiation as a perfect fluid. As the universe expands, the radiation density (think in terms of photons) decreases like it does in the case of matter. However, the energy density for radiation decreases faster because photons are redshifted as the universe expands, and hence loose energy.

## VACUUM DENSITY

Recent evidence indicates that the expansion of the universe is accelerating, and this is consistent with a nonzero cosmological constant. The cosmological constant represents vacuum energy that is modeled as a perfect fluid with the condition that $\rho = -p$.

## MATTER-DOMINATED AND RADIATION-DOMINATED UNIVERSES

The evolution of the scale factor $a\,(t)$ is influenced by whether or not the universe is *matter dominated* or *radiation dominated*. The ratio of $\rho_M/\rho_R$ tells us whether the universe is matter or radiation dominated. In our time, observation indicates that the ratio $\rho_M/\rho_R \sim 10^3$ and therefore the universe is matter dominated. In the early history of the universe, it was radiation dominated. We will see later that as the universe ages, it will eventually become dominated by vacuum energy.

## THE HUBBLE PARAMETER

The Hubble constant or Hubble parameter indicates the rate of expansion of the universe. It is defined in terms of the time derivative of the scale factor as

$$H = \frac{\dot{a}}{a} \tag{12.7}$$

Remember that the scale factor is a measure of the size of the universe. From the equation, the units of the Hubble constant can be seen to be $sec^{-1}$. However, for convenience in astronomical applications the units are often stated as km/sec/Mpc, where Mpc are *megaparsecs*, which is a measure of distance (see below). In addition, there is a great deal of uncertainty about the actual value of the Hubble constant; therefore, it is often defined as $H_0 = 100h \, km/sec/Mpc$ where present evidence indicates $h \sim 0.7$.

## THE HUBBLE TIME

The inverse of the Hubble constant is the *Hubble time,* which is a rough estimate of the age of the universe. At the Hubble time, all galaxies in the universe were located at the same point. The Hubble length is a measure of cosmological scales, and is given by $d = c/H_0$.

## HUBBLE'S LAW

The distance of a galaxy from us is related to its velocity by Hubble's law

$$v = H_0 r$$

## THE DECELERATION PARAMETER

We can quantify the rate of change of the expansion with the *deceleration parameter.* It has a simple definition

$$q = -\frac{a\ddot{a}}{\dot{a}^2}$$

## OPEN UNIVERSE

An open universe is one that expands forever. The geometry of an open universe is one of negative curvature (think potato chip or saddle, as in Fig. 12-4) and it can in principle extend to infinity. For an open universe, $k = -1$.

## CLOSED UNIVERSE

A closed universe expands to some maximum size, then reverses, and then falls back on itself. Think of a closed universe as a sphere. From our discussion in the previous section, this corresponds to $k = +1$ and is illustrated in Fig. 12-2.

## FLAT UNIVERSE

A flat universe is described by Euclidean geometry on large scales and will expand forever. In this case, $k = 0$ (see Fig. 12-4).

## CRITICAL DENSITY

Whether or not the universe is open or closed is determined by the density of stuff in the universe. In other words, is there enough matter, and therefore enough gravity, to slow down the expansion enough so that it will stop and reverse? If so, we would live in a closed universe. The density required to have a closed universe is called the *critical density*. It can be defined in terms of the Hubble constant, Newton's gravitational constant, and the speed of light as

$$\rho_c = \frac{3H_0^2}{8\pi G} \tag{12.8}$$

At the present time, the ratio of the observed density to the critical density is very close to unity, indicating that the universe is not closed. However, keep in mind the uncertainty in the Hubble constant.

## THE DENSITY PARAMETER

This is the ratio of the observed density to the critical density.

$$\Omega = \frac{8\pi G}{3H_0^2}\rho = \frac{\rho}{\rho_c} \tag{12.9}$$

The density used here is obtained by adding contributions from all possible sources (matter, radiation, vacuum). If $\Omega < 1$, this corresponds to $k < 0$ and the universe is open. If $\Omega = 1$ then $k = 0$ and the universe is flat. Finally, if $\Omega > 1$ then we have $k > 0$ and a closed universe. As we indicated above, it appears that $\Omega \cong 1$.

# The Robertson-Walker Metric and the Friedmann Equations

To model the large-scale behavior of the universe such that Einstein's equations are satisfied, we begin by modeling the matter and energy in the universe by a perfect fluid. The particles in the fluid are galaxy clusters and the fluid is described by an average density $\rho$ and pressure $P$. Moreover, in co-moving coordinates, $\dot{t} = 1$ and $\dot{x}_i = 0$ giving $u^a = (1, 0, 0, 0)$. Therefore, we set

$$T^a{}_b = \begin{pmatrix} \rho & 0 & 0 & 0 \\ 0 & -P & 0 & 0 \\ 0 & 0 & -P & 0 \\ 0 & 0 & 0 & -P \end{pmatrix} \tag{12.10}$$

Using the metric

$$ds^2 = dt^2 - \frac{a^2(t)}{1 - kr^2} dr^2 - a^2(t)r^2 d\theta^2 - a^2(t)r^2 \sin^2\theta \, d\phi^2$$

we can use the metric to derive the components of the curvature tensor in the usual way. This was done in Example 5-3. From there we can work out the components of the Einstein tensor and use Einstein's equation to equate its components to the stress-energy tensor. We remind ourselves how Einstein's equation relates the curvature to the stress-energy tensor:

$$G_{ab} - \Lambda g_{ab} = 8\pi T_{ab}$$

The details were worked out in Example 7-3. Note we used a different signature of the metric in that example. For the signature we are using in this case, the result is found to be

$$\frac{3}{a^2} \left( k + \dot{a}^2 \right) - \Lambda = 8\pi\rho$$
$$2\frac{\ddot{a}}{a} + \frac{1}{a^2} \left( k + \dot{a}^2 \right) - \Lambda = -8\pi P \tag{12.11}$$

We can augment these equations by writing down the conservation of energy equation using the stress-energy tensor (see Chapter 7):

$$\nabla_a T^a{}_t = \partial_a T^a{}_t + \Gamma^a{}_{ab} T^b{}_t - \Gamma^b{}_{at} T^a{}_b = 0$$

Since the stress-energy tensor is diagonal, this simplifies to

$$\partial_a T^a{}_t + \Gamma^a{}_{ab} T^b{}_t - \Gamma^b{}_{at} T^a{}_b = \partial_t T^t{}_t$$
$$+ \Gamma^t{}_{tt} T^t{}_t + \Gamma^r{}_{rt} T^t{}_t + \Gamma^\theta{}_{\theta t} T^t{}_t + \Gamma^\phi{}_{\phi t} T^t{}_t$$
$$- \Gamma^t{}_{tt} T^t{}_t - \Gamma^r{}_{rt} T^r{}_r - \Gamma^\theta{}_{\theta t} T^\theta{}_\theta - \Gamma^\phi{}_{\phi t} T^\phi{}_\phi$$

In the chapter Quiz you will derive the Christoffel symbols for the Robertson-Walker metric. The terms showing up in this equation are given by

$$\Gamma^t{}_{tt} = 0$$
$$\Gamma^r{}_{rt} = \Gamma^\theta{}_{\theta t} = \Gamma^\phi{}_{\phi t} = \frac{\dot{a}}{a}$$

Using this together with $T^t{}_t = \rho$ and $T^r{}_r = T^\theta{}_\theta = T^\phi{}_\phi = -P$, we have

$$\Gamma^r{}_{rt} T^t{}_t = \Gamma^\theta{}_{\theta t} T^t{}_t = \Gamma^\phi{}_{\phi t} T^t{}_t = \frac{\dot{a}}{a}\rho$$
$$\Rightarrow \Gamma^r{}_{rt} T^t{}_t + \Gamma^\theta{}_{\theta t} T^t{}_t + \Gamma^\phi{}_{\phi t} T^t{}_t = 3\frac{\dot{a}}{a}\rho$$

and

$$-\Gamma^r{}_{rt} T^r{}_r - \Gamma^\theta{}_{\theta t} T^\theta{}_\theta - \Gamma^\phi{}_{\phi t} T^\phi{}_\phi = -\frac{\dot{a}}{a}(-P) - \frac{\dot{a}}{a}(-P) - \frac{\dot{a}}{a}(-P) = 3\frac{\dot{a}}{a}P$$

Therefore, the conservation equation becomes

$$\frac{\partial \rho}{\partial t} = -3\frac{\dot{a}}{a}(\rho + P) \tag{12.12}$$

This is nothing more than a statement of the first law of thermodynamics. The volume of a slice of space is given by $V \sim a^3(t)$ and the mass energy enclosed in the volume is $E = \rho V$. Then (12.12) is nothing more than the statement $dE + P\,dV = 0$.

In the present matter-dominated universe, we can model the matter content of the universe as dust and set the pressure $P = 0$. In this case, the second equation of (12.11) can be written as

$$2\frac{\ddot{a}}{a} + \frac{1}{a^2}\left(k + \dot{a}^2\right) - \Lambda = 0 \tag{12.13}$$

Meanwhile, the conservation equation can be written as

$$\frac{\partial \rho}{\partial t} = -3\frac{\dot{a}}{a}\rho$$

This can be rearranged to give

$$\frac{d\rho}{\rho} = -3\frac{da}{a}$$

$$\Rightarrow \ln\rho = -3\ln a = \ln a^{-3}$$

where we are ignoring constants of integration to get a qualitative answer. The result is that for a matter-dominated universe

$$\rho \propto a^{-3}$$

we set $\rho = \frac{1}{a^3}$ in the first of the Friedmann equations listed in (12.11). This gives

$$\frac{3}{a^2}\left(k + \dot{a}^2\right) - \Lambda = \frac{8\pi}{a^3}$$

Now multiply through by $a^3$ and divide by 3 to obtain

$$a\left(k + \dot{a}^2\right) - \frac{1}{3}\Lambda a^3 = \frac{8\pi}{3}$$

We can rearrange this equation to obtain a relation giving the time variation of the scale factor with zero pressure:

$$\dot{a}^2 = \frac{1}{3}\Lambda a^2 - k + \frac{8\pi}{3a} \tag{12.14}$$

The solutions of this equation give rise to different Friedmann models of the universe. Before we move on to consider these models in the next section, we briefly return to the conservation equation (12.12). We have just worked out a reasonable approximation to the present universe by considering the modeling of a matter-dominated universe by dust. Now let's consider the early and possible future states of the universe by considering a radiation-dominated universe and a vacuum-dominated universe, respectively.

As the universe expands, photons get redshifted and therefore loose energy. Using the electromagnetic field tensor, it can be shown that the equation of state (which relates the pressure to the density for a fluid) for radiation is given by

$$\rho = 3P$$

We can use this to replace $P$ in (12.12) to obtain

$$\frac{\partial \rho}{\partial t} = -3\frac{\dot{a}}{a}(\rho + P) = -3\frac{\dot{a}}{a}\left(\rho + \frac{1}{3}\rho\right) = -4\frac{\dot{a}}{a}\rho$$

Rearranging terms, we obtain

$$\frac{\mathrm{d}\rho}{\rho} = -4\frac{\mathrm{d}a}{a}$$

Following the procedure used for matter density, we integrate and ignore any constants of integration, which gives

$$\ln \rho = -4 \ln a = \ln a^{-4}$$

Therefore, we conclude that the energy density $\rho \propto a^{-4}$ in the case of radiation. The density falls off faster than matter precisely because of the redshift we mentioned earlier.

To close this section, we consider the vacuum energy density. The vacuum energy density is a constant, and therefore $\rho$ remains the same at all times. Since matter density and radiation energy density are decreasing as the universe expands but the vacuum energy density remains the same, eventually the universe will become dominated by vacuum energy.

# Different Models of the Universe

We now turn to the problem of considering the evolution of the universe in time, which amounts to solving for the scale factor $a(t)$. In this section we will be a bit sloppy from time to time, because we are interested only in the qualitative behavior of the solutions. Therefore, we may ignore integration constants etc.

First we consider the very early universe which was dominated by radiation. In that case we use $\rho = 3P$. For simplicity, we set the cosmological constant to zero and the Friedmann equations can be written as

$$\frac{3}{a^2}\left(k + \dot{a}^2\right) = 8\pi \rho$$

$$2\frac{\ddot{a}}{a} + \frac{1}{a^2}\left(k + \dot{a}^2\right) = -\frac{8\pi \rho}{3}$$

Using the first equation to replace $\rho$ in the second, we obtain

$$2\frac{\ddot{a}}{a} + \frac{1}{a^2}\left(k + \dot{a}^2\right) = -\frac{1}{a^2}\left(k + \dot{a}^2\right)$$

Rearranging terms, we have

$$\ddot{a} + \frac{1}{a}\left(k + \dot{a}^2\right) = 0$$

In the very early universe we can neglect the $k/a$ term, which gives

$$\ddot{a} + \frac{\dot{a}^2}{a} = 0$$

This can be rewritten as $a\ddot{a} + \dot{a}^2 = 0$. Notice that

$$\frac{\mathrm{d}}{\mathrm{d}t}(a\dot{a}) = a\ddot{a} + \dot{a}^2$$

and so we can conclude that $a\dot{a} = C$, where $C$ is a constant. Writing this out, we find

$$a\frac{\mathrm{d}a}{\mathrm{d}t} = C$$
$$\Rightarrow\ a\,\mathrm{d}a = C\,\mathrm{d}t$$

Integrating both sides (and ignoring the second integration constant), we find

$$\frac{a^2}{2} = Ct$$
$$\Rightarrow\ a(t) \propto \sqrt{t}$$

As we will see later, this expansion is more rapid than the later one driven by matter (as dust). This is due to the radiation pressure that dominates early in the universe.

For a complete examination of the behavior of the universe from start to finish, we begin by considering a simple case, the *de Sitter model*. This is a flat model devoid of any content (i.e., it contains no matter or radiation). Therefore, we set $k = 0$ and the line element can be written as

$$\mathrm{d}s^2 = \mathrm{d}t^2 - a^2(t)\,\mathrm{d}r^2 - a^2(t)r^2\,\mathrm{d}\theta^2 - a^2(t)r^2\sin^2\theta\,\mathrm{d}\phi^2$$

The fact that this models a universe that contains no matter or radiation means that we set $P = \rho = 0$. However, we leave the cosmological constant in the equations. So while this is a toy model, it can give an idea of behavior in the very late history of the universe. Since the expansion of the universe appears to be expanding and the matter and radiation density will eventually drop to negligible levels, in the distant future the universe may be a de Sitter universe.

With these considerations, (12.11) becomes

$$\frac{3}{a^2}\dot{a}^2 - \Lambda = 0$$

$$2\frac{\ddot{a}}{a} + \frac{1}{a^2}\dot{a}^2 - \Lambda = 0$$

Obtaining a solution using the first equation is easy. We move the cosmological constant to the other side and divide by 3 to get

$$\frac{\dot{a}^2}{a^2} = \frac{\Lambda}{3}$$

Now we take the square root of both sides

$$\frac{1}{a}\frac{da}{dt} = \sqrt{\frac{\Lambda}{3}}$$

This can be integrated immediately to give

$$a(t) = C\,e^{\sqrt{\Lambda/3}\,t} \tag{12.15}$$

where $C$ is a constant of integration that we won't worry about. We are interested only in the qualitative features of the solution, which can be seen by plotting.

For the de Sitter universe, (see Fig. 12-5) the line element becomes

$$ds^2 = dt^2 - a^2(t)\,dr^2 - e^{\Lambda/3t}r^2\,d\theta^2 - a^2(t)r^2\sin^2\theta\,d\phi^2$$

Now let's move on to a universe that contains matter. Direct solution of the Friedmann equations is in general difficult, and basically requires numerical analysis. Since recent observations indicate that the universe is flat (therefore $k = 0$), we don't lose anything by dropping $k$. With this in mind, let's consider a universe that contains matter, but we set $k = 0 = \Lambda$. We can obtain a solution

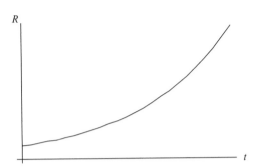

**Fig. 12-5.** The de Sitter solution represents a universe without matter. As time increases the universe expands exponentially.

by using (12.14). Setting $k = 0 = \Lambda$, we have

$$\dot{a}^2 = \frac{8\pi}{3a}$$

Rearranging terms and taking the square root of both sides, we have

$$\sqrt{a}\,\frac{da}{dt} = \sqrt{\frac{8\pi}{3}}$$

$$\Rightarrow \sqrt{a}\,da = \sqrt{\frac{8\pi}{3}}\,dt$$

Integrating on the left side, we have

$$\int \sqrt{a}\,da = \frac{2}{3}a^{3/2}$$

while on the right, ignoring the integration constant (we are interested only in the qualitative features), we have $\sqrt{8\pi/3}\,t$. This leads to

$$a\,(t) \propto t^{2/3} \tag{12.16}$$

A plot of this case is shown in Fig. 12-6. This describes a universe that evolves with a continuous expansion and a deceleration parameter given by $q = 1/2$. Remember we found that the radiation-dominated early universe had $a\,(t) \propto \sqrt{t}$.

We can also consider cases with positive and negative curvature. Adding more nonzero terms makes things difficult, so we proceed with the positive curvature

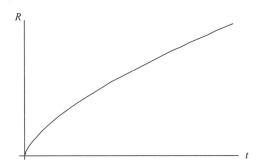

**Fig. 12-6.** A flat universe that contains matter but with zero cosmological constant.

case and leave the cosmological constant zero. This means we set $k = +1$ this describes a universe which collapses in on itself, as shown in Fig. 12-7. In this case, we have

$$\dot{a}^2 = \frac{8\pi}{3a} - 1 = \frac{C - a}{a} \tag{12.17}$$

where we have called $8\pi/3 = C$ for convenience. This equation can be solved parametrically. To obtain a solution, we define

$$a = C \sin^2 \tau \tag{12.18}$$

where $\tau = \tau(t)$. Then we have

$$\frac{da}{d\tau} = 2C \sin \tau \cos \tau \frac{d\tau}{dt}$$

Squaring, we find that

$$\dot{a}^2 = 4C^2 \sin^2 \tau \cos^2 \tau \left( \frac{d\tau}{dt} \right)^2$$

Using these results in (12.17), we obtain

$$4C^2 \sin^2 \tau \cos^2 \tau \left( \frac{d\tau}{dt} \right)^2 = \frac{C - C \sin^2 \tau}{C \sin^2 \tau} = \frac{\cos^2 \tau}{\sin^2 \tau}$$

Canceling terms on both sides, we arrive at the following:

$$2C \sin^2 \tau \, d\tau = dt$$

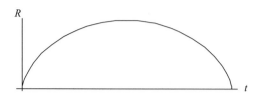

**Fig. 12-7.** Dust-filled universe with zero cosmological constant, and positive curvature. After expanding to a maximum size, the universe reverses and collapses in on itself.

Integrating, the right side becomes

$$\int 2C \sin^2 \tau \, d\tau = 2C \int \frac{1 - \cos 2\tau}{2} \, d\tau = C \left( \tau - \frac{1}{2} \sin 2\tau \right)$$

$$= \frac{C}{2} (2\tau - \sin 2\tau)$$

We can use the same trig identity used in the integral to write (12.18) as

$$a = \frac{C}{2} (1 - \cos 2\tau)$$

These equations allow us to obtain $a(t)$ parametrically. At $t = \tau = 0$, the universe begins with size $a(0) = 0$, i.e., at the "big bang" with zero size. The radius then increases to a maximum, and then contracts again to $a(0) = 0$. The maximum radius is the Schwarzschild radius determined by the constant $C$.

We leave the case $k = -1$ as an exercise.

Unfortunately, because of limited space our coverage of cosmology is very limited. We are leaving out a discussion of the big bang and inflation, for example. For a detailed overview of the different cosmological models, consider D'Inverno (1992). To review a discussion of recent observational evidence and an elementary but thorough description of cosmology, examine Hartle (2002). For a complete discussion of cosmology, consult Peebles (1993).

# Quiz

The following exercise will demonstrate that the cosmological constant has no effect over the scale of the solar system. Start with the general form of the

Schwarzschild metric

$$ds^2 = e^{2\nu(r)}\, dt^2 - e^{2\lambda(r)}\, dr^2 - r^2 \left( d\theta^2 + \sin^2 \theta\, d\phi^2 \right)$$

1.  With a nonzero cosmological constant, the Ricci scalar for this metric satisfies
    (a) $R = \Lambda$
    (b) $R = 0$
    (c) $R = 4\Lambda$
    (d) $R = -\Lambda$

2.  It can be shown that $\lambda(r) = \ln k - \nu(r)$. If $A$ and $B$ are constants of integration, then
    (a) $r\, e^\nu = A + Br - \Lambda \frac{kr^3}{3}$
    (b) $r\, e^\nu = A + Br$
    (c) $r\, e^\nu = A + Br - \Lambda r^2$

3.  By considering the field equation $R_{\theta\theta} = \Lambda g_{\theta\theta}$,
    (a) $B = 0$
    (b) $B = k$
    (c) $B = 4k$

4.  With the previous results in mind, choosing $k = 1$ we can write the spatial part of the line element as
    (a) $dl^2 = \dfrac{dr^2}{1 - \frac{2m}{r} - \frac{1}{3}\Lambda r^3} + r^2\, d\theta^2 + r^2\, \sin^2 \theta\, d\phi^2$
    (b) $dl^2 = \dfrac{dr^2}{1 - \frac{1}{3}\Lambda r^3} + r^2\, d\theta^2 + r^2\, \sin^2 \theta\, d\phi^2$
    (c) $dl^2 = \dfrac{dr^2}{1 + \frac{1}{3}\Lambda r^3} + r^2\, d\theta^2 + r^2\, \sin^2 \theta\, d\phi^2$

5.  If the cosmological constant is proportional to the size of the universe, say $a = \sqrt{\frac{3}{\Lambda}}$,
    (a) the Ricci scalar vanishes
    (b) particles would experience additional accelerations as revealed in tidal effects
    (c) the presence of the cosmological constant could not be detected by observations within the solar system

6. Consider a flat universe with positive cosmological constant. Starting with $\dot{a}^2 = \frac{C}{a} + \frac{\Lambda}{3}a^2$ use a change of variables $u = \frac{2\Lambda}{3C}a^3$, it can be shown that

   (a) $a^3 = \frac{3C}{2\Lambda} \left[ \cosh(3\Lambda)^{1/2} t - 1 \right]$

   (b) $a = \frac{3C}{2\Lambda} \left[ \cosh(3\Lambda)^{1/2} t - 1 \right]$

   (c) $a = \frac{3C}{2\Lambda} \left[ \sinh(3\Lambda)^{1/2} t - 1 \right]$

# Gravitational Waves

An important characteristic of gravity within the framework of general relativity is that the theory is nonlinear. Mathematically, this means that if $g_{ab}$ and $\gamma_{ab}$ are two solutions of the field equations, then $ag_{ab} + b\gamma_{ab}$ (where $a$, $b$ are scalars) may not be a solution. This fact manifests itself physically in two ways. First, since a linear combination may not be a solution, we cannot take the overall gravitational field of two bodies to be the summation of the individual gravitational fields of each body. In addition, the fact that the gravitational field has energy and special relativity tells us that energy and mass are equivalent leads to the remarkable fact that the gravitational field is its own source. We don't notice these effects in the solar system because we live in a region of weak gravitational fields, and so the linear newtonian theory is a very good approximation. But fundamentally these effects are there, and they are one more way that Einstein's theory differs from Newton's.

A common mathematical technique when faced with such a situation is to study a linearized version of the theory to gain some insight. In this chapter that is exactly what we will do. We consider a study of the linearized field equations and will make the astonishing discovery that gravitational effects can propagate as waves traveling at the speed of light. This will require a study of weak gravitational fields.

From here we will develop the Brinkmann metric that is used to represent pp-wave spacetimes. We will study this metric and the collision of gravitational waves using the Newman-Penrose formalism. We conclude the chapter with a very brief look at gravitational wave spacetimes with nonzero cosmological constant.

In this chapter we will use the shorthand notation $k_{a,b}$ to represent the covariant derivative of $k_a$ in place of the usual notation $\nabla_b k_a$. Therefore $k_{a,b} = \partial_b k_a - \Gamma^c{}_{ab} k_c$.

# The Linearized Metric

We begin by considering a metric that differs from the flat Minkowski metric by a small perturbation. If we take $\varepsilon$ to be some small constant parameter ($|\varepsilon| \ll 1$), then we can write the metric tensor as

$$g_{ab} = \eta_{ab} + \varepsilon h_{ab} \tag{13.1}$$

where we neglect all terms of order $\varepsilon^2$ or higher since $\varepsilon$ is small. Our first step in the analysis will be to write down the form of the various quantities such as the Christoffel symbols, the Riemann tensor, and the Ricci tensor when we write the metric in this form. Since we will drop all terms that are order $\varepsilon^2$ or higher, these quantities will assume fairly simple forms. Ultimately, we will show that this will allow us to write the Einstein field equations in the form of a wave equation.

We might as well take things in order and so we begin with the Christoffel symbols. We will work in a coordinate basis and so we compute the following:

$$\Gamma^a{}_{bc} = \frac{1}{2} g^{ad} \left( \frac{\partial g_{bc}}{\partial x^d} + \frac{\partial g_{cd}}{\partial x^b} - \frac{\partial g_{db}}{\partial x^c} \right)$$

To see how this works out, let's consider one term

$$\frac{\partial g_{bc}}{\partial x^d} = \frac{\partial}{\partial x^d} (\eta_{bc} + \varepsilon h_{bc}) = \frac{\partial \eta_{bc}}{\partial x^d} + \frac{\partial \varepsilon h_{bc}}{\partial x^d}$$

Now the Minkowski metric is given by $\eta_{ab} = \text{diag}(1, -1, -1, -1)$ and so $\frac{\partial \eta_{bc}}{\partial x^d} = 0$. We can pull the constant $\varepsilon$ outside of the derivative and so

$$\frac{\partial g_{bc}}{\partial x^d} = \varepsilon \frac{\partial h_{bc}}{\partial x^d}$$

To obtain the form of the Christoffel symbols, we need to know the form of $g^{ab}$. We begin by observing that we can raise indices with the Minkowski metric; i.e.,

$$h^{ab} = \eta^{ac} \eta^{bd} h_{cd}$$

We also recall that the metric tensor satisfies $g_{ab} g^{bc} = \delta_a^c$ (note this is also true for the Minkowski metric). The linearized form of the metric with raised indices will be similar but we assume it can be written as $g^{ab} = \eta^{ab} + \varepsilon \, a h^{ab}$, where $a$ is a constant to be determined. Now ignoring terms of order $\varepsilon^2$, we find

$$\delta_a^c = (\eta_{ab} + \varepsilon h_{ab}) \left( \eta^{bc} + a \varepsilon h^{bc} \right)$$

$$= \eta_{ab} \eta^{bc} + \varepsilon \eta^{bc} h_{ab} + a \varepsilon \eta_{ab} h^{bc} + a \varepsilon^2 h_{ab} h^{bc}$$

$$= \eta_{ab} \eta^{bc} + \varepsilon \eta^{bc} h_{ab} + a \varepsilon \eta_{ab} h^{bc}$$

$$= \delta_a^c + \varepsilon \left( \eta^{bc} h_{ab} + a \eta_{ab} h^{bc} \right)$$

For this to be true, the relation inside the parentheses must vanish. Therefore, we have

$$a \eta_{ab} h^{bc} = -\eta^{bc} h_{ab}$$

Let's work on the left-hand side:

$$a \eta_{ab} h^{bc} = a \eta_{ab} \eta^{be} \eta^{cf} h_{ef}$$

$$= a \delta_a^e \eta^{cf} h_{ef}$$

$$= a \eta^{cf} h_{af}$$

Now, notice that the index $f$ is repeated and is therefore a dummy index. We rename it $b$ and stick the result back into $a \eta_{ab} h^{bc} = -\eta^{bc} h_{ab}$ to obtain

$$a \eta^{bc} h_{ab} = -\eta^{bc} h_{ab}$$

Note that we used the fact that the metric is symmetric to write $\eta^{cb} = \eta^{bc}$. We conclude that $a = -1$ and we can write

$$g^{ab} = \eta^{ab} - \varepsilon h^{ab} \tag{13.2}$$

We now return to the Christoffel symbols. Using (13.2) together with $\frac{\partial g_{bc}}{\partial x^d} = \varepsilon \frac{\partial h_{bc}}{\partial x^d}$ and ignoring terms of second order in $\varepsilon$, we find

$$
\begin{aligned}
\Gamma^a{}_{bc} &= \frac{1}{2} g^{ad} \left( \frac{\partial g_{bc}}{\partial x^d} + \frac{\partial g_{cd}}{\partial x^b} - \frac{\partial g_{db}}{\partial x^c} \right) \\
&= \frac{\varepsilon}{2} g^{ad} \left( \frac{\partial h_{bc}}{\partial x^d} + \frac{\partial h_{cd}}{\partial x^b} - \frac{\partial h_{db}}{\partial x^c} \right) \\
&= \frac{\varepsilon}{2} \left( \eta^{ad} - \varepsilon h^{ad} \right) \left( \frac{\partial h_{bc}}{\partial x^d} + \frac{\partial h_{cd}}{\partial x^b} - \frac{\partial h_{db}}{\partial x^c} \right) \\
&= \frac{1}{2} \left( \varepsilon \eta^{ad} - \varepsilon^2 h^{ad} \right) \left( \frac{\partial h_{bc}}{\partial x^d} + \frac{\partial h_{cd}}{\partial x^b} - \frac{\partial h_{db}}{\partial x^c} \right)
\end{aligned}
$$

Dropping the second-order terms, we conclude that in the linearized theory we have

$$\Gamma^a{}_{bc} = \frac{1}{2} \varepsilon \eta^{ad} \left( \frac{\partial h_{bc}}{\partial x^d} + \frac{\partial h_{cd}}{\partial x^b} - \frac{\partial h_{db}}{\partial x^c} \right) \tag{13.3}$$

We can use this expression to write down the Riemann tensor and the Ricci tensor. The Riemann tensor is given by

$$R^a{}_{bcd} = \partial_c \Gamma^a{}_{bd} - \partial_d \Gamma^a{}_{bc} + \Gamma^e{}_{bd} \Gamma^a{}_{ec} - \Gamma^e{}_{bc} \Gamma^a{}_{ed}$$

Looking at the last two terms, $\Gamma^e{}_{bd} \Gamma^a{}_{ec}$ and $\Gamma^e{}_{bc} \Gamma^a{}_{ed}$, in comparison with (13.3) shows that these terms will result in terms involving $\varepsilon^2$, and so we obtain the simplified expression

$$R^a{}_{bcd} = \partial_c \Gamma^a{}_{bd} - \partial_d \Gamma^a{}_{bc}$$

Using (13.3), we find

$$
R^a{}_{bcd} = \partial_c \Gamma^a{}_{bd} - \partial_d \Gamma^a{}_{bc}
$$

$$
= \partial_c \left[ \frac{1}{2} \varepsilon \eta^{ae} \left( \frac{\partial h_{bd}}{\partial x^e} + \frac{\partial h_{de}}{\partial x^b} - \frac{\partial h_{eb}}{\partial x^d} \right) \right]
$$

$$
- \partial_d \left[ \frac{1}{2} \varepsilon \eta^{af} \left( \frac{\partial h_{bc}}{\partial x^f} + \frac{\partial h_{cf}}{\partial x^b} - \frac{\partial h_{fb}}{\partial x^c} \right) \right]
$$

$$
= \frac{1}{2} \varepsilon \left( \eta^{ae} \frac{\partial^2 h_{bd}}{\partial x^c \partial x^e} + \eta^{ae} \frac{\partial^2 h_{de}}{\partial x^c \partial x^b} - \eta^{ae} \frac{\partial^2 h_{eb}}{\partial x^c \partial x^d} \right)
$$

$$
- \frac{1}{2} \varepsilon \left( \eta^{af} \frac{\partial^2 h_{bc}}{\partial x^d \partial x^f} + \eta^{af} \frac{\partial^2 h_{cf}}{\partial x^d \partial x^b} - \eta^{af} \frac{\partial^2 h_{fb}}{\partial x^d \partial x^c} \right)
$$

Now the index $e$ in the first expression is a dummy index. Let's relabel it as $d$ in each term

$$
\eta^{ae} \frac{\partial^2 h_{bd}}{\partial x^c \partial x^e} = \eta^{af} \frac{\partial^2 h_{bd}}{\partial x^c \partial x^f}
$$

$$
\eta^{ae} \frac{\partial^2 h_{de}}{\partial x^c \partial x^b} = \eta^{af} \frac{\partial^2 h_{df}}{\partial x^c \partial x^b}
$$

$$
\eta^{ae} \frac{\partial^2 h_{eb}}{\partial x^c \partial x^d} = \eta^{af} \frac{\partial^2 h_{fb}}{\partial x^c \partial x^d}
$$

The last term will cancel, allowing us to write

$$
R^a{}_{bcd} = \frac{1}{2} \varepsilon \eta^{af} \left( \frac{\partial^2 h_{bd}}{\partial x^c \partial x^f} + \frac{\partial^2 h_{df}}{\partial x^c \partial x^b} - \frac{\partial^2 h_{bc}}{\partial x^d \partial x^f} - \frac{\partial^2 h_{cf}}{\partial x^d \partial x^b} \right) \qquad (13.4)
$$

Now the Ricci tensor is found using $R_{ab} = R^c{}_{acb}$. Using $R^a{}_{bcd} = \partial_c \Gamma^a{}_{bd} - \partial_d \Gamma^a{}_{bc}$, we have $R^a{}_b = \partial_c \Gamma^c{}_{bd} - \partial_d \Gamma^c{}_{bc}$. If we define the d'Alembertian operator

$$
W = \frac{\partial^2}{\partial t^2} - \left( \frac{\partial^2}{\partial x^2} + \frac{\partial^2}{\partial y^2} + \frac{\partial^2}{\partial z^2} \right) = \eta^{ab} \partial_a \partial_b
$$

along with $h = \eta^{cd} h_{cd}$, then the Ricci tensor can be written as

$$R_{ab} = \frac{1}{2}\varepsilon \left( \frac{\partial^2 h^c{}_a}{\partial x^b \partial x^c} + \frac{\partial^2 h^c{}_b}{\partial x^a \partial x^c} - W h_{ab} - \frac{\partial^2 h}{\partial x^a \partial x^b} \right) \tag{13.5}$$

The Ricci scalar is given by

$$R = \varepsilon \left( \frac{\partial^2 h^{cd}}{\partial x^c \, \partial x^d} - W h \right) \tag{13.6}$$

We can put these results together to write down the Einstein tensor for the linearized theory. To simplify notation, we will represent partial derivatives using commas; i.e.,

$$\frac{\partial^2 h^{cd}}{\partial x^c \, \partial x^d} = h^{cd}{}_{,cd} \quad \text{and} \quad \frac{\partial^2 h}{\partial x^a \, \partial x^b} = h_{,ab}$$

Using this notation, the Einstein tensor is

$$G_{ab} = \frac{1}{2}\varepsilon \left( h^c{}_{a,bc} + h^c{}_{b,ac} - W h_{ab} - h_{,ab} - \eta_{ab} h^{cd}{}_{,cd} + \eta_{ab} W h \right) \tag{13.7}$$

# Traveling Wave Solutions

We now define the following function, known as the *trace reverse* in terms of $h_{ab}$:

$$\psi_{ab} = h_{ab} - \frac{1}{2}\eta_{ab} h \tag{13.8}$$

The trace of this function is

$$\psi_{ab} = \eta_{ac} \psi^c{}_b \quad \text{and} \quad h_{ab} = \eta_{ac} h^c{}_b$$

Now note that

$$\eta_{ab} h = \eta_{ac} \delta^c_b h$$

Putting these results together, we obtain

$$\eta_{ac}\psi^c{}_b = \eta_{ac}h^c{}_b - \frac{1}{2}\eta_{ac}\delta^c_b h$$

$$\Rightarrow \psi^c{}_b = h^c{}_b - \frac{1}{2}\delta^c_b h$$

Now we take the trace by setting $b \to c$:

$$\psi^c{}_c = h^c{}_c - \frac{1}{2}\delta^c_c h$$

We are working in four dimensions. Therefore the trace of the Kronecker delta is $\delta^c_c = 4$. Setting $\psi = \psi^c{}_c$ and $h = h^c{}_c$, we find the trace to be

$$\psi = h - \frac{1}{2}\text{Tr}\left(\delta^c_c\right)h = h - \frac{1}{2}(4)h = h - 2h = -h \qquad (13.9)$$

This is why $\psi_{ab}$ is known as the trace reverse of $h_{ab}$. Now let's replace $h_{ab}$ by the trace reverse in Einstein's equation. Using (13.7) together with (13.8) and (13.9), we have

$$G_{ab} = \frac{1}{2}\varepsilon\left(h^c{}_{a,bc} + h^c{}_{b,ac} - Wh_{ab} - h_{,ab} - \eta_{ab}h^{cd}{}_{,cd} + \eta_{ab}Wh\right)$$

$$= \frac{1}{2}\varepsilon\left[\psi^c{}_{a,bc} + \frac{1}{2}\delta^c_a h_{,bc} + \psi^c{}_{b,ac} + \frac{1}{2}\delta^c_b h_{,ac} - W\left(\psi_{ab} + \frac{1}{2}\eta_{ab}h\right) - h_{,ab}\right.$$

$$\left. - \eta_{ab}\psi^{cd}{}_{,cd} - \frac{1}{2}\eta_{ab}\eta^{cd}h_{,cd} + \eta_{ab}Wh\right]$$

Let's rearrange the terms in this expression

$$G_{ab} = \frac{1}{2}\varepsilon\left[\psi^c{}_{a,bc} + \psi^c{}_{b,ac} - W\psi_{ab} - \eta_{ab}\psi^{cd}{}_{,cd} + \frac{1}{2}\delta^c_a h_{,bc} + \frac{1}{2}\delta^c_b h_{,ac}\right.$$

$$\left. - W\left(\frac{1}{2}\eta_{ab}h\right) - h_{,ab} - \frac{1}{2}\eta_{ab}\eta^{cd}h_{,cd} + \eta_{ab}Wh\right]$$

Notice that

$$\frac{1}{2}\delta_a^c h_{,bc} = \frac{1}{2}h_{,ba}$$

$$\frac{1}{2}\delta_b^c h_{,ac} = \frac{1}{2}h_{,ab}$$

Now partial derivatives commute, so $\frac{1}{2}h_{,ba} = \frac{1}{2}h_{,ab}$ and

$$\frac{1}{2}\delta_a^c h_{,bc} + \frac{1}{2}\delta_b^c h_{,ac} = h_{,ab}$$

Inserting this into the Einstein tensor, we can cancel a term reducing it to

$$G_{ab} = \frac{1}{2}\varepsilon\left[\psi^c{}_{a,bc} + \psi^c{}_{b,ac} - W\psi_{ab} - \eta_{ab}\psi^{cd}{}_{,cd} - W\left(\frac{1}{2}\eta_{ab}h\right)\right.$$
$$\left. - \frac{1}{2}\eta_{ab}\eta^{cd}h_{,cd} + \eta_{ab}Wh\right]$$
$$= \frac{1}{2}\varepsilon\left[\psi^c{}_{a,bc} + \psi^c{}_{b,ac} - W\psi_{ab} - \eta_{ab}\psi^{cd}{}_{,cd} + \frac{1}{2}\eta_{ab}Wh - \frac{1}{2}\eta_{ab}\eta^{cd}h_{,cd}\right]$$

Now $\eta_{ab}Wh = \eta_{ab}\eta^{cd}h_{,cd}$. This cancels the last two terms and so using the trace reverse the Einstein tensor becomes

$$G_{ab} = \frac{1}{2}\varepsilon\left(\psi^c{}_{a,bc} + \psi^c{}_{b,ac} - W\psi_{ab} - \eta_{ab}\psi^{cd}{}_{,cd}\right) \tag{13.10}$$

To obtain the wave equations, we perform a *gauge transformation*. This is a coordinate transformation that leaves $R^a{}_{bcd}$, $R_{ab}$, and $R$ unchanged if we consider terms only to first order in $\varepsilon$. The coordinate transformation that will do this is

$$x^a \rightarrow x^{a'} = x^a + \varepsilon\phi^a \tag{13.11}$$

where $\phi^a$ is a function of position and $|\phi^a{}_{,b}| \ll 1$. It can be shown that this coordinate transformation changes $h_{ab}$ as

$$h'_{ab} = h_{ab} - \phi_{a,b} - \phi_{b,a}$$

(prime does not represent differentiation, this is just a new label). Furthermore, we find that the derivative of $\psi_{ab}$ changes as $\psi_{b,a}^{\prime a} = \psi^a{}_{b,a} - \Box\phi$. We are free to choose $\phi$ as long as the Riemann tensor keeps the same form. Therefore we demand that $W\phi = \psi^b{}_{a,b}$, which leads to $\psi_{b,a}^{\prime a} = 0$. When we substitute this into the Einstein tensor, we can write the full field equations as

$$\frac{1}{2}\varepsilon W\psi_{ab} = -\kappa T_{ab} \qquad (13.12)$$

In vacuum, we obtain

$$W\psi_{ab} = 0 \qquad (13.13)$$

Recalling the definition of the d'Alembertian operator, this is nothing more than the wave equation for waves traveling at $c$. This choice of gauge using the coordinate transformation goes under several names. Two frequently used names are the de Donder or Einstein gauge. We can write (13.13) in terms of (13.8), which gives

$$W\psi_{ab} = W\left(h_{ab} - \frac{1}{2}\eta_{ab}h\right) = Wh_{ab} - \frac{1}{2}\eta_{ab}Wh = 0$$

However, recalling (13.9) we can multiply (13.13) by $\eta^{ab}$ to obtain

$$0 = W\psi_{ab} = \eta^{ab}W\psi_{ab} = W\left(\eta^{ab}\psi_{ab}\right) = W\left(\psi^b{}_b\right) = W\psi = -Wh$$

So we can drop $Wh$ in the expansion of $W\psi_{ab}$, and the study of gravitational waves reduces to a study of the equation

$$Wh_{ab} = 0 \qquad (13.14)$$

# The Canonical Form and Plane Waves

A *plane wave* is characterized by uniform fields that are perpendicular to the direction of propagation. More specifically, if the wave is traveling in the $z$ direction, then the condition of uniformity means that the fields have no $x$ and $y$ dependence.

Another way to think of plane waves that are familiar from electrodynamics is as follows: Plane waves are typically visualized as a wave whose surfaces of constant phase are infinite planes that are perpendicular to the direction of

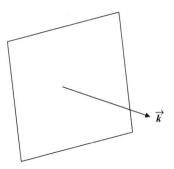

**Fig. 13-1.** A schematic representation of a plane wave. The wave vector $\vec{k}$ gives the direction of propagation of the wave. It is perpendicular to the wavefront, which is a plane and is a surface of constant phase.

propagation. If we define a vector $\vec{k}$ that gives the direction of propagation of the wave, then the wavefront is defined by the equation $k \cdot r = \text{const}$ (see Fig. 13-1). We call $\vec{k}$ the *wave vector*.

We will now consider plane gravitational waves traveling in the $z$ direction. In this case $h_{ab} = h_{ab}(t - z)$ and the condition that the field does not depend on $x$ and $y$ means that

$$\frac{\partial h_{ab}}{\partial x} = \frac{\partial h_{ab}}{\partial y} = 0$$

We will find that the Riemann tensor is greatly simplified in this case. We won't go into the details (see Adler et al., 1975) but under these conditions the Riemann tensor can be shown to be a function of $h_{xx}$, $h_{xy}$, $h_{yx}$, and $h_{yy}$ alone. The metric perturbation can then be split into two parts $h_{ab} = h_{ab}^{(1)} + h_{ab}^{(2)}$, where

$$h_{ab}^{(1)} = \begin{pmatrix} 0 & 0 & 0 & 0 \\ 0 & h_{xx} & h_{xy} & 0 \\ 0 & h_{yx} & h_{yy} & 0 \\ 0 & 0 & 0 & 0 \end{pmatrix} \quad \text{and} \quad h_{ab}^{(2)} = \begin{pmatrix} h_{tt} & h_{tx} & h_{ty} & h_{tz} \\ h_{xt} & 0 & 0 & h_{xz} \\ h_{yt} & 0 & 0 & h_{yz} \\ h_{zt} & h_{zx} & h_{zy} & h_{zz} \end{pmatrix} \quad (13.15)$$

It can be shown that we can find a coordinate system such that the components in $h_{ab}^{(2)}$ vanish and $h_{ab}^{(1)}$ represent the entire perturbation. First, noting that the Einstein gauge condition requires that $h^a{}_{b,a} - \frac{1}{2}h_{,b} = 0$, we can obtain further simplification of the perturbation components. Applying the gauge condition, we find that (remember, derivatives with respect to $y$ and $z$ vanish)

$$h_{tt,t} - h_{tz,z} - \frac{1}{2}h_{,t} = 0$$

$$h_{zt,t} - h_{zz,z} - \frac{1}{2}h_{,z} = 0$$

$$h_{xt,t} - h_{xz,z} = 0$$

$$h_{yt,t} - h_{yz,z} = 0$$

Now we define the new variable $u = t - z$. Then

$$\frac{\partial h_{ab}}{\partial t} = \frac{\partial h_{ab}}{\partial u}\frac{\partial u}{\partial t} = \frac{\partial h_{ab}}{\partial u} = h'_{ab}$$

$$\frac{\partial h_{ab}}{\partial z} = \frac{\partial h_{ab}}{\partial u}\frac{\partial u}{\partial z} = -\frac{\partial h_{ab}}{\partial u} = -h'_{ab}$$

and so writing the derivatives in terms of the new variable, we have

$$h_{tt}' + h_{tz}' - \frac{1}{2}h' = 0$$

$$h_{zt}' + h_{zz}' + \frac{1}{2}h' = 0$$

$$h_{xt}' + h_{xz}' = 0$$

$$h_{yt}' + h_{yz}' = 0$$

Let's take the last equation. We have

$$h_{yt}' + h_{yz}' = \left(h_{yt} + h_{yz}\right)' = 0$$

This can be true only if $h_{yt} + h_{yz}$ is a constant. We have an additional physical requirement: $h_{ab}$ must vanish at infinity. Therefore we must choose the constant to be zero. We then find that

$$h_{yt} = -h_{yz}$$

In addition, we find that $h_{xt} = -h_{zx}$. We are left with

$$h_{tt} + h_{tz} - \frac{1}{2}h = 0$$

$$h_{zt} + h_{zz} + \frac{1}{2}h = 0$$

Adding these equations give $h_{tz} = -\frac{1}{2}(h_{tt} + h_{zz})$. Now subtract the first equation from the second one to get $h_{zt} + h_{zz} + \frac{1}{2}h - \left(h_{tt} + h_{tz} - \frac{1}{2}h\right) = h - h_{tt} + h_{zz} = 0$. Now, writing out the trace explicitly, we have $h = h_{tt} - h_{xx} - h_{yy} - h_{zz}$. The end result is

$$h - h_{tt} + h_{zz} = h_{tt} - h_{xx} - h_{yy} - h_{zz} - h_{tt} + h_{zz} = -h_{xx} - h_{yy}$$

Since this term vanishes, we conclude that $h_{yy} = -h_{xx}$. The complete metric perturbation has now been simplified to

$$h_{ab} = \begin{pmatrix} h_{tt} & h_{tx} & h_{ty} & -\frac{1}{2}(h_{tt} + h_{zz}) \\ h_{tx} & h_{xx} & h_{xy} & -h_{tx} \\ h_{ty} & h_{xy} & -h_{xx} & -h_{ty} \\ -\frac{1}{2}(h_{tt} + h_{zz}) & -h_{tx} & -h_{ty} & h_{zz} \end{pmatrix} \tag{13.16}$$

We can go further with our choice of gauge so that most of the remaining terms vanish (see Adler et al., 1975, or D'Inverno, 1992, for details). We simply state the end result that we will then study. A coordinate transformation can always be found to put the perturbation into the *canonical form*, which means that we need to consider only $h_{ab}^{(1)}$ in (13.15) with the metric written as in (13.16). That is, we take

$$h_{ab} = \begin{pmatrix} 0 & 0 & 0 & 0 \\ 0 & h_{xx} & h_{xy} & 0 \\ 0 & h_{xy} & -h_{xx} & 0 \\ 0 & 0 & 0 & 0 \end{pmatrix} \tag{13.17}$$

Two polarizations result for gravity waves in the canonical form. In particular, we can take $h_{xx} \neq 0$ and $h_{xy} = 0$, which lead to $+$-*polarization*, or we can set $h_{xx} = 0$ and $h_{xy} \neq 0$, which gives $\times$-polarization. We examine both cases in detail in the next section.

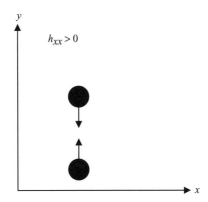

**Fig. 13-2.** When $h_{xx} > 0$, the relative distance of two particles separated along the
$y$-axis decreases.

# The Behavior of Particles as a Gravitational Wave Passes

To study the behavior of massive particles as a gravitational wave passes, we
consider the two cases of polarization which are given by $h_{xx} \neq 0$, $h_{xy} = 0$ and
$h_{xx} = 0$, $h_{xy} \neq 0$. Taking the former case first with $h_{xy} = 0$, we use (13.17)
together with $g_{ab} = \eta_{ab} + \varepsilon h_{ab}$ to write down the line element, which becomes

$$ds^2 = dt^2 - (1 - \varepsilon h_{xx})\, dx^2 - (1 + \varepsilon h_{xx})\, dy^2 - dz^2 \qquad (13.18)$$

As a gravitational wave passes, this metric tells us that the relative distances
between the particles will change. The wave will have oscillatory behavior and
so we need to consider the form of (13.18) as $h_{xx}$ changes from $h_{xx} > 0$ to zero
and then to $h_{xx} < 0$.

For simplicity we imagine particles lying in the $x$–$y$ plane. Furthermore,
suppose that the separation between the particles lies on a line that is parallel
with the $y$-axis. Then $dx$ vanishes and at a fixed time, we can write

$$ds^2 = -(1 + \varepsilon h_{xx})\, dy^2$$

This tells us that when $h_{xx} > 0$ the distance along $y$-axis between the particles
decreases because $ds^2$ becomes more negative. This is illustrated in Fig. 13-2.

On the other hand, when $h_{xx} < 0$, we can see that the relative distance between
the particles will increase. This is shown in Fig. 13-3.

When the separation of the particles is along the $x$-axis, we can see from
the line element that the behavior will be the opposite. In particular, the proper

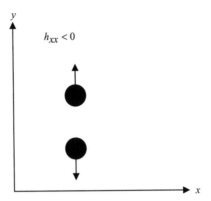

**Fig. 13-3.** When $h_{xx} < 0$, the relative distance between particles separated along the $y$-axis increases.

distance is given by

$$ds^2 = -(1 - \varepsilon h_{xx})\, dx^2$$

First we consider $h_{xx} < 0$. The form of the line element shown here indicates that the relative distance between the two particles will decrease. This behavior is shown in Fig. 13-4.

On the other hand, when $h_{xx} > 0$, the line element becomes more positive and therefore the relative distances between particles will increase. This is shown in Fig. 13-5.

The behavior of the particles discussed in these special cases allows us to extrapolate to a more general situation. It is common to consider a ring of

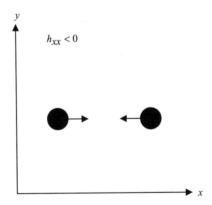

**Fig. 13-4.** Particles separated along the $x$-axis. When $h_{xx} < 0$, the relative physical displacements between the particles decrease.

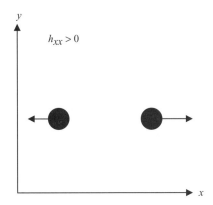

**Fig. 13-5.** Particles separated along the $x$-axis with $h_{xx} > 0$. The relative physical displacements between the particles increase.

particles lying in the plane and show how the ring is distorted by a passing gravitational wave. In particular, imagine that the ring starts off as a perfect circle. As the wave passes, $h_{xx}$ will oscillate between positive, zero, and negative values, causing the relative distances between particles to change in the manner just described. A transverse wave with $h_{xx} \neq 0$ and $h_{xy} = 0$ is referred to as one with +-polarization.

We now examine the other polarization case, by setting $h_{xx} = 0$. This time the line element is given by

$$ds^2 = dt^2 - dx^2 - dy^2 - dz^2 - 2\varepsilon h_{xy}\, dx\, dy \tag{13.19}$$

Consider the following transformation, which can be obtained by a rotation of $\pi/4$ :

$$dx' = \frac{dx - dy}{\sqrt{2}} \quad \text{and} \quad dy' = \frac{dx + dy}{\sqrt{2}}$$

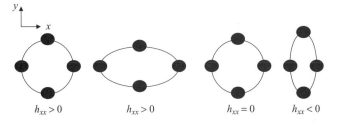

**Fig. 13-6.** The effect of a passing gravity wave with +-polarization on a ring of particles. The ring pulsates as the wave passes.

Writing the line element with these coordinates, we obtain

$$ds^2 = dt^2 - \left(1 - \varepsilon h_{xy}\right) dx'^2 - \left(1 + \varepsilon h_{xy}\right) dy'^2 - dz^2 \qquad (13.20)$$

Now doesn't that look familiar? It looks just like the line element we examined in (13.18). The behavior induced by the wave will be identical to that in the last case; however, this time everything is rotated by $\pi/4$. This polarization is known as $\times$-polarization.

In general, a plane gravitational wave will be a superposition of these two polarizations.

# The Weyl Scalars

In this section we review the Weyl scalars and briefly describe their meaning. They are calculated using the spin coefficients given in (9.15) in combination with a set of equations known as the Newman-Penrose identities. In all, there are five Weyl scalars which have the following interpretations:

$$
\begin{array}{ll}
\Psi_0 & \text{ingoing transverse wave} \\
\Psi_1 & \text{ingoing longitudinal wave} \\
\Psi_2 & \text{electromagnetic radiation} \\
\Psi_3 & \text{outgoing longitudinal wave} \\
\Psi_4 & \text{outgoing transverse wave}
\end{array}
\qquad (13.21)
$$

In most cases of interest we shall be concerned with transverse waves, and therefore with the Weyl scalars $\Psi_0$ and $\Psi_4$. The following Newman-Penrose identities can be used to calculate each of the Weyl scalars:

$$\Psi_0 = D\sigma - \delta\kappa - \sigma\left(\rho + \bar{\rho}\right) - \sigma\left(3\varepsilon - \bar{\varepsilon}\right) + \kappa\left(\pi - \bar{\pi} + \bar{\alpha} + 3\beta\right) \qquad (13.22)$$

$$\Psi_1 = D\beta - \delta\varepsilon - \sigma\left(\alpha + \pi\right) - \beta\left(\bar{\rho} - \bar{\varepsilon}\right) + \kappa\left(\mu + \gamma\right) + \varepsilon\left(\bar{\alpha} - \bar{\pi}\right) \qquad (13.23)$$

$$\Psi_2 = \bar{\delta}\tau - \Delta\rho - \rho\bar{\mu} - \sigma\lambda + \tau\left(\bar{\beta} - \alpha - \bar{\tau}\right)$$
$$\qquad + \rho\left(\gamma + \bar{\gamma}\right) + \kappa\nu - 2\Lambda \qquad (13.24)$$

$$\Psi_3 = \bar{\delta}\gamma - \Delta\alpha + \nu\left(\rho + \varepsilon\right) - \lambda\left(\tau + \beta\right) + \alpha\left(\bar{\gamma} - \bar{\mu}\right) + \gamma\left(\bar{\beta} - \bar{\tau}\right) \qquad (13.25)$$

$$\Psi_4 = \bar{\delta}\nu - \Delta\lambda - \lambda\left(\mu + \bar{\mu}\right) - \lambda\left(3\gamma - \bar{\gamma}\right) + \nu\left(3\alpha + \bar{\beta} + \pi - \bar{\tau}\right) \qquad (13.26)$$

# Review: Petrov Types and the Optical Scalars

It is very useful to study gravitational waves using the formalism introduced in Chapter 9. First we give a quick summary of the Petrov classification in relation to the Weyl scalars discussed in Chapter 9. The Petrov type of a spacetime indicates the number of principal null directions that spacetime has, and how many times each null direction is repeated. We can summarize the Petrov classifications that are primarily of interest in this chapter in relation to the Weyl scalars in the following way:

- *Petrov Type N:* There is a single principal null direction, repeated four times. If $l^a$ is aligned with the principal null direction, then $\Psi_0 = 0$ and $\Psi_4$ is the only nonzero Weyl scalar. If $n^a$ is aligned with the principal null direction, then $\Psi_4 = 0$ and $\Psi_0$ is the only nonzero Weyl scalar.
- *Petrov Type III:* There are two principal null directions, one of multiplicity one and one repeated three times. The nonzero Weyl scalars are $\Psi_3$ and $\Psi_4$.
- *Petrov Type II:* There is one doubly repeated principal null direction and one two distinct null directions. The nonzero Weyl scalars are $\Psi_2$, $\Psi_3$, and $\Psi_4$.
- *Petrov Type D:* There are two principal null directions, each doubly repeated. In this case $\Psi_2$ is the only nonzero Weyl scalar.

In particular, we recall three quantities defined in terms of the spin coefficients given in (9.15). These are the *optical scalars*, which describe the expansion,

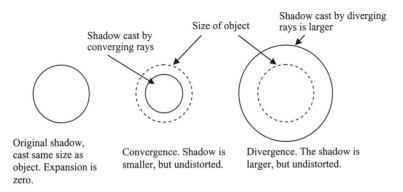

Shadow cast by converging rays

Size of object

Shadow cast by diverging rays is larger

Original shadow, cast same size as object. Expansion is zero.

Convergence. Shadow is smaller, but undistorted.

Divergence. The shadow is larger, but undistorted.

**Fig. 13-7.** An illustration of the expansion which is calculated from $-\mathrm{Re}(\rho)$. The first shadow shows no expansion, while the second illustrates convergence, and last one on the right divergence.

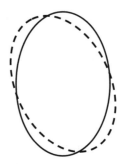

**Fig. 13-8.** The shadow of the object with no twist shown with the solid line. The dashed line shows the effect of twist on the null rays that result in the cast shadow.

twist, and shear of a null congruence:

$$-\text{Re}\,(\rho) \quad \text{expansion}$$
$$\text{Im}\,(\rho) \quad \text{twist} \qquad\qquad (13.27)$$
$$|\sigma| \quad \text{shear}$$

These quantities are interpreted in the following way. For the sake of understanding, we think of the null congruence as a set of light rays. Now imagine that an object is in the path of the light rays and it casts a shadow on a nearby screen. (see Fig. 13-8)

The *expansion* can be understood as seeing the cast shadow either larger or smaller than the object. That is, if the shadow is larger, the light rays are diverging while a smaller shadow indicates that the light rays are converging.

The *twist* is described in this thought experiment by a rotation of the shadow.

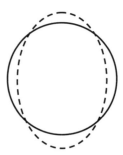

**Fig. 13-9.** Shear distorts the shadow of the object. We show the shadow with zero shear as a circle with the solid line. Shear distorts the circle into the ellipse, shown with the dashed line.

To understand shear, we consider an object that casts a perfect circle as a shadow if no shear is present. If the shear is nonzero, the shadow will be cast as an ellipse (see Fig. 13-9). The spin coefficient $\sigma$ used to define shear is a complex number. The magnitude $|\sigma|$ determines the amount of stretching or shrinking of the axes that define the ellipse while the phase of $\sigma$ defines the orientation of the axes.

# pp Gravity Waves

We now consider a more formal and generalized study of gravitational waves abstracted from sources and propagating at the speed of light. In particular, we study vacuum *pp-waves* where "pp" means *plane fronted* waves with *parallel* rays. Considered to travel in a flat background spacetime described by the Minkowski metric, a pp-wave is one that admits a *covariantly constant* null vector field. We now consider this definition.

Let $k^a$ be a null vector such that the covariant derivative vanishes; i.e.,

$$k_{a,b} = 0 \tag{13.28}$$

We say that a vector $k^a$ that satisfies (13.28) is covariantly constant. Recalling that we can define a null tetrad for a given metric by the vectors $(l^a, n^a, m^a, \overline{m}^a)$, in the case of a pp-wave spacetime it is possible to take $l^a$ as the covariantly constant null vector. This null vector field is taken to correspond to the rays of gravitational waves.

In Chapter 9 we learned that the covariant derivative of $l^a$ shows up in the definitions of several spin coefficients. Specifically, we have

$$\rho = l_{a,b}m^a\overline{m}^b \quad \text{and} \quad \sigma = l_{a,b}m^a m^b \tag{13.29}$$

This tells us that if $l^a$ is covariantly constant, then the gravitational wave of a pp-wave spacetime

- has no expansion or twist, and
- the shear vanishes.

Physically, the fact that null congruence has zero expansion means that the wave surfaces are planes. Furthermore, noting that the spin coefficient $\tau$ is given by

$$\tau = l_{a,b}m^a n^b \tag{13.30}$$

we see that in this case we also have $\tau = 0$. This implies that the null rays are parallel.

*Plane waves* are only a special class of the more general pp-waves. However, for simplicity we will focus on plane waves for the time being. First we will introduce two null coordinates $U$ and $V$, which are defined as follows:

$$U = t - z \quad \text{and} \quad V = t + z \tag{13.31}$$

Inverting these relations, we have

$$t = \frac{U + V}{2} \quad \text{and} \quad z = \frac{-(U - V)}{2}$$

Therefore we have $dt = \frac{1}{2}(dU + dV)$ and $dz = -\frac{1}{2}(dU - dV)$. Squaring, we find

$$dt^2 = \frac{1}{4}\left(dU^2 + 2\,dU\,dV + dV^2\right) \tag{13.32}$$

$$dz^2 = \frac{1}{4}\left(dU^2 - 2dU\,dV + dV^2\right) \tag{13.33}$$

We will now proceed to write the plane wave metric given by (13.18) in terms of $U$ and $V$ to find

$$ds^2 = dt^2 - (1 - \varepsilon h_{xx})\,dx^2 - (1 + \varepsilon h_{xx})\,dy^2 - dz^2$$

$$= \frac{1}{4}\left(dU^2 + 2dU\,dV + dV^2\right) - (1 - \varepsilon h_{xx})\,dx^2$$

$$\quad - (1 + \varepsilon h_{xx})\,dy^2 - \frac{1}{4}\left(dU^2 - 2dU\,dV + dV^2\right)$$

$$= dU\,dV - (1 - \varepsilon h_{xx})\,dx^2 - (1 + \varepsilon h_{xx})\,dy^2$$

Now, recalling that $h_{xx} = h_{xx}(t - z) = h_{xx}(U)$, we define

$$a^2(U) = 1 - \varepsilon h_{xx} \quad \text{and} \quad b^2(U) = 1 + \varepsilon h_{xx}$$

and we obtain the Rosen line element

$$ds^2 = dU\,dV - a^2(U)\,dx^2 - b^2(U)\,dy^2 \tag{13.34}$$

We choose the following basis of one forms for this metric:

$$\omega^{\hat{0}} = \frac{1}{2}(\mathrm{d}U + \mathrm{d}V), \quad \omega^{\hat{1}} = \frac{1}{2}(\mathrm{d}U - \mathrm{d}V), \quad \omega^{\hat{2}} = a(u)\,\mathrm{d}x, \quad \omega^{\hat{3}} = b(u)\,\mathrm{d}y$$

$$(13.35)$$

with

$$\eta_{\hat{a}\hat{b}} = \begin{pmatrix} 1 & 0 & 0 & 0 \\ 0 & -1 & 0 & 0 \\ 0 & 0 & -1 & 0 \\ 0 & 0 & 0 & -1 \end{pmatrix}$$

An exercise (see Quiz) shows that the nonzero components of the Ricci tensor in this basis are

$$R_{\hat{0}\hat{0}} = R_{\hat{0}\hat{1}} = R_{\hat{1}\hat{0}} = R_{\hat{1}\hat{1}} = -\left(\frac{1}{a}\frac{\mathrm{d}^2 a}{\mathrm{d}U^2} + \frac{1}{b}\frac{\mathrm{d}^2 b}{\mathrm{d}U^2}\right)$$

The vacuum equations $R_{\hat{a}\hat{b}} = 0$ give

$$\frac{1}{a}\frac{\mathrm{d}^2 a}{\mathrm{d}U^2} + \frac{1}{b}\frac{\mathrm{d}^2 b}{\mathrm{d}U^2} = 0$$

Letting $h(U) = \frac{1}{a}\frac{\mathrm{d}^2 a}{\mathrm{d}U^2}$, we see that this equation implies that $\frac{1}{b}\frac{\mathrm{d}^2 b}{\mathrm{d}U^2} = -h(U)$ and so we can write the metric in terms of the single function $h$. Therefore (13.34) becomes

$$\mathrm{d}s^2 = \mathrm{d}U\,\mathrm{d}V - h^2(U)\,\mathrm{d}x^2 - h^2(U)\,\mathrm{d}y^2$$

We now apply the following coordinate transformation. Let

$$u = U, \qquad v = V + x^2 aa' + y^2 bb', \qquad X = ax, \qquad Y = by$$

From the first two equations, we obtain

$$\mathrm{d}u = \mathrm{d}U$$

and

$$V = v - x^2 aa' - y^2 bb'$$

$$= v - \frac{a'}{a}X^2 - \frac{b'}{b}Y^2$$

Inverting the last two equations for $x$ and $y$ gives

$$x = \frac{1}{a}X, \quad y = \frac{1}{b}Y$$

$$\Rightarrow dx = \frac{1}{a}dX - \frac{a'}{a^2}X\,du$$

Now we use these relations along with the definition $h(u) = a''/a = -b''/b$ and obtain

$$dV = dv - 2\frac{a'}{a}X\,dX - \frac{a''}{a}X^2\,du + \frac{(a')^2}{a^2}X^2\,du - 2\frac{b'}{b}Y\,dY - \frac{b''}{b}Y^2\,du$$

$$+ \frac{(b')^2}{b^2}Y^2\,du$$

$$= dv - 2\frac{a'}{a}X\,dX - h(u)X^2\,du + \frac{(a')^2}{a^2}X^2\,du - 2\frac{b'}{b}Y\,dY + h(u)Y^2\,du$$

$$+ \frac{(b')^2}{b^2}Y^2\,du$$

We also have

$$a^2 dx^2 = a^2\left(-\frac{a'}{a^2}X\,du + \frac{1}{a}dX\right)^2 = \frac{(a')^2}{a^2}X^2\,du^2 - 2\frac{a'}{a}du\,dX + dX^2$$

$$b^2 dy^2 = b^2\left(-\frac{b'}{b^2}X\,du + \frac{1}{b}dX\right)^2 = \frac{(b')^2}{b^2}Y^2\,du^2 - 2\frac{b'}{b}du\,dY + dY^2$$

Substitution of these results into $ds^2 = dU\,dV - a^2(U)\,dx^2 - b^2(U)\,dy^2$ gives the *Brinkmann metric*

$$ds^2 = h(u)\left(Y^2 - X^2\right)du^2 + 2\,du\,dv - dX^2 - dY^2 \tag{13.36}$$

From here on we will drop the uppercase labels and let $X \to x$ and $Y \to y$ (just be aware of the relationship of these variables to the coordinates used in the original form of the plane wave metric given in (13.18)). More generally, if we define an arbitrary coefficient function $H(u, x, y)$, we can write this metric as

$$ds^2 = H(u, x, y)\,du^2 + 2\,du\,dv - dx^2 - dy^2 \tag{13.37}$$

This metric represents a pp-wave spacetime; however, it does not necessarily represent plane waves, which are a special case. We will describe the form of $H$ in the case of plane waves below.

In Example 9-5, we found the Weyl and Ricci scalars for this metric. In that problem we choose $l_a = (1, 0, 0, 0)$ as the covariantly constant null vector where the coordinates are given by $(u, v, x, y)$. The principal null direction is along $l_a$, which means that this vector defines the direction along which the rays of the gravity wave are coincident. The coordinate $v$ is another null coordinate while the coordinates $x, y$ define the surface of the plane wave.

It is obviously true that $l_{a,b} = 0$. To understand the implication of this requirement, we state some of the spin coefficients here:

$$\rho = l_{a,b} m^a \overline{m}^b, \qquad \sigma = l_{a,b} m^a m^b, \qquad \tau = l_{a,b} m^a n^b$$

We see that $l_{a,b} = 0$ implies that $\rho = \sigma = \tau = 0$. Therefore, as we explained in the beginning of this section, a pp-wave has no expansion, twist, or shear. In addition, the fact that $\tau = 0$ tells us that the null rays defined by $l_a$ are parallel. This shows that this metric is a pp-wave spacetime.

The form of the function $H$ in the Brinkmann metric can be studied further. A *generalized* pp-wave is one for which $H$ can be written in the form

$$H = Ax^2 + By^2 + Cxy + Dx + Ey + F$$

where $A, B, C, D, E$, and $F$ are real valued functions of $u$.

# Plane Waves

A pp-wave is a plane wave if we can write $H$ in the form

$$H(u, x, y) = a(u)(x^2 - y^2) + 2b(u)xy + c(u)(x^2 + y^2)$$
$$\text{(general plane wave)} \qquad\qquad (13.38)$$

(compare with our derivation of (13.36)). The functions $a$ and $b$ describe the polarization states of the gravitational wave, while $c$ represents waves of other types of radiation. To represent a plane gravitational wave in the vacuum, we let $c$ vanish; that is, $H(u, x, y)$ becomes

$$H(u, x, y) = a(u)(x^2 - y^2) + 2b(u)xy \quad \text{(plane wave in vacuum)}$$
$$(13.39)$$

Since these are plane waves, we expect that we can write the wave as an expression of the form $A e^{i\alpha}$, where $A$ is the amplitude of the wave. In fact we can, and it turns out that the wave is described by the Weyl scalar $\Psi_4$. That is, we write the gravity wave as

$$\Psi_4 = A e^{i\alpha}$$

In this case $\alpha$ represents the polarization of the wave. A linearly polarized wave is one for which $\alpha$ is a constant.

In Example 9-5, we found that

$$\Psi_4 = \frac{1}{4}\left(\frac{\partial^2 H}{\partial x^2} - \frac{\partial^2 H}{\partial y^2} + 2i\frac{\partial^2 H}{\partial x\,\partial y}\right)$$

Let's consider this with the form of $H$ given for a plane wave in vacuum (13.39). We write the metric as

$$ds^2 = \left(h_{11}x^2 + h_{22}y^2 + 2h_{12}xy\right)du^2 + 2\,du\,dv - dx^2 - dy^2$$

The form of the Weyl scalar for this metric will lead to a plane wave solution. We will also demonstrate that the vacuum equation using the Ricci scalar we found in Example 9-5 leads to the relations among $h_{11}$, $h_{22}$, and $h_{12}$ for plane waves that we found in the previous section.

To begin let's write down the Weyl scalar when $H = h_{11}x^2 + h_{22}y^2 + 2h_{12}xy$. We have

$$\begin{aligned}
\Psi_4 &= \frac{1}{4}\left(\frac{\partial^2 H}{\partial x^2} - \frac{\partial^2 H}{\partial y^2} + 2i\frac{\partial^2 H}{\partial x\partial y}\right) \\
&= \frac{1}{4}(2h_{11} - 2h_{22} + 4ih_{12}) \\
&= \frac{1}{2}(h_{11} - h_{22} + 2ih_{12})
\end{aligned}$$

In Chapter 9 we learned that the only nonzero component of the Ricci tensor was

$$\Phi_{22} = \frac{1}{4}\left(\frac{\partial^2 H}{\partial x^2} + \frac{\partial^2 H}{\partial y^2}\right)$$

In this case taking $H = h_{11}x^2 + h_{22}y^2 + 2h_{12}xy$, we have

$$\Phi_{22} = \frac{1}{2}(h_{11} + h_{22})$$

Now consider the vacuum field equations, which are defined by $R_{ab} = 0$ or $\Phi_{22} = 0$. This leads us to conclude that $h_{22} = -h_{11}$. The curvature tensor then becomes

$$\Psi_4 = \frac{1}{2}(2h_{11} + 2ih_{12}) = h_{11} + ih_{12}$$

Now let's consider the case of a plane wave with linear polarization, which means that $h_{12}$ is proportional to $h_{11}$. In that case we expect that we can write $\Psi_4 = h(u)e^{i\alpha}$. Using Euler's formula to expand $\Psi_4 = h(u)e^{i\alpha}$, we have

$$\Psi_4 = h(u)e^{i\alpha} = h(u)(\cos\alpha + i\sin\alpha) = h(u)\cos\alpha + ih(u)\sin\alpha$$

Comparison with $\Psi_4 = h_{11} + ih_{12}$ tells us that we can write $h_{11} = h(u)\cos\alpha$ and $h_{12} = h(u)\sin\alpha$. The meaning of $\alpha$ is taken to be that the polarization vector of the wave is at an angle $\alpha$ with the $x$-axis.

Summarizing, in the case of constant linear polarization in the vacuum we can write $H$ as

$$\begin{aligned}
H &= h_{11}x^2 + h_{22}y^2 + 2h_{12}xy \\
&= h(u)\cos\alpha x^2 - h(u)\cos\alpha y^2 + 2\sin\alpha xy \qquad (13.40) \\
&= h(u)\left[\cos\alpha(x^2 - y^2) + 2\sin\alpha xy\right]
\end{aligned}$$

# The Aichelburg-Sexl Solution

An interesting solution involving a black hole passing nearby was studied by Aichelburg and Sexl. The metric is given by

$$ds^2 = 4\mu\log(x^2 + y^2)du^2 + 2dudr - dx^2 - dy^2 \qquad (13.41)$$

This metric is clearly in the form of the Brinkmann metric and so represents a pp-wave spacetime. However, in this case $H = 4\mu\log(x^2 + y^2)$, which is not in the form given by (13.38); therefore, it does not represent plane waves.

# Colliding Gravity Waves

In this section we consider the collision of two gravitational waves. Such collisions lead to many interesting effects such as the introduction of nonzero expansion and shear. To begin the study of this phenomenon, we consider the simplest case possible, the collision of two impulsive plane gravitational waves.

An impulsive wave is a *shock wave* where the propagating disturbance is described by a *Dirac delta* function. That is, we imagine a highly localized disturbance propagating along the $z$ direction. Since $u = t - z$, we can create an idealized model of such a wave by taking $h(u) = \delta(u)$. This type of gravitational wave can be described with the metric

$$ds^2 = \delta(u)\left(Y^2 - X^2\right) du^2 + 2\, du\, dr - dX^2 - dY^2 \tag{13.42}$$

In this metric, $r$ is a spacelike coordinate. We can use a coordinate transformation (see Problem 7) to write this line element in terms of the null coordinates $u$ and $v$, which gives

$$ds^2 = 2\, du\, dv - [1 - u\Theta(u)]^2\, dx^2 - [1 + u\Theta(u)]^2\, dy^2 \tag{13.43}$$

where $\Theta(u)$ is the Heaviside step function. This function is defined by

$$\Theta(u) = \begin{cases} 0 \text{ for } & u < 0 \\ 1 \text{ for } & u \geq 0 \end{cases}$$

Therefore, in the region $u \geq 0$, we can write the line element as

$$ds^2 = 2\, du\, dv - (1 - u)^2\, dx^2 - (1 + u)^2\, dy^2$$

As an aside, note that the derivative of the Heaviside step function is the Dirac delta; i.e.,

$$\frac{d\Theta}{du} = \delta(u) \tag{13.44}$$

By considering the other null coordinate $v$, we can describe an opposing wave using

$$ds^2 = 2\, du\, dv - [1 - v\Theta(v)]^2\, dx^2 - [1 + v\Theta(v)]^2\, dy^2 \tag{13.45}$$

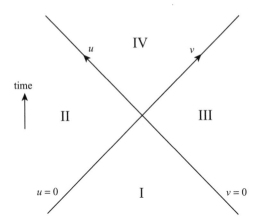

**Fig. 13-10.** Dividing spacetime into four regions to study colliding waves. [*Courtesy of J.B. Griffiths (1991).*]

and so for $v \geq 0$, (13.45) assumes the form

$$ds^2 = 2\,du\,dv - (1 - v)^2\,dx^2 - (1 + v)^2\,dy^2$$

Note that in either (13.43) or (13.45) if we set the step function $\Theta = 0$, we obtain the flat space line element

$$ds^2 = 2du\,dv - dx^2 - dy^2 \qquad (13.46)$$

This line element holds in the region $u, v < 0$.

With these observations in place, we see that we can divide spacetime into four regions, as shown in Fig. 13-10. Region I, where both $u, v < 0$, is the flat background spacetime described by (13.46). In Region II, $v < 0$ and $u \geq 0$, and so this region contains the approaching wave characterized by $\delta(u)$. It is described by the line element (13.43). An analogous result holds for Region III, which contains the approaching wave $\delta(v)$ and is described by the line element (13.46). Finally, Region IV, where both $u \geq 0$ and $v \geq 0$, is where the collision of the two waves takes place.

As an exercise in working with gravitational wave spacetimes using the Newman-Penrose formalism, we begin by considering a simple metric. We consider the case of the metric in Region III.

**EXAMPLE 13-1**
Show that the metric in Region III describes an impulsive gravitational wave characterized by $\delta(v)$. Find the nonzero Weyl scalars and determine the Petrov type.

**SOLUTION 13-1**

First let's write down some preliminaries we will need in calculations. The line element is given by

$$ds^2 = 2\,du\,dv - (1 - v\Theta(v))^2\,dx^2 - (1 + v\Theta(v))^2\,dy^2$$

The nonzero components of the metric tensor are

$$g^{uv} = g^{vu} = 1$$

$$g^{xx} = -\frac{1}{(1 - v\Theta(v))^2}, \quad g^{yy} = -\frac{1}{(1 + v\Theta(v))^2} \tag{13.47}$$

$$g_{uv} = g_{vu} = 1$$

$$g_{xx} = -(1 - v\Theta(v))^2, \quad g_{yy} = -(1 + v\Theta(v))^2$$

We can calculate the nonzero Christoffel symbols for this metric using $\Gamma^a{}_{bc} = \frac{1}{2}g^{ad}\left(\partial_b g_{dc} + \partial_c g_{db} - \partial_d g_{bc}\right)$. For example,

$$\Gamma^u{}_{xx} = \frac{1}{2}g^{ud}\left(\partial_x g_{dx} + \partial_x g_{dx} - \partial_d g_{xx}\right)$$

$$= \frac{1}{2}\partial_v g_{xx}$$

$$= \frac{1}{2}\frac{d}{dv}(1 - v\Theta(v))^2$$

$$= (1 - v\Theta(v))\frac{d}{dv}(-v\Theta(v))$$

$$= (1 - v\Theta(v))\left(-\Theta(v) - v\frac{d\Theta}{dv}\right)$$

$$= -(1 - v\Theta(v))(\Theta(v) + v\delta(v))$$

$$= -(1 - v\Theta(v))\Theta(v)$$

To move to the last line, we used the fact that $f(v)\,\delta(v) = f(0)$, so $v\delta(v) = 0$. It is a simple exercise to find all of the nonzero Christoffel symbols using equation (4.16), which turn out to be

$$\Gamma^u{}_{xx} = -(1 - v\Theta(v))\,\Theta(v), \quad \Gamma^u{}_{yy} = (1 + v\Theta(v))\,\Theta(v)$$

$$\Gamma^x{}_{xv} = -\frac{\Theta(v)}{1 - v\Theta(v)}, \quad \Gamma^y{}_{yv} = \frac{\Theta(v)}{1 + v\Theta(v)} \tag{13.48}$$

The wave is propagating along $v$, and so choosing $l_a$ to be along this direction we set $l_a = (0, 1, 0, 0)$. We can find the other basis vectors in the tetrad using

$$g_{ab} = l_a n_b + l_b n_a - m_a \overline{m}_b - m_b \overline{m}_a \qquad (13.49)$$

The vector $m_a$ is spacelike, and so considering $g_{uv}$ we can write

$$g_{uv} = l_u n_v + l_v n_u = l_v n_u = n_u$$
$$\Rightarrow n_u = 1$$

Since $n_u$ is null, we can take this to be the only nonzero component and write $n_a = (0, 1, 0, 0)$. Moving to the spacelike vector, we take $m_x$ to be real and so using (13.49) we have

$$g_{xx} = l_x n_x + l_x n_x - m_x \overline{m}_x - m_x \overline{m}_x$$
$$= -2m_x^2$$

Using $g_{xx} = -(1 - v\Theta(v))^2$ this gives

$$m_x = \frac{1 - v\Theta(v)}{\sqrt{2}} = \overline{m}_x$$

A similar exercise, taking $m_y$ to be complex, leads us to choose

$$m_y = i\left(\frac{1 + v\Theta(v)}{\sqrt{2}}\right) \quad \text{and} \quad \overline{m}_y = -i\left(\frac{1 + v\Theta(v)}{\sqrt{2}}\right)$$

Summarizing, for this metric we choose the following null tetrad:

$$l_a = (0, 1, 0, 0), \quad n_a = (1, 0, 0, 0)$$
$$m_a = \left(0, 0, \frac{1 - v\Theta(v)}{\sqrt{2}}, i\frac{1 + v\Theta(v)}{\sqrt{2}}\right),$$
$$\overline{m}_a = \left(0, 0, \frac{1 - v\Theta(v)}{\sqrt{2}}, -i\frac{1 + v\Theta(v)}{\sqrt{2}}\right) \qquad (13.50)$$

Raising indices with the metric, we find

$$l^a = (1, 0, 0, 0), \quad n^a = (0, 1, 0, 0)$$

$$m^a = \left( 0, 0, \frac{1}{\sqrt{2}\,(1 - v\Theta(v))}, -i\,\frac{1}{\sqrt{2}\,(1 + v\Theta(v))} \right) \tag{13.51}$$

$$\overline{m}^a = \left( 0, 0, \frac{1}{\sqrt{2}\,(1 - v\Theta(v))}, i\,\frac{1}{\sqrt{2}\,(1 + v\Theta(v))} \right)$$

Now that we have the null tetrad, we can compute the spin coefficients. First noting that the only nonzero component of the first null vector is the constant $l_v = 1$, looking at the Christoffel symbols (13.48) we can see that $l_{a;b} = \partial_b l_a - \Gamma^v{}_{ab} l_v = 0$. This is, of course, what we expect for a pp-wave spacetime. This requirement dictates that several of the spin coefficients will vanish, in particular, $\kappa = \sigma = \rho = v = 0$.

An exercise also shows that $\tau = \pi = \varepsilon = \gamma = \alpha = \beta = 0$. Let's compute the two remaining spin coefficients. Starting with $\lambda = -n_{a;b} \overline{m}^a \overline{m}^b$, looking at (13.51) we observe that only terms involving $\overline{m}^x$ and $\overline{m}^y$ will be nonzero. Expanding the sum with this in mind, we find

$$\lambda = -n_{a;b} \overline{m}^a \overline{m}^b = -\left( n_{x;x} \overline{m}^x \overline{m}^x + n_{x;y} \overline{m}^x \overline{m}^y + n_{y;x} \overline{m}^y \overline{m}^x + n_{y;y} \overline{m}^y \overline{m}^y \right)$$

Recall that the covariant derivative is

$$n_{a;b} = \partial_b n_a - \Gamma^d{}_{ab} n_d$$

The analysis is greatly simplified by the fact that $n_a$ only has a $u$ component that is a constant. So this reduces to $n_{a;b} = -\Gamma^u{}_{ab} n_u = -\Gamma^u{}_{ab}$. Considering the mixed terms first, we find

$$n_{x;y} = -\Gamma^u{}_{xy} = 0$$

$$n_{y;x} = -\Gamma^u{}_{yx} = 0$$

For the other two terms, we have

$$n_{x;x} = -\Gamma^u{}_{xx} = (1 - v\Theta(v))\,\Theta(v)$$

$$n_{y;y} = -\Gamma^u{}_{yy} = -(1 + v\Theta(v))\,\Theta(v)$$

Now we write down the products $\overline{m}^x \overline{m}^x$ and $\overline{m}^y \overline{m}^y$.

$$\overline{m}^x \overline{m}^x = \left( \frac{1}{\sqrt{2}\,(1 - v\Theta(v))} \right) \left( \frac{1}{\sqrt{2}\,(1 - v\Theta(v))} \right) = \frac{1}{2\,(1 - v\Theta(v))^2}$$

$$\overline{m}^y \overline{m}^y = \left( i\,\frac{1}{\sqrt{2}\,(1 + v\Theta(v))} \right) \left( i\,\frac{1}{\sqrt{2}\,(1 + v\Theta(v))} \right) = \frac{-1}{2\,(1 + v\Theta(v))^2}$$

and so the calculation for the spin coefficient $\lambda$ becomes

$$\lambda = - \left( n_{x;x} \overline{m}^x \overline{m}^x + n_{y;y} \overline{m}^y \overline{m}^y \right)$$

$$= - \Bigg[ (1 - v\Theta(v))\,\Theta(v) \left( \frac{1}{2\,(1 - v\Theta(v))^2} \right)$$

$$- (1 + v\Theta(v))\,\Theta(v) \left( \frac{-1}{2\,(1 + v\Theta(v))^2} \right) \Bigg]$$

$$= - \left( \frac{\Theta(v)}{2\,(1 - v\Theta(v))} + \frac{\Theta(v)}{2\,(1 + v\Theta(v))} \right)$$

$$= - \frac{\Theta(v)}{(1 - v\Theta(v))\,(1 + v\Theta(v))}$$

The calculation for the remaining spin coefficient is similar. The only difference this time is that we use terms like $\overline{m}^x m^x$ instead of $\overline{m}^x \overline{m}^x$. We have

$$\mu = - \left( n_{x;x} \overline{m}^x m^x + n_{y;y} \overline{m}^y m^y \right)$$

$$= - \Bigg[ (1 - v\Theta(v))\,\Theta(v) \left( \frac{1}{2\,(1 - v\Theta(v))^2} \right)$$

$$- (1 + v\Theta(v))\,\Theta(v) \left( \frac{-1}{2\,(1 + v\Theta(v))^2} \right) \Bigg]$$

$$= - \left( \frac{\Theta(v)}{2\,(1 - v\Theta(v))} - \frac{\Theta(v)}{2\,(1 + v\Theta(v))} \right)$$

$$= - \frac{v\Theta(v)}{(1 - v\Theta(v))\,(1 + v\Theta(v))}$$

Looking at the expressions for the Weyl scalars, we can see that the only nonzero term we have in this case is going to be $\Psi_4$. Since most of the spin coefficients vanish, the calculation of this term is relatively painless. First we need to compute the directional derivative of $\lambda$. Remember that $\lambda$ is a scalar, and so the directional derivative is just $\Delta\lambda = n^a \nabla_a \lambda = n^a \partial_a \lambda$ because the covariant derivative reduces to an ordinary partial derivative when applied to a scalar. Looking at (13.51), the only nonzero component is the $v$ component. Calculating the derivative, we find

$$\frac{\partial\lambda}{\partial v} = \frac{\partial}{\partial v}\left(-\frac{\Theta(v)}{1 - v^2\Theta(v)}\right)$$

We compute this derivative using $(f/g)' = \frac{f'g - g'f}{g^2}$. We take

$$f = \Theta(v) \Rightarrow f' = \delta(v)$$
$$g = 1 - v^2\Theta(v) \Rightarrow g' = -2v\Theta(v) - v^2\delta(v) = -2v\Theta(v)$$

Noting the overall minus sign, we obtain

$$\partial_v\lambda = -\frac{\delta(v)(1 - v^2\Theta(v)) + (2v\Theta(v))\Theta(v)}{(1 - v^2\Theta(v))^2} = -\frac{\delta(v) + 2v\Theta(v)}{(1 - v^2\Theta(v))^2}$$

Using $f(v)\delta(v) = f(0)\delta(v)$, we can simplify this term because

$$\frac{\delta(v)}{(1 - v^2\Theta(v))^2} = \delta(v)$$

This allows us to write the derivative as

$$\partial_v\lambda = -\delta(v) - \frac{2v\Theta(v)}{(1 - v^2\Theta(v))^2}$$

All together, the Weyl scalar turns out to be

$$\Psi_4 = \bar{\delta}v - \Delta\lambda - \lambda(\mu + \bar{\mu}) - \lambda(3\gamma - \bar{\gamma}) + v(3\alpha + \bar{\beta} + \pi - \bar{\tau})$$
$$= -\Delta\lambda - \lambda(\mu + \bar{\mu})$$
$$= -\Delta\lambda - 2\lambda\mu$$

$$= \delta(v) + \frac{2v\Theta(v)}{(1-v^2\Theta(v))^2} - 2\left(-\frac{\Theta(v)}{(1-v\Theta(v))(1+v\Theta(v))}\right)$$

$$\left(-\frac{v\Theta(v)}{(1-v\Theta(v))(1+v\Theta(v))}\right)$$

$$= \delta(v) + \frac{2v\Theta(v)}{(1-v^2\Theta(v))^2} - \frac{2v\Theta(v)}{(1-v^2\Theta(v))^2}$$

$$= \delta(v)$$

As expected, we obtain a Dirac delta function. Since $\Psi_4$ is the only nonzero Weyl scalar, we conclude that this spacetime is Petrov Type N.

# The Effects of Collision

Referring once again to Fig. 13-10, Regions I, II, and III are flat. In Region IV, which represents the interaction region of the two waves, spacetime is curved. Of specific interest: is the interaction of the two waves causes a focusing effect that does two things. Since focusing means that the wave no longer has zero convergence, the waves are no longer plane waves in Region IV. More interesting is the fact that the focusing in this case results in a singularity described by $u^2 + v^2 = 1$

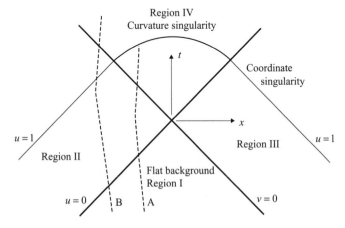

**Fig. 13-11.** The collision of two impulsive plane gravitational waves. Focusing effects induce a curvature singularity which is inevitable for Particle A, which crosses both wavefronts. [*Courtesy of J.B. Griffiths (1991).*]

In Fig. 13-11, the world lines of two particles are portrayed. Noting that the wave approaching from Region II is characterized by the Dirac delta function $\delta(u)$, we note that a particle will cross the wave if it passes the line $u = 0$ and analogously for a particle crossing $v = 0$. Looking at the figure, we see that Particle B crosses the wave approaching from Region II, but it encounters the line $u^2 + v^2 = 1$ before encountering the second wave, which is characterized by $\delta(v)$. Therefore Particle B avoids the curvature singularity in Region IV. For Particle B, the singularity, which comes across in Region II, is just a coordinate singularity.

Particle A, meanwhile, has a different fate. The world line of this particle indicates that it encounters both wavefronts before encountering the singularity. Unfortunately for Particle A it encounters a real curvature singularity in Region IV.

# More General Collisions

We now leave impulsive waves behind and consider more general types of collision. We will again consider the collision of two waves. First we imagine a null congruence in vacuum and review its characteristics. The geodesics of the congruence are parallel and $\rho = \sigma = 0$, meaning that the contraction, twist, and shear vanish for the congruence.

For a more general type of collision than that considered in the last section, we can imagine that the gravity wave encounters either an electromagnetic wave (which has nonzero energy density and therefore is a source of gravitational field) or a collision with a gravitational wave. The interaction can be described by the two Newman-Penrose equations

$$\mathrm{D}\rho = \rho^2 + \sigma\overline{\sigma} + \Phi_{00}$$
$$\mathrm{D}\sigma = \sigma\left(\rho + \overline{\rho}\right) + \Psi_0$$

The terms $\Phi_{00}$ and $\Psi_0$ can represent an opposing electromagnetic and gravitational wave, respectively. Initially, as the wave travels through a region with no other waves present, $\Phi_{00} = 0$ and $\Psi_0 = 0$. Since the wave has zero expansion, shear, and twist, this situation is described by

$$\mathrm{D}\rho = 0$$
$$\mathrm{D}\sigma = 0$$

If the wave encounters an electromagnetic wave, which means that $\Phi_{00} > 0$, then initially

$$D\rho = \Phi_{00}$$

$$D\sigma = 0$$

This causes $\rho$ to increase, causing the congruence to converge since $-\text{Re}\rho$ gives the expansion of the wave. Therefore as $\rho$ gets larger the expansion gets smaller.

On the other hand, if the wave encounters another gravitational wave, then initially

$$D\sigma = \Psi_0$$

$$\Rightarrow \ D\rho = \sigma\overline{\sigma}$$

and so we can see that shear $\sigma$ caused by the collision induces a contraction via the first equation.

In other words, these equations represent the following effects:

- If a congruence passes through a region with nonzero energy density (which means that $\Phi_{00}$ is nonzero), it will *focus*.
- If a gravitational wave collides with another gravitational wave, it will begin to shear. This induces a contraction in the congruence and it will therefore begin to focus. Taking these effects together, we see that the opposing gravity wave causes an *astigmatic* focusing effect.

In the next example, we imagine that a null congruence begins in vacuum. We take the region $v < 0$ to be a flat region of spacetime. Defining a plane wave by $v = \text{const}$, we choose the null vector $l^a$ to point along $v$. The null hypersurface is given by $v = 0$. In the region past $v = 0$, an opposing wave is encountered (see Fig. 13-12). In the next example, we consider the line element in the region where the two waves interact and illustrate that the shear and convergence become nonzero.

**EXAMPLE 13-2**
The region $v > 0$ is described by the line element

$$ds^2 = 2\,du\,dv - \cos^2 av\,dx^2 - \cosh^2 av\,dy^2 \qquad (13.52)$$

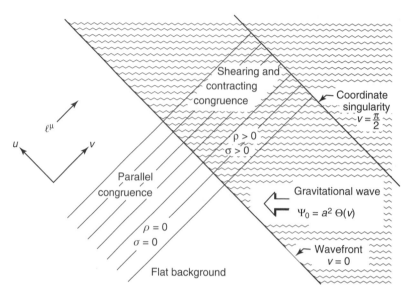

**Fig. 13-12.** An illustration of two colliding gravity waves. In the flat region, the null congruence has parallel rays with no shear or contraction. In the region where the two gravity waves collide, there is shear and congruence. [*Courtesy of J.B. Griffiths (1991).*]

Show that the congruence contracts and shears after passing the gravitational wavefront. Determine the Petrov type and interpret. Describe the focusing effect and determine the alignment of the shear axes.

**SOLUTION 13-2**

The components of the metric tensor are given by

$$g^{uv} = g^{vu} = 1$$

$$g^{xx} = -\frac{1}{\cos^2{(av)}}, \quad g^{yy} = -\frac{1}{\cosh^2{(av)}} \tag{13.53}$$

$$g_{uv} = g_{vu} = 1$$

$$g_{xx} = -\cos^2{(av)}, \quad g_{yy} = -\cosh^2{(av)}$$

Using (4.16), one can show that the nonzero Christoffel symbols are given by

$$\Gamma^u{}_{xx} = -a\cos av \sin av, \quad \Gamma^u{}_{yy} = a\cosh av \sinh av$$

$$\Gamma^x{}_{xv} = -a\tan av, \quad \Gamma^y{}_{yv} = a\tanh av \tag{13.54}$$

We define the null tetrad such that $l^a$ points along $v$, giving

$$l^a = (0, 1, 0, 0), \quad n^a = (1, 0, 0, 0)$$

$$m^a = \left(0, 0, -\frac{1}{\sqrt{2}\cos av}, -\frac{i}{\sqrt{2}\cosh av}\right) \qquad (13.55)$$

Lowering indices with the metric tensor (i.e., $l_a = g_{ab}l^b$, etc.), we find

$$l_a = (1, 0, 0, 0), \quad n_a = (0, 1, 0, 0)$$

$$m_a = \left(0, 0, \frac{\cos av}{\sqrt{2}}, -\frac{i\cosh av}{\sqrt{2}}\right) \qquad (13.56)$$

To show that the congruence contracts and shears, we must show that $\rho$ and $\sigma$ are nonzero. Now $l_{a;b} = \partial_b l_a - \Gamma^c{}_{ab}l_c$. The simplicity of $l_a$ means that this expression will take a very simple form. In fact, since $l_a$ has only a $u$ component which is constant (and therefore $\partial_b l_a = 0$ for all $a,b$), we can write

$$l_{a;b} = -\Gamma^u{}_{ab}l_u = -\Gamma^u{}_{ab}$$

We can calculate the spin coefficient representing contraction using $\rho = l_{a;b}m^a\overline{m}^b$, where $\overline{m}^b$ is the complex conjugate of $m^a$ as given in (13.55). There are only two nonzero terms in the sum. We calculate each of these individually:

$$l_{x;x} = -\Gamma^u{}_{xx} = a\cos av\sin av$$

$$l_{y;y} = -\Gamma^u{}_{yy} = a\cosh av\sinh av$$

and so we have

$$\rho = l_{a;b}m^a\overline{m}^b$$

$$= l_{x;x}m^x\overline{m}^x + l_{y;y}m^y\overline{m}^y$$

$$= (a\cos av\sin av)\left(-\frac{1}{\sqrt{2}\cos av}\right)\left(-\frac{1}{\sqrt{2}\cos av}\right)$$

$$+ (-a\cosh av\sinh av)\left(-\frac{i}{\sqrt{2}\cosh av}\right)\left(\frac{i}{\sqrt{2}\cosh av}\right)$$

$$= \frac{a}{2}\left(\frac{\cos av \sin av}{\cos av \cos av}\right) - \frac{a}{2}\left(\frac{\cosh av \sinh av}{\cosh av \cosh av}\right)$$

$$= \frac{a}{2}\left(\frac{\sin av}{\cos av}\right) - \frac{a}{2}\left(\frac{\sinh av}{\cosh av}\right)$$

$$\Rightarrow \rho = \frac{a}{2}(\tan av - \tanh av)$$

Next we compute the shear. This can be done by computing $\sigma = l_{a;b}m^a m^b$. Again, the only nonzero terms in the sum are those with $l_{x;x}, l_{y;y}$. And so we obtain

$$\sigma = l_{x;x}m^x m^x + l_{y;y}m^y m^y$$

$$= (a \cos av \sin av)\left(-\frac{1}{\sqrt{2}\cos av}\right)\left(-\frac{1}{\sqrt{2}\cos av}\right)$$

$$+ (-a \cosh av \sinh av)\left(-\frac{i}{\sqrt{2}\cosh av}\right)\left(\frac{i}{\sqrt{2}\cosh av}\right)$$

$$= \frac{a}{2}\left(\frac{\cos av \sin av}{\cos av \cos av}\right) + \frac{a}{2}\left(\frac{\cosh av \sinh av}{\cosh av \cosh av}\right)$$

$$\Rightarrow \sigma = \frac{a}{2}(\tan av + \tanh av)$$

An exercise shows that the remaining spin coefficients vanish. Let's make a quick qualitative sketch of the shear.

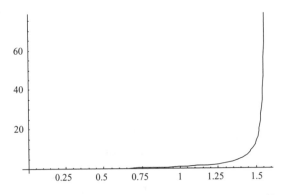

**Fig. 13-13.** The tan function blows up at $av = \pi/2$.

As we can see from $\sigma = \frac{a}{2}(\tan av - \tanh av)$ and by looking at the sketch (see Fig. 13-13), the shear blows up at $av = \pi/2$. In other words, there is a singularity.

Looking at the Newman-Penrose identities, we see that the only nonzero Weyl scalar is given by $\Psi_0$. The only nonzero spin coefficients are $\rho, \sigma$, and so we have

$$\Psi_0 = -D\sigma + 2\sigma\rho \qquad (13.57)$$

where $D$ is the directional derivative along $l^a$. Since $\sigma$ is a scalar, we need to compute only $l^a\partial_a\sigma$. Once again, with only one component this becomes relatively simple. We find

$$Do = l^a\partial_a\sigma = \partial_v\sigma$$

$$= \frac{\partial}{\partial v}\left[\frac{a}{2}(\tan av - \tanh av)\right]$$

$$= \frac{a}{2\cos^2 av}$$

Therefore we find

$$\Psi_0 = -D\sigma + 2\sigma\rho$$

$$= \frac{a^2}{2\cos^2 av} + 2\left[\frac{a}{2}(\tan av + \tanh av)\right]\left[\frac{a}{2}(\tan av - \tanh av)\right]$$

$$= -\frac{a^2}{2\cos^2 av} + \frac{a^2}{2}\left[\tan^2 av - \tanh^2 av\right]$$

$$= -\frac{a^2}{2\cos^2 av} + \frac{a^2\sin^2 av}{2\cos^2 av} - \frac{a^2}{2}\tanh^2 av$$

$$= -\frac{a^2}{2}(1 + \tanh^2 av)$$

We can make the following observations of our calculated results. First the fact that $\Psi_4 = 0$ and $\Psi_0 \neq 0$ tells us that $n^a$ is aligned with the principal null direction. Since all other Weyl scalars vanish, the null direction is repeated four times and therefore the Petrov Type is N.

# Nonzero Cosmological Constant

An ongoing area of research involves the investigation of gravitational radiation with a nonzero cosmological constant. Unfortunately, limited space prevents us from covering this interesting topic in much detail.

Here we will simply examine an example to give the reader a final "how-to" demonstration on using Newman-Penrose methods.

### EXAMPLE 13-3

The *Narai spacetime* is a solution to the vacuum field equations with positive cosmological constant; that is, $R_{ab} = \Lambda g_{ab}$. The line element is given by

$$ds^2 = -\Lambda v^2 \, du^2 + 2 \, du \, dv - \frac{1}{\Omega^2} \left( dx^2 + dy^2 \right)$$

where $\Omega = 1 + \frac{\Lambda}{2}(x^2 + y^2)$ and $\Lambda$ is the cosmological constant that we take to be positive.

### SOLUTION 13-3

The components of the metric tensor are given by

$$g_{uu} = -\Lambda v^2, \quad g_{uv} = g_{vu} = 1, \quad g_{xx} = g_{yy} = -\frac{1}{\Omega^2}$$

$$g^{vv} = -\Lambda v^2, \quad g^{uv} = g^{vu} = 1, \quad g^{xx} = g^{yy} = -\Omega^2$$

The nonzero Christoffel symbols are

$$\Gamma^u{}_{uu} = -\Lambda v, \quad \Gamma^v{}_{uu} = -\Lambda^2 v^3, \quad \Gamma^v{}_{vu} = -\Lambda v$$

To construct a null tetrad, we begin by taking $l_a = (1, 0, 0, 0)$. Then using $g_{uu} = 2l_u n_u$ we find that $n_u = -\frac{1}{2}\Lambda v^2$. Setting $g_{uv} = l_u n_v + l_v n_u$ we conclude that $n_v = 1$. All together this procedure leads to the tetrad

$$l_a = (1, 0, 0, 0), \qquad n_a = \left(-\frac{1}{2}\Lambda v^2, \, 1, \, 0, \, 0\right)$$

$$m_a = \left(0, \, 0, \, -\frac{i}{\sqrt{2}\Omega}, \, -\frac{1}{\sqrt{2}\Omega}\right), \qquad \overline{m}_a = \left(0, \, 0, \, \frac{i}{\sqrt{2}\Omega}, \, -\frac{1}{\sqrt{2}\Omega}\right)$$

$$l_a = (1, 0, 0, 0), \qquad n_a = \left(-\frac{1}{2}\Lambda v^2, \, 1, \, 0, \, 0\right)$$

$$m_a = \left(0, \, 0, \, -\frac{i}{\sqrt{2}\Omega}, \, -\frac{1}{\sqrt{2}\Omega}\right), \qquad \overline{m}_a = \left(0, \, 0, \, \frac{i}{\sqrt{2}\Omega}, \, -\frac{1}{\sqrt{2}\Omega}\right)$$

We can raise indices with the metric tensor; i.e., $l^a = g^{ab}l_b$, $n^a = g^{ab}n_b$, $m^a = g^{ab}m_b$. This gives

$$l^a = (0,\ 1,\ 0,\ 0), \qquad n^a = \left(1,\ -\frac{1}{2}\Lambda v^2,\ 0,\ 0\right)$$

$$m^a = \left(0,\ 0,\ \frac{i}{\sqrt{2}}\Omega,\ \frac{1}{\sqrt{2}}\Omega\right), \quad \overline{m}^a = \left(0,\ 0,\ -\frac{i}{\sqrt{2}}\Omega,\ \frac{1}{\sqrt{2}}\Omega\right)$$

The only nonzero spin coefficient is $\gamma$. From (9.15) we have

$$\gamma = \frac{1}{2}(l_{a;b}n^a n^b - m_{a;b}\overline{m}^a n^b)$$

looking at the first term, since $l_a = (1, 0, 0, 0)$ the solution is pretty simple. With only the $u$ component to consider we have

$$l_{a;b} = \partial_b l_a - \Gamma^c{}_{ab}l_c = -\Gamma^u{}_{ab}$$

Looking at the Christoffel symbols, the only nonzero term is $l_{u;u} = -\Gamma^u{}_{uu} = \Lambda v$. And so the first sum gives

$$l_{a;b}n^a n^b = l_{u;u}n^u n^u = \Lambda v$$

Considering the nonzero terms of the other members of the null tetrad, the second sum can be written as

$$m_{a;b}\overline{m}^a n^b = m_{x;u}\overline{m}^x n^u + m_{x;v}\overline{m}^x n^v + m_{y;u}\overline{m}^y n^u + m_{y;v}\overline{m}^y n^v$$

However, all of these terms vanish. For example, consider

$$m_{x;u} = \partial_u m_x - \Gamma^c{}_{xu}m_c$$

All the Christoffel symbols of the form $\Gamma^c{}_{xu}$ are zero for this spacetime, and since $m_x = -\frac{i}{\sqrt{2}\Omega}$ but $\Omega = 1 + \frac{\Lambda}{2}(x^2 + y^2)$, the derivative with respect to $u$ vanishes, and so $m_{x;u}$ vanishes as well. The same argument holds for each term. Therefore we conclude that

$$\gamma = \frac{1}{2}\left(l_{u;u}n^u n^u\right) = \frac{1}{2}\Lambda v$$

An exercise shows that all the remaining spin coefficients vanish. With only one nonzero spin coefficient to consider, the Newman-Penrose identities become rather simple. There are three identities we can use that contain $\gamma$

$$D\gamma - \Delta\varepsilon = \alpha(\tau + \overline{\pi}) + \beta(\overline{\tau} + \pi) - \gamma(\varepsilon + \overline{\varepsilon})$$
$$+ \tau\pi - \nu\kappa + \Psi_2 - \Lambda_{NP} + \Phi_{11} \tag{13.58}$$

$$\delta\alpha - \overline{\delta}\beta = \rho\mu - \sigma\lambda + \alpha\overline{\alpha} + \beta\overline{\beta} - 2\alpha\beta + \gamma(\rho - \overline{\rho})$$
$$+ \epsilon(\mu - \overline{\mu}) - \Psi_2 + \Lambda_{NP} + \Phi_{11} \tag{13.59}$$

$$\Delta\rho - \overline{\delta}\tau = -\rho\overline{\mu} - \sigma\lambda + \tau(\overline{\beta} - \alpha - \overline{\tau}) + \rho(\gamma + \overline{\gamma}) + \kappa\nu - \Psi_2 - 2\Lambda_{NP} \tag{13.60}$$

Here the Newton-Penrose scalar is related to the Ricci scalar via $\Lambda_{NP} = \frac{1}{24}R$. Looking at the left side of (13.58), the only nonzero term is

$$D\gamma = l^a \partial_a \gamma = \partial_v \gamma = \partial_v \left(\frac{1}{2}\Lambda v\right) = \frac{1}{2}\Lambda$$

Putting this together with the nonzero terms on the right side gives

$$\Psi_2 - \Lambda_{NP} + \Phi_{11} = \frac{1}{2}\Lambda \tag{13.61}$$

Moving to the next equation, everything vanishes except the last three unknown terms. Therefore (13.59) becomes

$$-\Psi_2 + \Lambda_{NP} + \Phi_{11} = 0$$
$$\Rightarrow \Phi_{11} = \Psi_2 - \Lambda_{NP} \tag{13.62}$$

Finally, all terms vanish in (13.60) with the exception of the last two, giving us

$$\Psi_2 = -2\Lambda_{NP} \tag{13.63}$$

Using (13.62) and (13.63) in (13.61), we find

$$\frac{1}{2}\Lambda = \Psi_2 - \Lambda_{NP} + \Phi_{11} = -2\Lambda_{NP} - \Lambda_{NP} - 3\Lambda_{NP} = -6\Lambda_{NP}$$
$$\Rightarrow \Lambda_{NP} = -\frac{1}{12}\Lambda \tag{13.64}$$

Back substitution of this result into (13.62) and (13.63) gives

$$\Psi_2 = \frac{1}{6}\Lambda \quad \text{and} \quad \Phi_{11} = \frac{1}{4}\Lambda \qquad (13.65)$$

It can be shown that the other Weyl scalars vanish. Therefore, with $\Psi_2 \neq 0$, we conclude that this spacetime is *Petrov Type D*. This means there are two principal null directions, each doubly repeated. The fact that the spacetime contains $\Psi_2$ and not $\Psi_4$ or $\Psi_0$ indicates that this spacetime describes electromagnetic fields and not gravitational radiation. This spacetime represents a vacuum universe that contains electromagnetic fields with no matter.

# Further Reading

The study of gravitational radiation is an active and exciting area. With LIGO coming into operation, exciting experimental results will soon complement the theory. Unfortunately we can scratch only the surface in this brief treatment. Limited space precluded us from covering experimental detection of gravitational waves, and energy and power carried by the waves. The interested reader is encouraged to consult Misner et al. (1973) for an in-depth treatment, or Schutz (1985) for a more elementary but thorough presentation. The Bondi metric is important for the analysis of radiating sources (see D'Inverno, 1992). Hartle (2002) has up-to-date information on the detection of gravity waves, and active area of research in current physics. The section on the collision of gravitational waves relied on *Colliding Plane Waves in General Relativity* by J.B. Griffiths (1991), which, while out of print, is available for free download at http://www-staff.lboro.ac.uk/~majbg/jbg/book.html. The reader is encouraged to examine that text for a thorough discussion of gravitational wave collisions.

In addition, this chapter also relied on *Generalized pp-Waves* by J.D. Steele, which the mathematically advanced reader may find interesting. It is available at http://web.maths.unsw.edu.au/~jds/Papers/gppwaves.pdf. The reader interested in gravitational waves and the cosmological constant should consult "Generalized Kundt waves and their physical interpretation," J.B. Griffiths, P. Docherty, and J. Podolský, *Class. Quantum Grav.*, 21, 207–222, 2004 (gr-qc/0310083), or on which Example 13-3 was based.

# Quiz

1. Following the procedure used in Example 9-2, consider the collision of a gravitational wave with an electromagnetic wave. The line element in the region $v \geq 0$ is given by

$$ds^2 = 2 \, du \, dv - \cos^2 av \left( dx^2 + dy^2 \right)$$

   By calculating the nonzero spin coefficients, one finds that
   (a)  there is pure focusing
   (b)  the Weyl tensor vanishes
   (c)  there is twist and shear

2. Consider the Aichelburg-Sexl metric given in (13.41). The only nonzero spin coefficient is given by
   (a)  $\nu = -2\sqrt{2} \frac{x}{x^2+y^2}$
   (b)  $\nu = -2\sqrt{2} \frac{(x+iy)}{x^2+y^2}$
   (c)  $\pi = \sqrt{2} \frac{(x+iy)}{x^2+y^2}$

3. Compute the Ricci scalar $\Phi_{22}$ for the metric used in Example 9-1. You will find
   (a)  $\Phi_{22} = \mu \left( \delta + \bar{\delta} \right) + \bar{\nu}\pi - \nu \left( \tau - 3\beta - \bar{\alpha} \right)$
   (b)  $\Phi_{22} = \delta \nu - \Delta \mu - \mu^2 - \lambda \bar{\lambda} - \mu \left( \delta + \bar{\delta} \right) + \bar{\nu}\pi - \nu \left( \tau - 3\beta - \bar{\alpha} \right)$

   $$= -\Delta \mu - \mu^2 - \lambda^2$$

   (c)  $\Phi_{22} = \delta \nu - \Delta \mu - \mu^2$

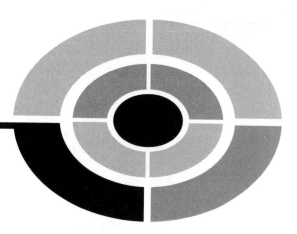

# Final Exam

1. In flat space, $\Delta s$ for the following pairs of events $E_1 = (3, -1, 2, 4)$ and $E_2 = (0, 4, -1, 1)$ is given by
   (a) 1
   (b) 4
   (c) $-2$
   (d) 6

2. Consider two events that are *simultaneous* in some rest frame. The interval between these events is
   (a) spacelike
   (b) timelike
   (c) null
   (d) cannot be determined

3. A rocket ship with rest, length 20 m, approaches a barn with open ends. An observer named Sally is at rest with respect to the barn. She sees that the ship is traveling at $v/c = 0.95$ and her measurements indicate that the barn is 13 m long. The ship

(a) will not fit in the barn, because it is nearly 64 m long, as measured by Sally

(b) does not fit in the barn, because the ship is 20 m long according to all observers

(c) fits in the barn because Sally sees the length of the ship as about 6.2 m

(d) fits in the barn because from Sally's point of view, the ship is 10 m long

4. The four force is defined to be $K^a = dp^a/d\tau$, where $p^a = (p^0, p)$ is the four momentum and $\tau$ is proper time. If $u^a = dx^a/d\tau$ is the four velocity, then

(a) $u_a K^a = 0$

(b) $u_a K^a = -1$

(c) $u_a K^a = 1$

5. Let $V^a = (2, 1, 1, -1)$ and $W^a = (-1, 3, 0, 1)$. Then if $\eta_{ab} = \text{diag}(1, -1, -1, -1)$, $V_a W^a$ is

(a) 2

(b) −6

(c) −4

(d) 0

6. If a metric $g_{ab}$ is diagonal then

(a) $\Gamma_{aba} = \Gamma_{baa} = -\frac{1}{4}\frac{\partial g_{aa}}{\partial x^b}$

(b) $\Gamma_{aba} = \Gamma_{baa} = \frac{1}{2}\frac{\partial g_{ab}}{\partial x^b}$

(c) $\Gamma_{aba} = -\Gamma_{baa} = \frac{1}{2}\frac{\partial g_{aa}}{\partial x^b}$

(d) $\Gamma_{aba} = \Gamma_{baa} = \frac{1}{2}\frac{\partial g_{aa}}{\partial x^b}$

Consider the metric given by

$$ds^2 = (u^2 + v^2)(du^2 + dv^2) + u^2 v^2 d\theta^2$$

7. Two nonzero Christoffel symbols for this metric are

(a) $\Gamma^u_{vv} = \dfrac{v}{u^2 + v^2}$, $\Gamma^u_{\theta\theta} = \dfrac{uv^2}{u^2 + v^2}$

(b) $\Gamma^u{}_{vv} = \frac{u}{u^2-v^2}$, $\Gamma^u{}_{\theta\theta} = \dfrac{uv^2}{u^2+v^2}$

(c) $\Gamma^u{}_{vv} = \frac{u}{u^2+v^2}$, $\Gamma^u{}_{\theta\theta} = \dfrac{uv^2}{u^2+v^2}$

8. All components of the Riemann tensor are equal and are given by
   (a) $R_{abcd} = u$
   (b) $R_{abcd} = v$
   (c) $R_{abcd} = \dfrac{1}{u^2+v^2}$
   (d) $R_{abcd} = 0$

For the next two problems consider the metric

$$ds^2 = d\psi^2 + \sinh^2\psi\, d\theta^2 + \sinh^2\psi\, \sin^2\theta\, d\phi^2$$

9. If you calculate the Christoffel symbols, you find that
   (a) $\Gamma^\psi{}_{\theta\theta} = \sinh\psi\,\cosh\psi$, $\Gamma^\psi{}_{\phi\phi} = \sinh\psi\,\cosh\psi\,\sin^2\theta$
   (b) $\Gamma^\psi{}_{\theta\theta} = -\sinh\psi\,\cosh\psi$, $\Gamma^\psi{}_{\phi\phi} = \tanh\psi\,\sin^2\theta$
   (c) $\Gamma^\psi{}_{\theta\theta} = \sinh\psi\,\cosh\psi$, $\Gamma^\psi{}_{\phi\phi} = -\sinh\psi\,\cosh\psi\,\sin^2\theta$

10. The nonzero Ricci rotation coefficients are given by
    (a) $\Gamma_{\hat\psi\hat\theta\hat\theta} = \Gamma_{\hat\psi\hat\phi\hat\phi} = -\dfrac{\cosh\psi}{\sinh\psi}$, $\Gamma_{\hat\theta\hat\phi\hat\phi} = -\dfrac{\cot\theta}{\sinh\psi}$
    (b) $\Gamma_{\hat\psi\hat\theta\hat\theta} = \Gamma_{\hat\psi\hat\phi\hat\phi} = \dfrac{\cosh\psi}{\sinh\psi}$, $\Gamma_{\hat\theta\hat\phi\hat\phi} = -\frac{\cot\theta}{\sinh\psi}$
    (c) $\Gamma_{\hat\psi\hat\theta\hat\theta} = \Gamma_{\hat\psi\hat\phi\hat\phi} = -\dfrac{\cos\psi}{\sinh\psi}$, $\Gamma_{\hat\theta\hat\phi\hat\phi} = -\dfrac{\cot\theta}{\sinh\psi}$

11. If a space is conformally flat, i.e., $g_{ab}(x) = f(x)\eta_{ab}$, then
    (a) the Weyl tensor cannot be calculated
    (b) the Weyl tensor vanishes
    (c) the Ricci rotation coefficients vanish

12. Consider the Reissner-Nordström metric

$$ds^2 = \left(1 - \frac{2m}{r} + \frac{e^2}{r^2}\right) dt^2 - \left(1 - \frac{2m}{r} + \frac{e^2}{r^2}\right)^{-1}$$
$$\times\, dr^2 - r^2\, d\theta^2 - r^2\, \sin^2\theta\, d\phi^2$$

The only nonzero Weyl scalar is

(a) $\Psi_2 = \dfrac{e^2 - mr}{r^4}$

(b) $\Psi_1 = \dfrac{e^2 - mr}{r^4}$

(c) $\Psi_2 = \dfrac{e^2 + mr}{r^4}$

(d) $\Psi_2 = \dfrac{e^2 - mr}{r^2}$

13. Consider the Schwarzschild solution and suppose that the line element is given by

$$ds^2 = \left(1 - \frac{2m_1}{r}\right) dt^2 - \left(1 - \frac{2m_2}{r}\right)^{-1} dr^2 - r^2 \left(d\theta^2 + \sin^2 \theta \, d\phi^2\right)$$

The deflection of a light ray in this case is proportional to
(a) $m_1 - m_2$
(b) $m_1 m_2$
(c) $m_1^2 + m_2^2$
(d) $m_1 + m_2$

14. The Gödel metric, which is a physically unrealistic model that allows for the possibility of time travel, is given by

$$ds^2 = (dt + e^{\omega x} \, dy)^2 - dx^2 - \frac{1}{2} e^{2\omega x} \, dy^2 - dz^2$$

Suppose that the stress-energy tensor be given by

$$T_{ab} = \rho \begin{pmatrix} 1 & 0 & e^{\omega x} & 0 \\ 0 & 0 & 0 & 0 \\ e^{\omega x} & 0 & e^{2\omega x} & 0 \\ 0 & 0 & 0 & 0 \end{pmatrix}$$

If the cosmological constant is not zero, the Einstein field equations in this case require that
(a) $\Lambda = -\omega^2/2$
(b) $\Lambda = -\omega^4/2$

(c) $\Lambda = \omega^2$

(d) $\Lambda = \omega^4/2$

15. Rotating black holes are described by
    (a) the Reissner-Nordstrøm solution
    (b) the Kerr solution
    (c) the Schwarzschild metric

16. Consider two fixed observers near a Schwarzschild black hole. One observer at $r = 3m$ emits a pulse of ultraviolet light (wavelength about 400 nm) to a second fixed observer at $r = 5m$. The second observer finds that the light is redshifted, with a wavelength such that the light is
    (a) green
    (b) yellow
    (c) blue
    (d) orange

17. The best description of the coordinate singularity in the Schwarzschild metric is that
    (a) it has no effect whatsoever; it is just an artifact of our coordinate choice. Nothing happens here and particles that cross the coordinate singularity can return to the outside, given enough energy.
    (b) it is an artifact of our choice of coordinates; however, it pinpoints the surface of the event horizon. Once you pass this point you cannot return.
    (c) it is the location where the geometry "blows up."

18. A space with $k = 1$ can be said to have
    (a) zero curvature
    (b) positive curvature
    (c) negative curvature
    (d) embedded curvature

19. In a perfect fluid, the diagonal components of the stress-energy tensor are given by
    (a) $T^a{}_b = \text{diag}(-\rho, -P, -P, -P)$
    (b) They all vanish.
    (c) $T^a{}_b = \text{diag}(\rho, -P, -P, -P)$
    (d) $T^a{}_b = \text{diag}(\rho, 0, 0, 0)$

20. For a type N spacetime
    (a) the only nonzero Weyl scalar is $\Psi_0$ or $\Psi_4$, and there is only one principal null direction

(b) the Weyl scalars all vanish

(c) the only nonzero Weyl scalar is $\Psi_0$ or $\Psi_4$, and there are two principal null directions

(d) there is one doubly repeated null direction and the nonzero Weyl scalars are $\Psi_0$, $\Psi_2$, and $\Psi_4$

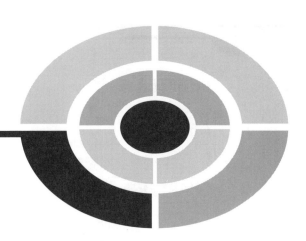

# Quiz and Exam
# Solutions

## CHAPTER 1
1. c      2. a      3. c      4. a      5. b

## CHAPTER 2
1. c      2. a      3. b      4. a      5. c

## CHAPTER 3
1. b      2. c      3. b      4. c      5. d

## CHAPTER 4

| | | | | |
|---|---|---|---|---|
| 1. c | 2. a | 3. c | 4. d | 5. b |
| 6. a | 7. d | 8. b | 9. c | 10. b |

## CHAPTER 5

| | | | | |
|---|---|---|---|---|
| 1. a | 2. c | 3. a | 4. d | 5. c |
| 6. a | 7. d | 8. a | | |

## CHAPTER 6

| | | | | |
|---|---|---|---|---|
| 1. c | 2. b | 3. a | 4. d | 5. a |
| 6. c | 7. a | 8. a | | |

## CHAPTER 7

| | | | | |
|---|---|---|---|---|
| 1. a | 2. b | 3. a | 4. a | 5. a |

## CHAPTER 8

| | | |
|---|---|---|
| 1. d | 2. b | 3. a |

## CHAPTER 9

| | | | | |
|---|---|---|---|---|
| 1. c | 2. a | 3. a | 4. d | 5. b |

## CHAPTER 10

| | | | | |
|---|---|---|---|---|
| 1. a | 2. b | 3. b | 4. a | 5. d |

## CHAPTER 11

| | | | | |
|---|---|---|---|---|
| 1. b | 2. b | 3. a | 4. c | 5. a |

## CHAPTER 12

| | | | | |
|---|---|---|---|---|
| 1. c | 2. a | 3. b | 4. a | 5. c |
| 6. a | | | | |

## CHAPTER 13

| | | |
|---|---|---|
| 1. a | 2. b | 3. b |

## FINAL EXAM

| | | | | |
|---|---|---|---|---|
| 1. c | 2. a | 3. c | 4. a | 5. b |
| 6. d | 7. c | 8. d | 9. a | 10. a |
| 11. b | 12. a | 13. d | 14. a | 15. b |
| 16. a | 17. b | 18. b | 19. c | 20. a |

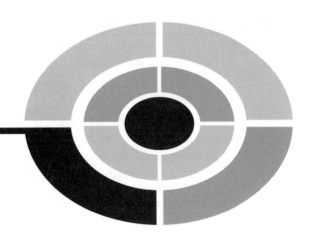

# References and Bibliography

## Books

Of course it is impossible to write a math or science book without relying on previous publications. Below is a list of the materials used in the preparation of the manuscript. Hopefully, we have not forgotten to list anyone. Since those studying the subject will likely have the desire to purchase additional books, I have included my comments.

Adler, R., Bazin, M., and Schiffer, M., *Introduction to General Relativity*, 2d ed., McGraw-Hill, New York, 1975.

We relied heavily on this book for the development of the manuscript. Unfortunately it is out of print, which is really too bad. Although the presentation is a bit old fashioned, it includes some nice discussions on the Schwarzschild solution and the Petrov classification. I also enjoy their use of Lagrangian methods throughout the text.

Carlip, S., *Quantum Gravity in 2 + 1 Dimensions*, Cambridge University Press, Cambridge, 1998.

Carroll, S.M., *An Introduction to General Relativity Spacetime and Geometry*, Addison-Wesley, San Francisco, 2004.

This is a nice edition to the books in the field that came out recently. Clearly written.

Chandrasekhar, S., *The Mathematical Theory of Black Holes*, Clarendon Press, Oxford, 1992.

A good reference book that includes description of the Newman-Penrose formalism. Trying to read it cover to cover might induce a headache.

D'Inverno, R., *Introducing Einstein's Relativity*, Oxford University Press, Oxford, 1992.

A nice concise presentation that's a good place to start building mathematical background for the subject. Chapters are relatively short.

Griffiths, J.B., *Colliding Plane Waves in General Relativity*, Clarendon Press, Oxford, 1991.

A specialized book that we referenced for discussion of Petrov classification and the meaning of the Weyl scalars. More on the advanced side, but good if you are interested in gravitational waves.

Hartle, J.B., *Gravity: An Introduction to Einstein's General Relativity*, Addison-Wesley, San Francisco, 2002.

Good for discussions on the physics with a slow and gentle introduction to the math. Also, some discussion of experimental/observational research like LIGO. Hartle would be a good compliment to this book since we took a more mathematical approach, and so you can find what's missing from our book here.

Hawking, S.W. and Ellis, G.F.R., *The Large-Scale Structure of Spacetime*, Cambridge University Press, Cambridge, 1973.

Great read if you are more mathematically inclined. This is a personal favorite.

Hughston, L.P. and Tod, K.P., *An Introduction to General Relativity*, Cambridge University Press, Cambridge, 1990.

Very slim volume, covers topics quickly in a few pages. It kind of has a mathematically formal prose.

Kay, D.C., *Schaum's Outline of Tensor Calculus*, McGraw-Hill, New York, 1988.

Good to use to learn about differential geometry.

Lightman, A.P., Press, W.H., Price, R.H., and Teukolsky, S.A., *Problem Book in Relativity and Gravitation*, Princeton University Press, Princeton, New Jersey, 1975.

Lots of solved relativity problems, but light on discussion.

Ludvigsen, M., *General Relativity: A Geometric Approach*, Cambridge University Press, Cambridge, 1999.

A really nice but slim volume. Good to read after you have mastered some of the others.

Misner, C., Thorne, K., and Wheeler, J., *Gravitation*, Freeman, San Francisco, 1973.

Certainly the most comprehensive book out there. Everything you need to master, the subject can be found here. This is a good place to read about the Cartan methods.

Peebles, P.J.E., *Principles of Physical Cosmology*, Princeton University Press, Princeton, New Jersey, 1993.

Schutz, B.F., *A First Course in Relativity*, Cambridge University Press, Cambridge, 1985.

A nice book that provides a gentle introduction to the subject. I like the fact that it discusses one forms and how they are introduced. Good book to get started.

Stephani, H., *Relativity: An Introduction to Special and General Relativity*, 3d ed., Cambridge University Press, Cambridge, 2004.

Taylor, E.F., and Wheeler, J.A., *Exploring Black Holes: An Introduction to General Relativity*, Addison-Wesley, San Francisco, 2000.

# Papers and Web Sites

We also relied on a few papers and Web sites. The papers referenced were downloaded online, so we are providing the Web references.

Alcubierre, M., *The Warp Drive: Hyper-Fast Travel Within General Relativity*. arXiv:gr-qc/0009013 v1 5 Sep 2000. Available at: http://arxiv.org/PS_cache/gr-qc/pdf/0009/0009013.pdf

Finley, D., *Physics 570 Course Notes*. Available at: http://panda.unm.edu/courses/finley/p570/handouts/connections.pdf

The class where I first learned relativity. Dr. Finley instilled my preference for using Cartan's equations and sparked my interest in null tetrads. In particular, you may find the notes on connections useful.

Goncalves, S.M.C.V., *Naked Singularities in Tolman-Bondi-de Sitter Collapse*. Available at: http://xxx.lanl.gov/PS_cache/gr-qc/pdf/0012/0012032.pdf

Gutti, S., *Gravitational Collapse of Inhomogeneous Dust in (2 + 1) Dimensions*. Available at: http://lanl.arxiv.org/abs/gr-qc/0506100

Gutti, S., Singh, T.P., Sundararajan, P., and Vaz, C., *Gravitational Collapse of an Infinite Dust Cylinder.*

Living Reviews is a nice Web site we used to do research on fluids and the stress-energy tensor. It has a lot of great information on relativity.
http://relativity.livingreviews.org/

Wikipedia is a great resource but should be used with caution. They allow anyone to post, so there is no telling what's true and what's not. In any case, there are many great articles covering many topics, including relativity. Just be sure to double check what you find.

Two we used were
http://en.wikipedia.org/wiki/Petrov_classification
http://en.wikipedia.org/wiki/Pp-wave_spacetimes

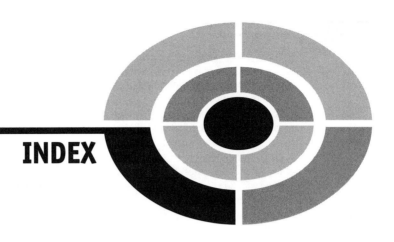

# INDEX

# Index

# Index

# Index

# Index

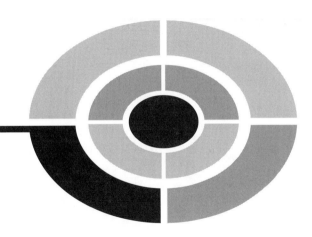

# About the Author

**David McMahon** works as a researcher in the national laboratories on nuclear energy. He has advanced degrees in physics and applied mathematics, and has written several titles for McGraw-Hill.